Application of Chromatography-Mass Spectrometry and Related Techniques, 2nd Edition

Application of Chromatography-Mass Spectrometry and Related Techniques, 2nd Edition

Topic Editors

Chao Kang
Ronald Beckett

MDPI

Basel • Beijing • Wuhan • Barcelona • Belgrade • Novi Sad • Cluj • Manchester

Topic Editors
Chao Kang
School of Chemistry and Chemical Engineering
Guizhou University
Guiyang
China

Ronald Beckett
School of Chemistry
Monash University
Melbourne
Australia

Editorial Office
MDPI AG
Grosspeteranlage 5
4052 Basel, Switzerland

This is a reprint of the Topic, published open access by the journals *Analytica* (ISSN 2673-4532), *Chemosensors* (ISSN 2227-9040), *Foods* (ISSN 2304-8158), *Molecules* (ISSN 1420-3049) and *Separations* (ISSN 2297-8739), freely accessible at: https://www.mdpi.com/topics/8PXED6W3YS.

For citation purposes, cite each article independently as indicated on the article page online and as indicated below:

Lastname, A.A.; Lastname, B.B. Article Title. *Journal Name* **Year**, *Volume Number*, Page Range.

ISBN 978-3-7258-8457-5 (Hbk)
ISBN 978-3-7258-8458-2 (PDF)
https://doi.org/10.3390/books978-3-7258-8458-2

Contents

About the Editors

Chao Kang

Chao Kang received his bachelor's degree and his doctorate in September 2006 and June 2015, respectively, from the College of Chemistry and Chemical Engineering of Hunan University and the State Key Laboratory of Chemo/Biosensing and Chemometrics; he also won the National Scholarship and other awards. From September 2015 to the present, he has taught chemistry and performed scientific research at the School of Chemistry and Chemical Engineering of Guizhou University. His research interests include chemometrics and multivariate analysis, as well as multiway analysis theory and its applications in chromatography and spectroscopy, including liquid biopsy, biological analysis, drug analysis, and food analysis. He has presided over scientific research projects such as the National Natural Science Foundation of China, the Guizhou Provincial Basic Research (Natural Science) Key Project, the Guizhou Provincial Basic Research (Natural Science) Project, the Guizhou Provincial Department of Education Yunian Science and Technology Talent Growth Project, and the Guizhou University Talent Introduction Science and Technology Fund. He has published more than 50 papers in academic journals such as *Theranostics*, *Biosensors and Bioelectronics*, *Industrial Crops and Products*, *Talanta*, *Analytica Chimica Acta*, and *Journal of Separation Science.*

Ronald Beckett

Ronald Beckett worked as an academic staff member in the Department of Chemistry at Caulfield Institute of Technology, Chisholm Institute of Technology, and Monash University in Melbourne, Australia, from 1970 to 2012. His main research interest was the characterization and role of suspended particles and natural organic matter in natural waters, particularly their role in the speciation and transport of pollutants. He pioneered the use of field-flow fractionation (FFF), a separation and sizing technique, to study these materials while working with FFF's inventor, Prof. Calvin Giddings from the University of Utah, USA.

foods

Article

Determination of 69 Pesticide Residues in 42 Batches of Platycodonis Radix and Dietary Risk Assessment Using Combined QuEChERS with GC-MS/MS and UHPLC-MS/MS

Jing Ma [1,2], Xinyue Qiu [1], Suiqing Chen [1,3,4], Haibo Wang [2,5,6,*] and Xiaoya Sun [1,3,*]

1 School of Pharmacy, Henan University of Chinese Medicine, Zhengzhou 450046, China; mj021017@163.com (J.M.); q_xinyue@163.com (X.Q.); suiqingchen0371@163.com (S.C.)
2 NMPA Key Laboratory for Quality Control of Traditional Chinese Medicine (Chinese Materia Medica and Prepared Slices), Zhengzhou 450018, China
3 Collaborative Innovation Center of Research and Development on the Whole Industry Chain of Yu-Yao, Henan University of Chinese Medicine, Zhengzhou 450046, China
4 Collaborative Innovation Center for Respiratory Disease Diagnosis and Treatment & Chinese Medicine Development of Henan Province, Henan University of Chinese Medicine, Zhengzhou 450046, China
5 Henan Institute for Drug and Medical Device Inspection, Zhengzhou 450018, China
6 Henan Province Engineering Research Center for Quality Control of Traditional Chinese Medicine, Zhengzhou 450018, China
* Correspondence: haibowang99@163.com (H.W.); sxy7336@163.com (X.S.)

Abstract

This study aimed to establish a rapid analytical method for the determination of 69 pesticide residues in Platycodonis Radix using GC-MS/MS and UHPLC-MS/MS, as well as carry out a dietary risk assessment on 42 batches of Platycodonis Radix samples collected from different geographical origins. Samples were prepared using the QuEChERS method, followed by high-speed centrifugation and membrane filtration, and the target pesticides were analyzed in selected reaction monitoring (SRM) mode via GC-MS/MS and in multiple reaction monitoring (MRM) mode via UHPLC-MS/MS. Among the 42 tested batches of Platycodonis Radix samples, 3 out of 27 pesticide compounds were detected via GC-MS/MS screening, while only 1 pesticide compound was positive from the 42 compounds determined via UHPLC-MS/MS, and the risk assessment results demonstrated that both chronic and acute dietary exposure risks of all detected pesticides were considerably lower than 1.

Keywords: pesticide residues; Platycodonis Radix; QuEChERS; GC-MS/MS; UHPLC-MS/MS; risk assessment

1. Introduction

Platycodon grandiflorus (Jacq.) A. DC. (PG) is predominantly distributed across East Asia, including China, Korea, and Japan, and has also been introduced into cultivation in certain regions of Europe and North America [1]. Originating in China, wild populations of PG are widely distributed in Guangdong, Guizhou, Yunnan, Sichuan, and other provinces. Platycodonis Radix is the dried root of PG [2]. This broad geographical distribution has led to the classification of two primary varieties: Northern and Southern Platycodonis Radix [3]. At present, the core producing areas are concentrated in Chifeng (Inner Mongolia), Bozhou (Anhui), Zibo (Shandong), Shangluo (Shaanxi), and Song County in Luoyang (Henan) [4]. As a classical medicinal and edible herb, Platycodonis Radix holds significant value in both clinical therapy and the production of health-promoting foods. Modern pharmacological studies have not only validated its traditional antitussive and expectorant effects [5] but also elucidated its multifarious pharmacological activities, including anti-inflammatory [6], antioxidant [7], immunomodulatory [8], hepatoprotective [9], and antitumor properties [10]. In clinical practice, it is frequently employed in compound formulations for the treatment of respiratory diseases such as chronic bronchitis, pharyngitis, and pneumonia. Moreover, it exhibits certain therapeutic potential in the management of digestive disorders, cardiovascular diseases, and as an adjuvant in cancer therapy [11]. Furthermore, owing to its distinctive flavor and nutritional profile, Platycodonis Radix has been developed into a variety of commercial products in the food industry, including health-promoting teas, functional candies, fermented pickles, and functional beverages [12]. Nevertheless, with escalating market demand and the depletion of wild resources, artificial cultivation has become the primary means of supplying this herb for commercial purposes.

Platycodonis Radix exhibits a relatively prolonged growth cycle and thrives in moist soil conditions, yet remains susceptible to various pests and diseases during cultivation. Predominant diseases include root rot caused by Fusarium solani, leaf spot induced by Alternaria spp., and powdery mildew resulting from Erysiphe cichoracearum. Common insect pests encompass aphids, cutworms, and wireworms, among others [13]. To effectively manage these biotic stressors and ensure yield stability, growers frequently employ chemical pesticides such as organophosphates, pyrethroids, and triazoles. Although chemical interventions can rapidly suppress pest and disease outbreaks, excessive or improper application often leads to elevated pesticide residues in the plant material. This not only compromises the quality of the medicinal herb but also poses potential health risks to consumers through dietary exposure [14]. Consequently, pesticide residue has emerged as a critical factor constraining its safe utilization and the sustainable development of the Platycodonis Radix industry.

In the field of pesticide residue analysis, major international pharmacopeias such as the European Pharmacopoeia (EP 11.0) and the United States Pharmacopoeia (USP 46) have systematically established maximum residue limit standards (MRLs) for over 120 pesticides, including organophosphates, organochlorines, pyrethroids, and neonicotinoids [15]. Globally recognized MRLs specifically applicable to Platycodonis Radix are still lacking. The 2025 edition of the Pharmacopoeia of the People's Republic of China sets general MRLs for 47 banned pesticides, including relevant metabolites and isomers, which are applicable to common herbal medicines. However, specific residue limits for routine quality supervision of Platycodonis Radix remain limited, underscoring the need for reliable multi-residue analytical methods to support comprehensive quality and safety evaluation.

The advancement of modern instrumental analysis technologies has significantly propelled the detection of pesticide residues in traditional Chinese medicines toward higher sensitivity, selectivity, and throughput. In terms of sample preparation, the QuEChERS method has emerged as the mainstream choice for multi-pesticide residue extraction and cleanup in complex matrices, owing to its simplicity, rapidity, and cost-effectiveness, making it particularly suitable for high-lipid and multi-interference herbal matrices [16]. At the detection level, GC-MS/MS and UHPLC-MS/MS have become core techniques for the quantitative analysis of pesticide residues in medicinal plants. Recent years have witnessed continuous improvements in their detection sensitivity and multi-component simultaneous analysis capabilities. For instance, Hu et al. employed UHPLC-MS/MS to successfully achieve the simultaneous separation and quantification of 30 pesticides in Astragalus [17]. In a study on Coix Seed, Huang et al. established a non-targeted screening method based on UHPLC-MS/MS, which, through matching with a database containing 1608 compounds, identified 39 potential pesticide residues, further validating the applicability of this technology for multi-residue screening in rhizome-based medicinal materials [18]. GC-MS/MS demonstrates distinct advantages in detecting non-polar and volatile pesticides and is often used complementarily with UHPLC-MS/MS. In research on pesticide residues in Ginseng, Xu et al. utilized a combined GC-MS/MS and UHPLC-MS/MS approach to systematically elucidate the migration and transformation patterns of 30 pesticides during processing, demonstrating that the combined use of both techniques enables comprehensive coverage of pesticide residue analysis in traditional Chinese medicines [19]. These analytical systems, refined through years of research and validation, have established standardized protocols for rhizome-based medicinal materials such as Astragalus, Ginseng, and Coix Seed, exhibiting good accuracy, reproducibility, and high technical maturity.

Previous studies on pesticide residue analysis in Platycodonis Radix have faced certain limitations. Conventional sample preparation techniques, such as gel permeation chromatography (GPC) and solid-phase extraction (SPE) [20,21], have been used for pesticide residue analysis in complex matrices. However, these procedures may require additional cleanup steps, longer processing time, or greater solvent and material consumption, which can limit their convenience for high-throughput multi-residue analysis. Many previous studies have relied on a single analytical platform, such as GC-MS/MS, GC-IT-MS, or GC-FPD, and often focused on a limited subset of pesticide classes, which may restrict the coverage of pesticides with diverse physicochemical properties [22]. These challenges can be further influenced by the complex chemical composition of Platycodonis Radix, which contains polysaccharides, saponins, inulin, and various secondary metabolites, potentially causing matrix effects during extraction and detection, and thereby affecting the performance of conventional sample preparation and single-platform analyses for multi-class pesticide residue determination.

Based on the aforementioned technological context and practical regulatory needs, this study systematically selected 69 prohibited pesticides as target analytes, covering multiple categories such as organochlorines and organophosphates, with reference to literature reports and pharmacopeial standards. A validated analytical workflow for multi-pesticide residues in Platycodonis Radix was established using a conventional QuEChERS sample preparation procedure. Previous studies have reported [23] the applicability of QuEChERS-based procedures for multi-residue pesticide analysis in various root and rhizome medicinal herbs; therefore, this procedure was selected in this study based on its reported applicability and operational convenience for complex plant matrices. This pretreatment method was further combined with GC-MS/MS and UHPLC-MS/MS. Notably, the two instrumental platforms were applied in a complementary manner, with distinct and non-overlapping target compounds, allowing full utilization of their respective applicability: GC-MS/MS for volatile and semi-volatile pesticides and UHPLC-MS/MS for polar and thermolabile compounds. The analytical method developed in this study can provide reliable technical means and data support for the scientific selection of high-quality producing areas, standardized cultivation management, and whole-process quality control of Platycodonis Radix. Furthermore, it offers a technical basis for quality and safety supervision of Platycodonis Radix and related medicinal and edible products, thereby contributing to the advancement of the traditional Chinese medicine industry toward standardization, regulation, and sustainable development.

2. Materials and Methods

2.1. Chemicals and Reagents

Acetonitrile and methanol (HPLC grade) were acquired from MREDA Inc. (Beijing, China). Formic acid (HPLC grade) was obtained from Macklin Biochemical Technology Co., Ltd. (Shanghai, China). Glacial acetic acid (analytical grade) was sourced from Shanghai

Yuanye Bio-Technology Co., Ltd. (Shanghai, China). Organic microporous membranes (0.22 μm) were acquired from ANPEL Laboratory Technologies Inc. (Shanghai, China).

The QuEChERS dSPE kits used in this study were specifically designed for pesticide residue analysis in traditional Chinese medicinal materials according to the Chinese Pharmacopoeia. The kits, purchased from ANPEL Laboratory Technologies (Shanghai, China), included the SBEQ-CA8115-BZ dSPE extraction kit (AOAC 2007.01), with each unit containing 6 g magnesium sulfate ($MgSO_4$) and 1.5 g sodium acetate (NaAc), and SBEQ-CA8773-25 Pharmacopoeia-oriented dSPE dispersive SPE purification tubes, each containing 900 mg $MgSO_4$, 300 mg PSA (primary secondary amine), 30 mg C18, 30 mg graphitized carbon black (GCB), and 300 mg silica (Si). A Platycodonis Radix blank matrix (P/N 63102) was purchased from Shanghai Standard Technology Service Co., Ltd. (Shanghai, China).

2.2. Sample Collection

Platycodonis Radix samples were collected from major cultivation bases in China. A total of 42 samples were obtained from 9 different origins, and the specific information is listed in Table S1. All collected medicinal materials were peeled, dried in a far-infrared drying oven at 50 °C, ground, sieved with a No. 3 sieve, and stored in a dry place for further use.

2.3. Sample Preparation

The extraction procedure was performed using the QuEChERS method. Specifically, 3 g of pulverized Platycodonis Radix was accurately weighed into a 50 mL polypropylene centrifuge tube. Thereafter, 15 mL of 1% (v/v) glacial acetic acid solution was added, and the mixture was vortex-mixed for 1 min and allowed to stand for 30 min. Subsequently, 15 mL of acetonitrile was introduced, followed by vortex mixing and mechanical shaking for 8 min at 300 strokes/min to ensure complete sample homogenization. A QuEChERS extraction salt packet (containing $MgSO_4$ and NaAc) was then added and manually dispersed, and the tube was vigorously shaken for 6 min under the same oscillation conditions.

The mixture was subsequently left to cool in an ice bath for 10 min and then centrifuged at 4000× g for 5 min at 4 °C. A 9 mL aliquot of the supernatant was transferred into a dSPE cleanup tube. The mixture was vortexed for 1 min and shaken vigorously for 8 min at 300 strokes/min. After centrifugation under the same conditions (4000× g, 5 min, 4 °C), 5 mL of the purified supernatant was transferred to a 10 mL glass tube and concentrated to approximately 0.4 mL under a gentle stream of nitrogen in a 40 °C water bath. The residue was reconstituted to 1 mL with acetonitrile and filtered through a 0.22 μm nylon syringe filter into an autosampler vial prior to chromatographic analysis.

2.4. Preparation of Stock and Working Solutions

A 1 mL aliquot of the mixed standard solution containing 69 pesticides (concentration range: 4–20 μg/mL) was accurately pipetted into a 20 mL volumetric flask. The solution was diluted to the mark with acetonitrile and thoroughly mixed. The resulting working standard solution was stored at −25 °C until further use.

2.5. GC-MS/MS Analyses

GC-amenable pesticides were determined using a Thermo Scientific TSQ 8000 triple quadrupole mass spectrometer (Thermo Fisher Scientific, Waltham, MA, USA) coupled with a Trace 1310 gas chromatograph and equipped with a TriPlus RSH autosampler (Thermo Fisher Scientific, Waltham, MA, USA). Chromatographic separation was achieved using a TG-17Sil MS capillary column (30 m × 0.25 mm, 0.25 μm, Thermo Scientific, USA). The column temperature was initially held at 60 °C for 1 min and then increased to 170 °C at a rate of 30 °C/min, followed by a gradual rise to 230 °C at 2 °C/min. Finally, the temperature was elevated to 300 °C at 15 °C/min and maintained for 6 min. The total run time was 45.33 min. High-purity helium (He) was used as the carrier gas, and the injector temperature was set at 250 °C. The sample was introduced in splitless mode with an injection volume of 1 μL.

The triple quadrupole MS was operated in selected reaction monitoring (SRM) mode and electron ionization (EI) mode, with the transfer line and ion source temperatures set to 250 °C.

2.6. UHPLC-MS/MS Analyses

An Agilent 1290 series II liquid chromatograph coupled with a 6475 triple quadrupole mass spectrometer (Agilent Technologies, Santa Clara, CA, USA) was employed for the analysis. Chromatographic separations were performed on a reversed-phase Agilent Zorbax Eclipse Plus C18 column (3.0 × 150 mm, 1.8 μm; Santa Clara, CA, USA). Mobile phase A consisted of 0.1% (v/v) formic acid aqueous solution containing 5 mmol/L ammonium formate. Mobile phase B was HPLC-grade methanol. A linear binary mobile phase solvent gradient was used as follows: 80% A at 0 min, 50% A at 2 min, 0% A at 12 min, 0% A at 15 min, and 80% A at 15.1 min. The flow rate, column temperature, and injection volume were 0.3 mL/min, 40 °C, and 1 μL, respectively.

The mass spectrometer was operated using electrospray ionization (ESI) in positive- and negative-ion modes. The ionization parameters were as follows: capillary voltage: 4 kV; drying gas temperature: 250 °C; drying gas flow rate: 7 L/min; sheath gas temperature: 350 °C; sheath gas flow rate: 12 L/min; nebulizer pressure: 350 kPa.

2.7. Pesticide Residue Risk Assessment

Pesticide residue risk was assessed using the hazard index method [24]. The estimated daily exposure was calculated according to the following equation:

$$\mathrm{EXP} = \mathrm{Ef} \times \mathrm{Ed} \times \mathrm{IR} \times \frac{\mathrm{C}}{\mathrm{AT} \times \mathrm{BW}} \tag{1}$$

where EXP represents the estimated daily pesticide exposure (mg/kg/bw); Ef (exposure frequency) was set as 90 days per year based on the P95 percentile from the national food consumption survey [25]; Ed (exposure duration) refers to the lifetime exposure to herbal medicine, assumed to be 20 years [26]; IR is the daily intake of herbal material (kg/d)—for short-term exposure assessment, the maximum daily intake at the P95 percentile (0.01 kg/d) was applied, while for long-term assessment, the mean daily intake (0.0065 kg/d) was used, both according to the Platycodonis Radix monograph in the Chinese Pharmacopoeia 2025 edition [2]; C denotes the measured concentration of pesticide residue in the herbal material (mg/kg); AT (averaging time) was set as 25,550 days, corresponding to a 70-year lifetime; and BW (body weight) was set at 63 kg.

$$\mathrm{HQ_a} = \frac{\mathrm{EXP_a} \times \mathrm{SF}}{\mathrm{ARfD}} \tag{2}$$

where HQ_a is the acute hazard quotient, where the subscript "a" refers to acute exposure; EXP_a represents the estimated short-term exposure (mg/kg); SF denotes the safety factor with a standard value of 100; and ARfD (acute reference dose) indicates the maximum acceptable daily intake for acute exposure (mg/kg).

$$\mathrm{HQ_c} = \frac{\mathrm{EXP_c} \times \mathrm{SF}}{\mathrm{ADI}} \tag{3}$$

https://doi.org/10.3390/foods15111835

where HQ_c is the chronic hazard quotient, where the subscript "c" refers to chronic exposure; EXP_c denotes the estimated long-term daily intake (mg/kg); and ADI (acceptable daily intake) refers to the maximum acceptable daily intake under chronic exposure (mg/kg).

$$HI = \sum_{i=1}^{n} HQ \quad (4)$$

where HQ (hazard quotient) denotes the risk quotient for an individual pesticide residue, and HI (hazard index) represents the cumulative risk index of multiple pesticide residues, both of which are utilized for assessing the long-term intake risk of pesticide residues in herbal medicines.

3. Results and Discussion

3.1. Optimizing the GC-MS/MS Parameters

During the development of the GC–MS/MS method, selected reaction monitoring mode was employed for analysis [27], in which one quantitative ion pair and two qualitative ion pairs were established for each of the 27 target pesticides; the optimized mass spectrometric parameters are listed in Table 1. The optimized total ion chromatogram (TIC) is shown in Figure 1. The results demonstrate that SRM mode enabled accurate identification of pesticides that were chromatographically co-eluted, thereby enhancing the specificity and reliability of the analytical method.

Table 1. GC-MS/MS parameters for the detection of 27 pesticide compounds.

NO.	Pesticides	Retention Time (min)	Quantitative Ion Pair and Collision Energy (V)	Qualitative Ion Pair and Collision Energy (V)	
				1	2
1	Hexachlorobenzene	9.06	283.9 > 213.9, 32	142.0 > 107.0, 24	285.9 > 250.9, 16
2	α-HCH	9.58	181.0 > 145, 14	111.0 > 75.1, 12	183.0 > 147.0, 14
3	δ-HCH	9.58	181.0 > 145.0, 14	111.0 > 75.1, 12	183.0 > 147.0, 14
4	Terbufos	9.84	231.0 > 128.9, 24	97.0 > 79.0, 14	57.1 > 29.2, 10
5	β-HCH	11.18	181.0 > 145.0, 14	111.0 > 75.1, 12	183.0 > 147.0, 14
6	Heptachlor	12.26	100.1 > 65.1, 12	65.1 > 39.1, 10	271.9 > 236.9, 16
7	γ-HCH	12.68	181.0 > 145.0, 14	111.0 > 75.1, 12	183.0 > 147.0, 14
8	Octachlorodipropyl ether	12.76	79.0 > 49.0, 12	130.0 > 95.0, 22	131.9 > 97.0, 22
9	Aldrin	13.73	66.1 > 65.1, 8	262.9 > 192.9, 30	264.9 > 229.9, 16
10	Parathion-methyl	15.02	109.1 > 79.0, 8	125.0 > 79.1, 8	263.0 > 109.2, 12
11	Dicofol	15.7	139.1 > 111.1, 12	141.0 > 113.0, 12	75.1 > 74.1, 14
12	Oxychlordane	16.35	149.0 > 121.0, 6	115.0 > 87.0, 10	185.0 > 121.0, 10
13	Heptachlor exo-epoxide	17.3	81.1 > 53.1, 10	139.1 > 111.0, 12	236.9 > 142.9, 22
14	trans-Chlordane (γ)	18.68	136.1 > 108.1, 14	121.1 > 93.0, 8	236.9 > 142.8, 24
15	cis-Chlordane (α)	19.41	372.9 > 266.0, 24	236.9 > 118.9, 22	65.2 > 39.1, 10
16	α-Endosulfan	19.65	195.0 > 159.1, 12	170.0 > 99.1, 28	236.9 > 165.0, 26
17	4,4′-DDE	21.72	246.0 > 176.1, 28	176.1 > 150.1, 20	248.0 > 176.0, 32
18	Dieldrin	21.81	79.1 > 77.1, 12	77.1 > 51.1, 12	81.1 > 53.1, 12
19	Endrin	24.26	262.9 > 193.0, 30	139.1 > 111.1, 12	81.1 > 53.1, 10
20	Chlorobenzilate	24.34	139.0 > 111.0, 12	111.0 > 75.1, 12	251.1 > 139.0, 14
21	2,4′-DDT	25.37	235.1 > 165.1, 22	165.1 > 164.2, 18	237.1 > 165.1, 20
22	Nitrofen	25.98	283.0 > 253.1, 10	139.1 > 113.0, 18	63.1 > 62.1, 12
23	4,4′-DDD	26.55	235.0 > 165.1, 22	165.1 > 164.2, 16	237.0 > 165.1, 22
24	β-Endosulfan	26.77	195.0 > 159.0, 10	170.0 > 99.2, 28	236.9 > 165.1, 22
25	4,4′-DDT	28.47	235.0 > 165.1, 22	165.1 > 164.2, 18	237.0 > 165.1, 22
26	Endosulfan sulfate	30.61	271.9 > 236.9, 12	238.9 > 204.1, 10	236.9 > 140.9, 16
27	Mirex	34.57	271.9 > 236.9, 12	236.9 > 118.9, 24	273.8 > 238.7, 16

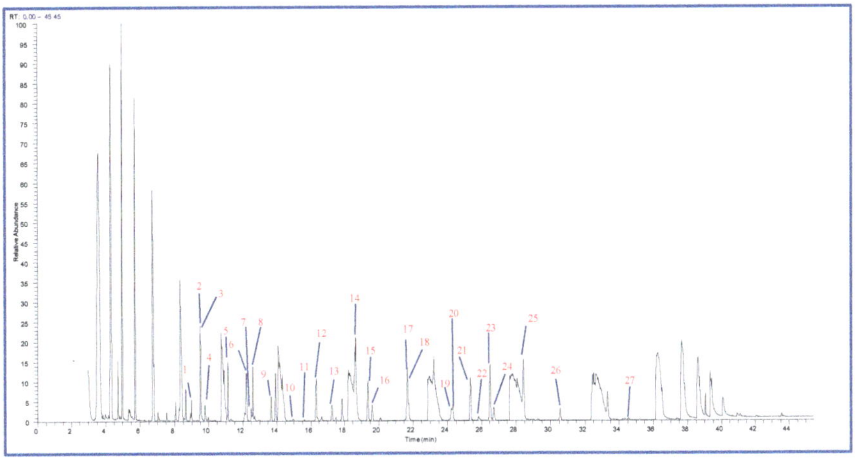

Figure 1. GC-MS/MS total ion chromatogram (TIC). The numbers 1–27 indicate the corresponding pesticide compounds listed in Table 1, and the blue leader lines indicate the corresponding chromatographic peaks.

3.2. Optimizing the UHPLC-MS/MS Parameters

Three columns with different specifications were compared: the ZORBAX Eclipse Plus C18 column (3.0 × 150 mm, 1.8 μm), Hypersil GOLD C18 column (2.1 × 100 mm, 1.9 μm), and Hypersil GOLD column (2.1 × 150 mm, 3 μm). The results indicate that the ZORBAX Eclipse Plus C18 column provided the highest separation efficiency with a sharp peak shape and no apparent tailing or fronting. The mobile phase system was optimized with reference to the fifth method for pesticide residue determination specified in the Chinese Pharmacopoeia (2025 Edition), General Chapter 2341. The mobile phase in the reference method consisted of mobile phase A (0.1% aqueous formic acid containing 5 mmol/L ammonium formate) and mobile phase B (methanol–0.1% aqueous formic acid containing 5 mmol/L ammonium formate, 95:5, v/v). In this study, we optimized the mobile phase system by changing mobile phase B to HPLC-grade pure methanol while maintaining mobile phase A as 0.1% (v/v) aqueous formic acid containing 5 mmol/L ammonium formate, and adjusting the time program of linear binary gradient elution (consistent with the gradient program described in Section 2.6).

Under consistent experimental conditions, the optimized mobile phase system (mobile phase A: 0.1% aqueous formic acid containing 5 mmol/L ammonium formate; mobile phase B: pure methanol) was evaluated in comparison with the mobile phase system in the reference method (methanol–0.1% aqueous formic acid containing 5 mmol/L ammonium formate, 95:5, v/v). The findings demonstrate that the optimized mobile phase system (with pure methanol as mobile phase B) offered better separation performance and simpler operation, as it eliminated the need for preparing mixed solutions containing acid and salts, thereby streamlining the experimental procedure. From the perspective of column maintenance, pure methanol as mobile phase B may also reduce the risk of salt precipitation in the organic phase and simplify routine preparation.

The mixed standard stock solution was diluted 20-fold with acetonitrile. Under the optimized chromatographic conditions described in Section 2.6, full-scan analyses were performed in both positive- and negative-ion modes to identify the precursor ions of the target compounds. Among them, fipronil and its three metabolites (Fipronil desulfinyl, Fipronil sulfone, Fipronil sulfide) consistently generated stable $[M–H]^-$ ions in full-scan mode and were therefore analyzed in negative-ion mode. The remaining 38 pesticides were analyzed in positive-ion mode. Product ion scanning was then employed to screen characteristic product ions and explore appropriate collision energy ranges. Further optimization of mass spectrometric parameters—including fragmentor voltage, collision energy, and product ion selection—was conducted using multiple reaction monitoring mode. The final optimized MS parameters are summarized in Table 2. The optimized total ion chromatogram (TIC) is shown in Figure 2.

Table 2. UHPLC-MS/MS parameters for the detection of 42 pesticide compounds.

NO.	Pesticides	Retention Time (min)	Quantitative Ion Pair and Collision Energy (V)	Qualitative Ion Pair and Collision Energy (V)	
				1	2
1	Methamidophos	3.31	142.0 > 94.0, 80	142.0 > 125.0, 80	/
2	Acephate	3.58	184.0 > 125.0, 60	184.0 > 143.0, 80	/
3	Omethoate	3.74	214.1 > 125.0, 80	214.1 > 183.0, 80	/
4	Aldicarb-sulfoxide	3.84	207.1 > 89.0, 80	207.1 > 132.0, 80	/
5	Aldicarb sulfone	4.11	223.1 > 76.0, 110	223.1 > 86.0, 110	240.1 > 86.0, 110
6	Methomyl	4.54	163.0 > 88.0, 80	163.0 > 106.0, 80	/
7	Monocrotophos	4.66	224.1 > 127.0, 100	224.1 > 193.0, 100	/
8	Phosfolan-methyl	4.78	228.0 > 109.0, 90	228.0 > 168.0, 90	/
9	Chlordimeform	4.85	197.1 > 46.0, 120	197.1 > 89.0, 120	197.1 > 117.0, 30
10	3-Hydroxycarbofuran	5.59	238.1 > 163.0, 80	238.1 > 181.0, 80	238.1 > 220.1, 80
11	Dimethoate	5.82	230.0 > 171.0, 80	230.0 > 199.0, 80	/
12	Phosfolan	6.61	256.0 > 140.0, 100	256.0 > 168.0, 100	256.0 > 228.0, 100
13	Phosphamidon	6.91	300.1 > 127.0, 120	300.1 > 174.0, 120	/
14	Aldicarb	7.00	208.1 > 89.0, 65	208.1 > 116.0, 65	/
15	Metsulfuron-methyl	7.60	382.0 > 167.1, 90	328.0 > 199.0, 90	/
16	Fenamiphos sulfoxide	7.67	320.1 > 171.0, 120	320.1 > 233.0, 120	320.1 > 292.0, 120
17	Fenamiphos sulfone	7.87	336.1 > 188.0, 125	336.1 > 266.0, 125	/
18	Carbofuran	7.89	222.1 > 123.1, 100	222.1 > 165.1, 100	/
19	Chlorsulfuron	8.12	358.0 > 141.0, 120	358.0 > 167.0, 120	/
20	Ethametsulfuron-methyl	8.36	411.1 > 168.0, 120	411.1 > 196.0, 120	/
21	Phorate sulfoxide	8.81	277.0 > 97.0, 80	277.0 > 143.0, 80	277.0 > 171.0, 80
22	Phorate sulfone	8.97	293.0 > 115.0, 60	293.0 > 171.0, 60	293.0 > 247.0, 60
23	Isocarbophos	9.26	312.0 > 236.0, 100	312.0 > 270.0, 100	/
24	Methidathion	9.66	303.0 > 85.0, 80	303.0 > 145.0, 80	/
25	Demeton	10.01	259.1 > 61.0, 100	259.2 > 89.0, 100	/
26	Terbufos sulfone	10.02	321.0 > 97.0, 120	321.0 > 171.0, 120	/
27	Terbufos sulfoxide	10.06	305.1 > 97.0, 60	305.1 > 187.0, 60	/
28	Isazafos	10.93	314.0 > 120.0, 100	314.0 > 162.0, 100	/
29	Fipronil desulfinyl	11.07	386.9 > 282.0, 100	386.9 > 351.0, 100	/
30	Fipronil	11.33	434.9 > 250.0, 120	434.9 > 330.0, 120	/
31	Ethoprophos	11.34	243.1 > 97.0, 100	243.1 > 131.0, 100	243.1 > 173.0, 100
32	Fipronil sulfide	11.44	418.9 > 262.0, 110	418.9 > 383.0, 110	/
33	Fenamiphos	11.48	304.1 > 202.0, 140	304.1 > 217.0, 140	304.1 > 234.1, 140
34	Parathion	11.66	292.0 > 236.0, 120	292.0 > 264.0, 120	/
35	Fipronil sulfone	11.67	450.9 > 282.0, 135	450.9 > 415.0, 135	/
36	Isocarbophos-methyl	11.75	332.0 > 121.0, 90	332.0 > 231.0, 90	332.0 > 273.0, 90
37	Sulfotep	11.77	323.0 > 97.0, 120	323.0 > 115.0, 120	323.0 > 171.0, 120
38	Coumaphos	12.01	363.0 > 227.0, 120	363.0 > 307.0, 120	/
39	Fonofos	12.13	247.0 > 109.1, 80	247.0 > 137.1, 80	/
40	Phorate	12.43	261.0 > 47.0, 70	261.0 > 75.0, 70	/
41	Cadusafos	12.67	271.1 > 97.0, 80	271.1 > 131.0, 80	271.1 > 159.0, 80
42	Sulfluramid	13.40	526.0 > 169.0, 160	526.0 > 219.0, 160	/

https://doi.org/10.3390/foods15111835

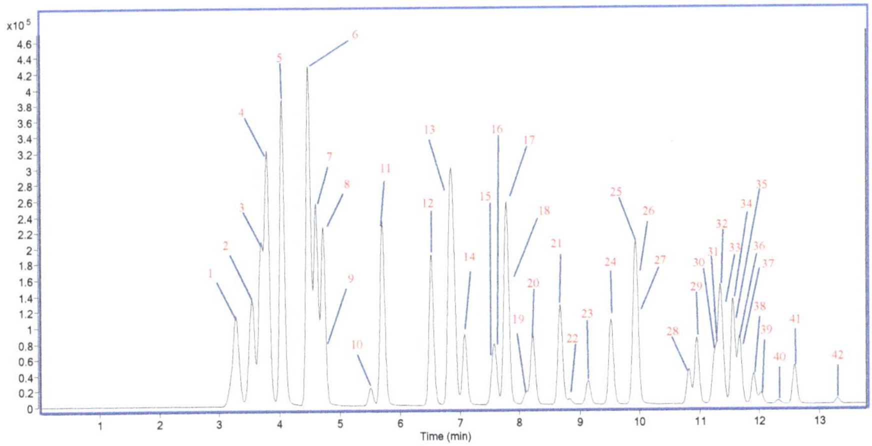

Figure 2. UHPLC-MS/MS total ion chromatogram (TIC). The numbers 1–42 indicate the corresponding pesticide compounds listed in Table 2, and the blue leader lines indicate the corresponding chromatographic peaks.

3.3. Method Validation

The method was validated following the SANTE 11312/2021 guideline document (SANTE/11312/2021, 2021) [28]. Although this guideline is primarily intended for pesticide analysis in food samples, it has been extensively adopted and proven applicable for similar multi-residue analysis in complex botanical matrices, including traditional Chinese medicines. Its validation protocols provide a comprehensive and accepted benchmark for ensuring method reliability. According to the SANTE guideline, a method must be tested for linearity, extraction recoveries, LODs and LOQs.

3.3.1. Linearity and Limits of Detection and Quantification

Under the optimized chromatographic–mass spectrometric conditions, a series of matrix-matched standard solutions were prepared by diluting 50, 100, 150, 200, and 500 μL of the 20-fold diluted mixed standard working solution to 1 mL with the blank matrix solution. Matrix-matched calibration curves were constructed by plotting the peak area (y) against the mass concentration of each target pesticide (x, μg/L), and the corresponding regression equations were derived.

The limit of detection (LOD) was determined at a signal-to-noise ratio (S/N) of 3, and the limit of quantification (LOQ) at $S/N \geq 10$. The results (Tables 3 and 4) indicate that all 69 pesticides exhibited good linearity within their respective concentration ranges, with correlation coefficients (R^2) greater than 0.99. The LODs ranged from 0.0058 to 2.7933 μg/kg, and the LOQs from 0.0196 to 9.3109 μg/kg, all of which were below the maximum residue limits specified in the current national standards. These findings demonstrate that the developed method possesses high sensitivity and is fully suitable for the trace-level determination of pesticide residues.

https://doi.org/10.3390/foods15111835

Table 3. Method validation results for 27 pesticides in GC-MS/MS analysis.

Pesticides	Calibration Curve	Linear r2	Linear Range (ng/mL)	LOD (μg/kg)	LOQ (μg/kg)	Recovery (%)			Repeatability, % RSD			ME (%)
						Low	Medium	High	Low	Medium	High	
Hexachlorobenzene	y = 1767.5x − 629.4	0.9961	4–100	0.2989	0.9963	93.10	90.92	82.71	7.5693	7.8146	8.7716	45.57
α-HCH	y = 339.1x + 4.2	0.9980	20–500	1.3190	4.3966	95.46	104.56	97.28	3.0184	6.1493	3.4101	3.63
δ-HCH	y = 2054.8x − 13,239.0	0.9981	20–500	1.0523	3.5075	104.19	107.74	97.24	4.0447	4.6947	8.1929	59.73
Terbufos	y = 2921.3x − 4885.2	0.9980	4–100	0.3622	1.2074	93.19	102.87	99.13	1.9742	3.2914	3.4123	37.57
β-HCH	y = 2558.2x − 16,950.0	0.9974	20–500	2.7586	9.1954	98.18	102.29	97.24	3.9500	3.3365	3.2164	29.63
Heptachlor	y = 3865.1x − 13,266.0	0.9985	10–250	0.6402	2.1340	92.64	95.63	91.09	2.7013	4.8958	3.8789	35.57
γ-HCH	y = 194.7x + 215.0	0.9960	20–500	0.8247	2.7491	94.42	104.23	100.27	2.2264	3.9292	3.7558	61.70
Octachlorodipropyl ether	y = 2093.6x − 3573.6	0.9981	4–100	0.4845	1.6149	95.09	107.73	95.30	2.1736	3.5528	6.0079	34.29
Aldrin	y = 574.5x − 1612.5	0.9995	10–250	1.8204	6.0680	95.55	77.80	108.19	9.8979	2.5176	3.4616	14.27
Parathion-methyl	y = 1683.3x − 5126.0	0.9972	4–100	0.7160	2.3866	91.83	103.28	120.44	2.5976	5.8234	7.3344	17.51
Dicofol	y = 165.9x + 2526.2	0.9881	4–100	0.5658	1.8859	97.65	107.15	101.10	2.0197	7.2231	3.8249	102.76
Oxychlordane	y = 1173.6x − 13,151.0	0.9961	20–500	1.2508	4.1693	86.88	88.49	92.48	5.4079	3.9166	3.4170	17.17
Heptachlor exo-epoxide	y = 555.9x − 4199.5	0.9962	10–250	1.6547	5.5157	112.32	108.26	69.13	4.7242	7.2043	7.2055	25.02
trans-Chlordane (γ)	y = 766.4x − 6184.4	0.9974	20–500	2.1622	7.2072	100.13	107.90	100.87	1.9672	3.4806	4.7888	9.70
cis-Chlordane (α)	y = 787.0x − 6531.1	0.9986	20–500	2.7015	9.0050	95.98	94.79	91.98	2.0306	4.6125	5.0554	31.59
α-Endosulfan	y = 447.3x − 2541.9	0.9980	10–250	2.7933	9.3110	96.34	95.60	109.02	9.9602	8.2000	4.9355	53.63
4,4′-DDE	y = 4297.6x − 6539.5	0.9979	20–500	0.5299	1.7665	97.32	88.38	90.94	2.2477	5.3730	3.9884	43.73
Dieldrin	y = 329.3x − 2391.0	0.9937	10–250	2.5084	8.3612	70.46	106.90	79.05	7.2211	9.1227	7.2179	1.83
Endrin	y = 398.7x − 3254.2	0.9994	10–250	2.7397	9.1324	98.86	100.01	95.18	5.0103	3.5718	3.0654	26.22
Chlorobenzilate	y = 5394.8x − 4961.1	0.9988	10–250	0.5201	1.7337	97.76	109.90	106.65	2.9734	5.5055	4.4307	58.33
2,4′-DDT	y = 5277.5x − 32,902.0	0.9965	20–500	0.4152	1.3841	98.39	87.24	89.08	3.2908	9.0795	5.3690	32.38
Nitrofen	y = 1179.9x − 10,358.0	0.9968	10–250	1.4218	4.7393	95.57	112.03	106.49	2.4172	5.0886	6.2379	69.27
4,4′-DDD	y = 9006.3x − 17,829.0	0.9960	20–500	0.3563	1.1878	97.68	105.33	99.32	1.7184	5.5988	3.6595	83.55
β-Endosulfan	y = 494.5x − 4324.4	0.9956	10–250	1.3793	4.5977	106.96	108.43	96.92	5.0383	4.9671	3.9683	52.06
4,4′-DDT	y = 3183.7x − 23,798.0	0.9968	20–500	0.1845	0.6151	102.22	96.15	90.20	8.3384	6.6958	7.4234	29.95
Endosulfan sulfate	y = 1614.3x − 4951.0	0.9982	10–250	0.6529	2.1763	101.52	109.96	96.49	3.1242	5.1162	6.5594	101.10
Mirex	y = 3867.2x − 4208.2	0.9980	2–50	0.4444	1.4815	95.85	70.06	69.35	3.0953	4.9632	6.8593	19.98

https://doi.org/10.3390/foods1

Table 4. Method validation results for 42 pesticides in UHPLC-MS/MS analysis.

Pesticides	Calibration Curve	Linear r2	Linear Range (ng/mL)	LOD (μg/kg)	LOQ (μg/kg)	Recovery (%)			Repeatability, %RSD			ME (%)
						Low	Medium	High	Low	Medium	High	
Methamidophos	y = 2099.2x − 2925.0	0.9988	10–250	0.1586	0.5288	62.03	65.85	60.87	1.1610	0.3931	0.9561	8.10
Acephate	y = 2678.6x − 6253.1	0.9984	10–250	0.6736	2.2452	63.17	66.00	62.50	0.8052	1.1499	1.5546	19.77
Omethoate	y = 2270.0x − 4264.7	0.9988	10–250	0.0059	0.0196	83.47	89.72	83.12	2.7526	0.7831	0.7813	15.66
Aldicarb sulfoxide	y = 1266.7x − 4001.8	0.9991	20–500	0.0257	0.0856	79.32	85.31	96.27	4.7910	0.4489	3.5245	16.51
Aldicarb sulfone	y = 1003.9x − 1368.0	0.9996	20–500	0.0365	0.1218	63.80	69.77	64.67	1.2925	1.1238	1.4495	16.99
Methomyl	y = 732.7x − 4071.9	0.9986	20–500	0.0479	0.1595	68.05	92.74	60.05	1.7282	2.8155	1.5290	−16.92
Monocrotophos	y = 2481.6x − 3567.2	0.9988	6–150	0.0100	0.0332	89.76	64.25	78.31	3.7893	0.6128	0.7295	11.98
Phosfolan-methyl	y = 2758.6x − 3819.9	0.9975	6–150	0.0438	0.1458	63.95	66.07	77.52	1.7462	1.0428	0.3421	11.18
Chlordimeform	y = 881.8x − 1007.6	0.9973	4–100	0.0142	0.0474	63.77	64.97	99.78	0.6325	1.0605	0.9600	5.84
3-Hydroxycarbofuran	y = 431.7x + 491.9	0.9971	4–100	0.0716	0.2388	89.15	87.00	78.52	2.0910	1.5404	0.6614	28.85
Dimethoate	y = 2465.2x − 6109.8	0.9985	10–250	0.0427	0.1423	63.53	64.53	61.07	0.9888	1.0495	0.6356	10.32
Phosfolan	y = 2408.9x − 3606.3	0.9990	6–150	0.0227	0.0757	105.01	90.43	99.67	1.5330	0.4929	0.9124	−24.40
Phosphamidon	y = 2211.5x − 5955.7	0.9954	10–250	0.0167	0.0555	86.32	87.77	85.35	1.0878	0.9524	0.6741	41.87
Aldicarb	y = 781.3x − 3695.5	0.9982	2–50	0.0263	0.0876	70.55	76.98	72.33	1.1152	1.2471	0.8045	36.01
Metsulfuron methyl	y = 1815.2x − 1943.4	0.9990	4–100	0.1449	0.4832	83.78	85.87	83.90	1.2908	2.1330	1.1605	−27.56
Fenamiphos sulfoxide	y = 1675.5x − 1549.3	0.9968	4–100	0.0268	0.0892	77.07	80.20	75.58	1.5390	1.0462	1.0398	29.52
Fenamiphos sulfone	y = 2150.9x − 2049.8	0.9972	4–100	0.0178	0.0592	82.05	87.12	82.28	1.4436	0.6429	1.2894	31.15
Carbofuran	y = 4479.1x − 4884.8	0.9976	4–100	0.0095	0.0318	76.43	81.32	77.13	0.4872	0.6480	0.6745	32.19
Chlorsulfuron	y = 487.2x − 1303.4	0.9984	10–250	0.0544	0.1814	73.22	72.75	71.00	3.6906	1.7832	0.5270	−23.76
Ethametsulfuron methyl	y = 4333.9x − 16,102.7	0.9987	10–250	0.0121	0.0402	66.37	72.37	70.77	1.1502	0.5504	0.7131	6.37
Phorate sulfoxide	y = 3234.8x − 967.8	0.9982	2–50	0.0159	0.0530	78.73	83.12	80.07	0.9628	0.5561	0.3225	34.68
Phorate sulfone	y = 337.6x − 100.1	0.9962	2–50	0.1155	0.3851	81.60	84.27	79.35	8.3120	5.1068	4.0223	37.45
Isocarbophos	y = 192.0x − 493.8	0.9995	10–250	0.1562	0.5208	71.52	81.08	75.50	1.2171	1.2392	0.8122	16.74
Methidathion	y = 1792.6x − 3626.7	0.9986	10–250	0.0436	0.1454	64.75	85.73	81.42	1.5682	0.8270	1.0259	38.50

https://doi.org/10.3390/foods1

Table 4. *Cont.*

Pesticides	Calibration Curve	Linear r2	Linear Range (ng/mL)	LOD (μg/kg)	LOQ (μg/kg)	Recovery (%)			Repeatability, %RSD			ME (%)
						Low	Medium	High	Low	Medium	High	
Demeton	y = 506.9x − 529.4	0.9979	4–100	0.0153	0.0509	74.10	74.87	71.98	1.7155	2.4841	0.7106	36.44
Terbufos sulfone	y = 1482.6x − 1323.3	0.9977	4–100	0.0185	0.0616	77.85	78.78	73.22	1.4042	0.3801	0.7649	37.84
Terbufos sulfoxide	y = 5450.0x − 5880.7	0.9967	4–100	0.0090	0.0298	80.55	82.57	78.30	0.4695	0.7724	0.7748	46.46
Isazafos	y = 2789.1x − 1511.6	0.9976	2–50	0.0191	0.0636	88.30	93.90	86.05	2.1666	0.9087	0.8009	43.06
Fipronil desulfinyl	y = 1862.9x − 1363.4	0.9981	4–100	0.0314	0.1046	103.38	105.63	103.18	1.4560	0.6844	0.3749	52.69
Fipronil	y = 181.4x − 158.1	0.9985	4–100	0.0220	0.0734	107.13	106.12	110.25	2.9409	3.0793	1.6643	68.98
Ethoprophos	y = 1566.3x − 1709.5	0.9987	4–100	0.0298	0.0993	81.43	89.97	84.45	1.7671	0.5520	1.7383	50.63
Fipronil sulfide	y = 1915.5x − 1565.5	0.9974	4–100	0.0115	0.0383	102.65	104.38	102.75	1.4668	1.0658	0.9669	55.35
Fenamiphos	y = 3612.6x − 3260.2	0.9995	4–100	0.0184	0.0613	82.77	87.80	83.82	1.8398	1.5732	0.9587	21.47
Parathion	y = 48.5x + 36.1	0.9997	4–100	0.0187	0.0622	97.70	107.52	102.92	9.6365	4.0675	4.8284	34.06
Fipronil sulfone	y = 2209.6x − 2320.4	0.9958	4–100	0.0186	0.0621	99.70	102.70	100.30	1.2876	1.2378	0.4965	53.51
Isocarbophos-methyl	y = 744.6x − 456.6	0.9993	4–100	0.0476	0.1587	89.10	91.72	84.63	3.2139	4.9962	3.5828	32.12
Sulfotep	y = 1667.7x − 1760.0	0.9986	4–100	0.0273	0.0909	81.80	89.45	82.80	1.3325	1.4451	0.7638	33.12
Coumaphos	y = 1537.5x − 3534.7	0.9981	10–250	0.0580	0.1932	73.52	86.07	82.57	1.9013	0.6997	1.1532	55.09
Fonofos	y = 323.5x − 304.5	0.9990	4–100	0.1046	0.3488	77.25	81.02	76.37	4.8950	4.8535	2.2386	22.54
Phorate	y = 276.7x − 64.6	0.9995	2–50	0.0286	0.0953	92.37	107.65	94.20	3.0543	2.3462	1.9860	−0.37
Cadusafos	y = 2726.6x − 2506.4	0.9997	4–100	0.0158	0.0527	73.63	93.43	87.22	1.1254	0.2845	0.4436	48.75
Sulfluramid	y = 87.5x + 70.2	0.9959	4–100	0.0212	0.0707	62.03	65.85	60.87	1.1610	0.3931	0.9561	69.12

https://doi.org/10.3390/foods1

3.3.2. Matrix Effect

In mass spectrometric analysis, the detection of target compounds in complex matrix samples is susceptible to matrix interference, often manifesting as signal enhancement or suppression—a phenomenon known as the matrix effect (ME) [29]. To evaluate its impact on detection outcomes, this study calculated the matrix effect by comparing the slope of the matrix-matched calibration curve, prepared using blank matrix extract, with that of the solvent-based (acetonitrile) calibration curve. The intensity of the matrix effect was classified as follows: an ME value between 20% and 50% indicated a moderate effect, below 20% a weak effect, and above 90% a strong effect [30].

The results demonstrate (Figures 3 and 4) that among the 27 pesticides analyzed via GC-MS/MS, the matrix effect ranged from 1.83% to 102.76%, with 2 exhibiting a strong effect, 18 a moderate effect, and 7 a weak effect. For the 42 pesticides analyzed via UHPLC-MS/MS, the matrix effect ranged from −24.58% to 69.12%, with 17 showing a weak effect and 25 a moderate effect. To effectively mitigate interference from the matrix effect, all calibration standards in this experiment were prepared using a blank Platycodonis Radix matrix free of the target analytes.

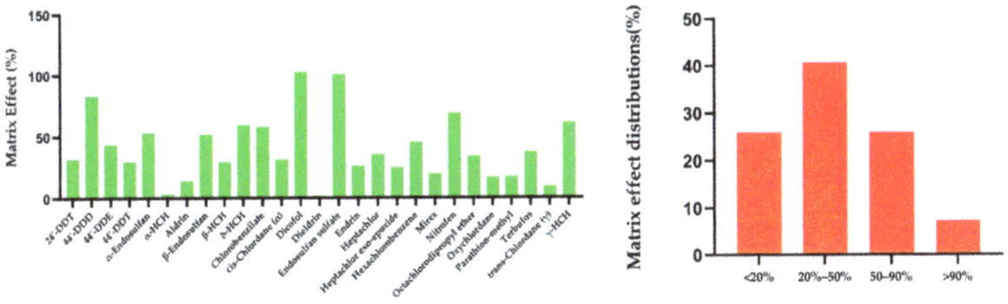

Figure 3. Matrix effects of the 27 pesticides detected via GC-MS/MS. The left panel shows the matrix effect (%) of each pesticide, and the right panel shows the distribution of pesticides across different matrix-effect ranges.

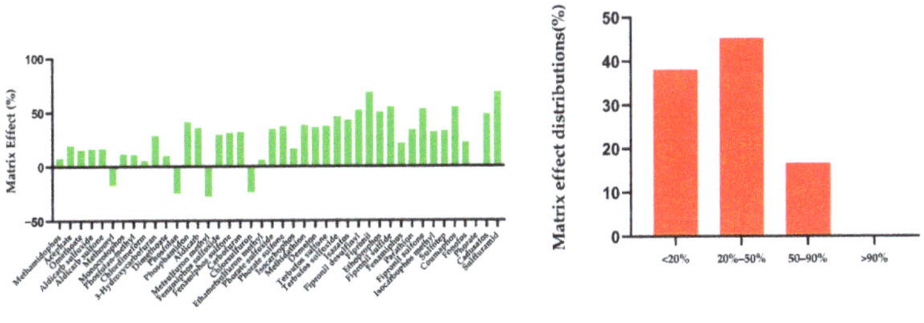

Figure 4. Matrix effects of the 42 pesticides detected via UHPLC-MS/MS. The left panel shows the matrix effect (%) of each pesticide, and the right panel shows the distribution of pesticides across different matrix-effect ranges.

3.3.3. Extraction Recoveries

The accuracy and precision of the method were evaluated through recovery experiments. A 3 g aliquot of blank Platycodonis Radix matrix powder was fortified at three levels (low, medium, and high), corresponding to 0.5-, 1-, and 2-fold the reporting limit (RL). For each fortification level, six replicates were prepared according to the procedure outlined in Section 2.3. As presented in Figures 5 and 6, the recovery rates for the majority of pesticides fell within the acceptable range of 70% to 120%. Specifically, the recoveries for all target compounds ranged from 60.05% to 120.44%, with associated relative standard deviations (RSDs) all below 10%. These data indicate that the method showed generally acceptable accuracy and good precision.

https://doi.org/10.3390/foods15111835

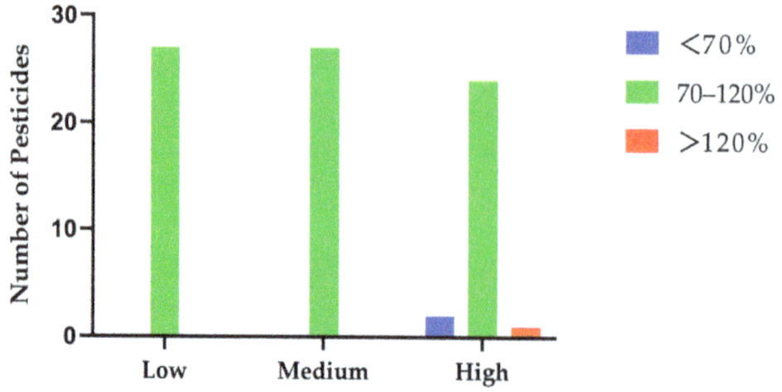

Figure 5. Recovery rates of the 27 pesticides (GC-MS/MS) at low, medium, and high spiked levels. Bars indicate the number of pesticides with recoveries of <70%, 70–120%, and >120%; absence of a bar indicates zero pesticides in that category.

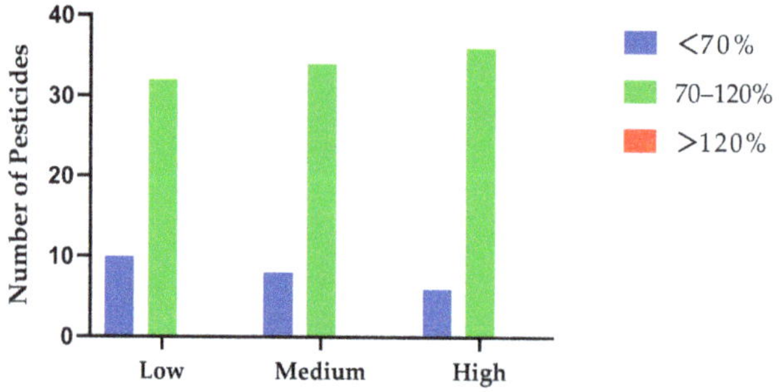

Figure 6. Recovery rates of the 42 pesticides (UHPLC-MS/MS) at low, medium, and high spiked levels. Bars indicate the number of pesticides with recoveries of <70%, 70–120%, and >120%; absence of a bar indicates zero pesticides in that category.

3.4. Application of the Analytical Method

The established method was applied to analyze 42 batches of Platycodonis Radix samples collected from different origins. The results are presented in Figure 7 and Table 5.

Among the 27 pesticide compounds monitored via GC-MS, 3 were detected: α-HCH was found in 35 batches at concentrations ranging from 0.035 to 0.14 mg/kg, 1 batch exceeded the ChP MRL (0.1 mg/kg), and 20 batches exceeded the GB 2763 MRL (0.05 mg/kg). No MRLs are specified by the EU or CAC; oxychlordane was detected in 32 batches within the range of 0.028–0.44 mg/kg, 12 batches exceeded the ChP MRL (0.1 mg/kg), and 32 batches exceeded the GB 2763 MRL (0.02 mg/kg). No MRLs are specified by the EU or CAC; parathion-methyl was identified in 24 batches at levels between 0.0048 and 0.051 mg/kg, 5 batches exceeded the MRLs set by ChP, GB 2763, and the EU (all at 0.02 mg/kg), while no batches exceeded the CAC MRL (3 mg/kg). Among the 42 pesticides analyzed via UHPLC-MS, only Parathion was detected, present in 19 batches at concentrations of 0.0076–0.054 mg/kg, 5 batches exceeded the MRLs of ChP and the EU (both 0.02 mg/kg), 7 batches exceeded the GB 2763 MRL (0.01 mg/kg), and no batches exceeded the CAC MRL (0.2 mg/kg).

https://doi.org/10.3390/foods15111835

Table 5. Determination results of pesticide residues in 42 batches of Platycodonis Radix.

Instrument	Pesticides	Detected (n) [(a)]	Concentration (mg/kg)	ChP MRL (mg/kg) [(b)]	GB 2763 MRL (mg/kg) [(c)]	EU MRL (mg/kg) [(d)]	CAC MRL (mg/kg) [(e)]	Exceed (ChP) [(f)]	Exceed (GB 2763) [(g)]	Exceed (EU) [(h)]	Exceed (CAC) [(i)]
GC-MS/MS	α-HCH	35	0.035–0.14	0.1	0.05	-	-	1	20	-	-
GC-MS/MS	Oxychlordane	32	0.028–0.44	0.1	0.02	-	-	12	32	-	-
GC-MS/MS	Parathion-methyl	24	0.0048–0.051	0.02	0.02	0.02	3	5	5	5	0
UHPLC-MS/MS	Parathion	19	0.0076–0.054	0.02	0.01	0.02	0.2	5	7	5	0

[(a)] Number of samples in which the pesticide was detected among the 42 batches of Platycodonis Radix samples. [(b)] Maximum residue limit specified in the Chinese Pharmacopoeia. [(c)] Maximum residue limit specified in GB 2763 National Food Safety Standard—Maximum Residue Limits for Pesticides in Food. [(d)] Maximum residue limit specified by the European Union (EU). [(e)] Maximum residue limit specified by the Codex Alimentarius Commission (CAC). [(f)] Number of samples exceeding the MRL set by the Chinese Pharmacopoeia. [(g)] Number of samples exceeding the MRL set by GB 2763. [(h)] Number of samples exceeding the EU MRL. [(i)] Number of samples exceeding the CAC MRL.

https://doi.org/10.3390/foods1

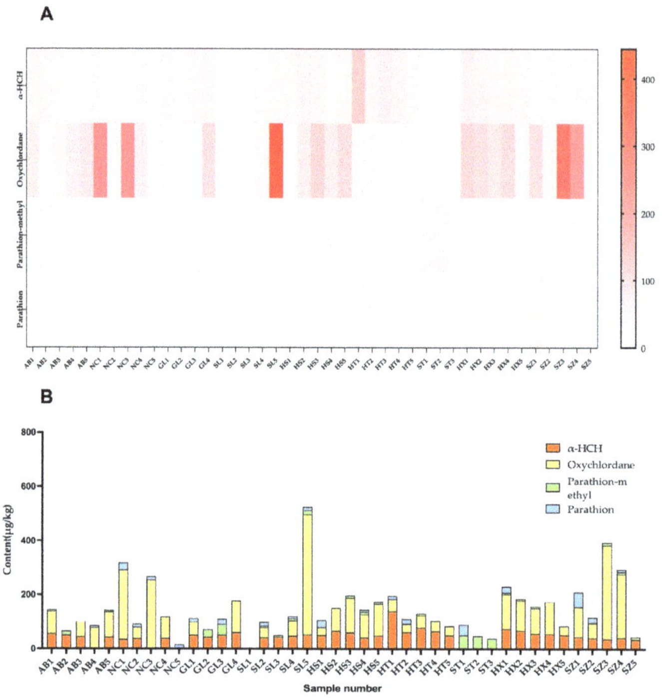

Figure 7. Residual concentrations of four pesticides in different samples. (**A**): Heatmap of residual concentrations. (**B**): Stacked bar chart of individual component concentrations.

These findings indicate that pesticide residues or pesticide-related environmental contamination remain present in some Platycodonis Radix samples and producing areas. Among the detected compounds, α-HCH and oxychlordane belong to the organochlorine pesticide class. China has historically been a major producer and user of organochlorine pesticides, which, while effectively improving crop yields, have also significantly contaminated soil and water systems [31,32]. In this study, α-HCH and oxychlordane were detected in the majority of samples, with particularly high concentrations observed in certain batches (Figure 7). This widespread occurrence suggests that their presence is likely derived from not only historical agricultural practices but also persistent residual contamination in the soil and water environments of major Platycodonis Radix-producing areas, rather than recent direct application.

Notably, the distribution patterns of these pesticides varied among different geographical origins. Oxychlordane exhibited notably high concentrations in samples from several origins, with some batches exceeding the regulatory MRL, indicating that the planting environments in these regions may have a higher legacy pollution burden from organochlorine pesticides. Similarly, α-HCH showed widespread but uneven distribution across samples, with a few batches reaching concentrations close to or above the limit, reflecting regional differences in historical pollution levels and environmental persistence. In contrast, methyl Parathion and Parathion were detected in fewer batches and at lower concentrations, but their presence in multiple origins suggests potential cross-contamination from adjacent agricultural fields or non-targeted pesticide use during cultivation.

Organochlorine pesticides are highly persistent and bioaccumulative, leading to long-term retention in human adipose tissue, liver, and other organs with limited metabolic clearance. Prolonged exposure may elevate the risks of hepatic injury and hematologi-

https://doi.org/10.3390/foods15111835

cal disorders [33,34]. Methyl Parathion and Parathion are organophosphorus pesticides, characterized by high acute toxicity but relatively rapid degradation [35]. Their health hazards are primarily associated with acute neurotoxic effects, including symptoms such as sweating, tremors, and confusion upon high-dose exposure, which can progress to respiratory paralysis and even death in severe cases [36].

The uneven distribution of these pesticide residues across different producing areas highlights the need for origin-specific quality control strategies. Excessive or irrational pesticide application not only harms the environment and ecosystems but also poses direct threats to human health, especially considering the long-term medicinal and dietary use of Platycodonis Radix. Therefore, strengthening, environmental monitoring of historical pollution in major producing areas, promoting standardized cultivation practices that minimize cross-contamination, and establishing targeted quality testing protocols for high-risk origins are crucial steps to ensure the safety of Platycodonis Radix. Concurrently, enhancing the quality and safety testing services for agricultural products and establishing a comprehensive quality and safety control system throughout the production chain have become urgent priorities to reduce dietary exposure risks and ensure the sustainable development of the Platycodonis Radix industry.

3.5. Risk Assessment

The results indicate (Tables 6 and 7) that the chronic/acute exposure risks for all pesticides detected were significantly below 1. Therefore, long-term exposure to these pesticides through the consumption of Platycodonis Radix poses no significant health risk to humans. Note that MRL exceedance and dietary risk assessment represent different evaluation perspectives. The former indicates potential non-compliance in quality control, whereas HI values below 1 suggest that the estimated dietary exposure risk under the assumed consumption scenarios remains low. This conclusion underscores the overall safety of the analyzed samples concerning pesticide residues under the current usage patterns. However, continuous monitoring and adherence to Good Agricultural Practices (GAPs) remain essential to ensure the sustained quality and safety of herbal materials. Future studies could expand the risk assessment to include combined exposure effects from multiple pesticide residues and different consumption scenarios.

Table 6. Acute pesticide residue hazard index.

Pesticides	AB	NC	GL	SL	HS	HT	ST	HX	SZ
α-HCH	0.0045	0.0025	0.0058	0.0042	0.0061	0.0089	0.0000	0.0069	0.0046
Oxychlordane	0.0137	0.0282	0.0074	0.0240	0.0198	0.0087	0.0000	0.0216	0.0334
Parathion-methyl	0.0051	0.0011	0.0168	0.0068	0.0050	0.0083	0.0473	0.0035	0.0062
Parathion	0.0006	0.0048	0.0030	0.0025	0.0026	0.0023	0.0029	0.0016	0.0059
HI_a [(a)]	0.0238	0.0367	0.0331	0.0376	0.0334	0.0282	0.0502	0.0337	0.0501

[(a)] HI_a, the acute hazard index, is the sum of the hazard quotients (HQs) of all detected pesticides per producing region.

Table 7. Chronic pesticide residue hazard index.

Pesticides	AB	NC	GL	SL	HS	HT	ST	HX	SZ
α-HCH	0.0053	0.0030	0.0069	0.0050	0.0072	0.0105	0.0000	0.0082	0.0054
Oxychlordane	0.0890	0.2824	0.0742	0.2405	0.1978	0.0872	0.0000	0.2163	0.3339
Parathion-methyl	0.0182	0.0038	0.0601	0.0244	0.0178	0.0297	0.1690	0.0123	0.0221
Parathion	0.0014	0.0118	0.0073	0.0061	0.0063	0.0055	0.0072	0.0040	0.0144
HI_c [(a)]	0.1139	0.3011	0.1485	0.2760	0.2291	0.1329	0.1762	0.2408	0.3758

[(a)] HI_c, the chronic hazard index, is the sum of the chronic hazard quotients (HQs) of all detected pesticides per producing region.

https://doi.org/10.3390/foods15111835

4. Conclusions

This study developed an applicable analytical method for the determination of 69 pesticide residues in Platycodonis Radix using a QuEChERS-based extraction procedure coupled with gas chromatography–tandem mass spectrometry (GC–MS/MS) and liquid chromatography–tandem mass spectrometry (UHPLC–MS/MS). The method demonstrated good linearity over the concentration range of 4–500 ng/mL, with correlation coefficients (R^2) exceeding 0.99 for all target pesticides. The mean recoveries at low, medium, and high spiked levels ranged from 60.05% to 120.44%, and the relative standard deviations (RSDs) were between 0.28% and 9.96%. The limits of detection (LODs) and quantification (LOQs) were 0.0058–2.7933 μg/kg and 0.0196–9.3109 μg/kg, respectively.

The proposed workflow exhibited satisfactory sensitivity, precision, and accuracy and was suitable for the qualitative and quantitative analysis of the target pesticide residues in Platycodonis Radix. In the analysis of 42 batches of samples, four pesticides were detected, and the dietary risk assessment showed that both the acute and chronic hazard index values were below 1, indicating that the estimated dietary exposure risks were low under the assumed consumption scenarios. Therefore, this validated workflow may provide a useful analytical approach for pesticide residue monitoring, quality control, and dietary risk assessment of Platycodonis Radix. However, several limitations should be acknowledged. First, the QuEChERS procedure was not systematically optimized in terms of extraction and cleanup conditions. Second, the present workflow was not directly compared with other sample preparation methods (e.g., SPE, GPC, and ASE) or other instrumental analytical methods (e.g., GC-IT-MS and GC-FPD). Further studies involving systematic method optimization and direct comparison with general/reference methods would be valuable for more comprehensive evaluations of the relative performance, robustness, and broader applicability of this workflow.

Supplementary Materials: The following supporting information can be downloaded at https://www.mdpi.com/article/10.3390/foods15111835/s1: Table S1: Platycodonis Radix sample information.

Author Contributions: J.M.: writing—original draft preparation, data curation, formal analysis, methodology, visualization, and conceptualization. X.Q.: data curation, formal analysis, methodology, and validation. S.C.: funding acquisition, resources, investigation, project administration, and supervision. H.W.: data curation and project administration. X.S.: writing—review and editing, conceptualization, investigation, methodology, project administration, and supervision. All authors have read and agreed to the published version of the manuscript.

Funding: This work was supported by the Open Project of the NMPA Key Laboratory for Quality Control of Traditional Chinese Medicine (Chinese Materia Medica and Prepared Slices) [grant number: KF202402], Science and Technology Tackling Project of Henan Province [grant number: 252102310530] and Major Science and Technology Project in Henan Province [grant number: 221100310400].

Institutional Review Board Statement: Not applicable.

Informed Consent Statement: Not applicable.

Data Availability Statement: The original contributions presented in the study are included in the article/Supplementary Materials; further inquiries can be directed to the corresponding authors.

Conflicts of Interest: The authors declare no conflicts of interest.

References

1. Zhang, L.; Wang, X.; Zhang, J.; Liu, D.; Bai, G. Ethnopharmacology, phytochemistry, pharmacology and product application of *Platycodon grandiflorum*: A review. *Chin. Herb. Med.* **2024**, *16*, 327–343. [CrossRef]
2. Chinese Pharmacopoeia Commission. *Pharmacopoeia of the People's Republic of China, Vol. 1*; Chinese Medical Science Press: Beijing, China, 2025; pp. 297–298.

3. Aliya, A.; Li, C.; Wei, N.; Sun, Q.; Xu, J.W.; Sun, X.Y.; Zhang, Y.Q.; Xie, J.B. Investigating the Bitter Substance Foundation of Platycodonis Radix Based on Taste Component Correlation Analysis. *J. Li-Shizhen Tradit. Chin. Med.* **2024**, *35*, 1767–1772.
4. Zhao, Q.; Wang, L.; Wu, Y.L.; Wang, P.; Peng, D.Y. Textual Research on Chinese Herb Platycodonis Radix. *J. Chin. Med. Mater.* **2022**, *45*, 1266–1272.
5. Lee, H.-Y.; Lee, G.-H.; Kim, H.-K.; Chae, H.-J. Platycodi Radix and its active compounds ameliorate against house dust mite-induced allergic airway inflammation and ER stress and ROS by enhancing anti-oxidation. *Food Chem. Toxicol.* **2019**, *123*, 412–423. [CrossRef]
6. Li, J.J.; Liu, M.L.; Lv, J.N.; Chen, R.L.; Ding, K.; He, J.Q. Polysaccharides from Platycodonis Radix ameliorated respiratory syncytial virus-induced epithelial cell apoptosis and inflammation through activation of miR-181a-mediated Hippo and SIRT1 pathways. *Int. Immunopharmacol.* **2022**, *104*, 108510. [CrossRef]
7. Xu, S.S.; Wang, N.; Yan, D.D.; Zhong, Y. Platycoside E alleviates allergic airway inflammation in obesity-related asthma mouse model. *Mol. Immunol.* **2023**, *162*, 74–83. [CrossRef]
8. Ma, X.T.; Shao, S.; Xiao, F.Q.; Zhang, H.Y.; Zhang, R.R.; Wang, M.; Li, G.Z.; Yan, M.M. Platycodon grandiflorum extract: Chemical composition and whitening, antioxidant, and anti-inflammatory effects. *RSC Adv.* **2021**, *11*, 10814–10826. [CrossRef] [PubMed]
9. Ke, W.; Wang, P.; Wang, X.; Zhou, X.; Hu, X.; Chen, F. Dietary *Platycodon grandiflorus* Attenuates Hepatic Insulin Resistance and Oxidative Stress in High-Fat-Diet Induced Non-Alcoholic Fatty Liver Disease. *Nutrients* **2020**, *12*, 480. [CrossRef]
10. Chen, S.; Wang, Q.; Ming, S.; Zheng, H.; Hua, B.; Yang, H.S. Platycodin D induces apoptosis through JNK1/AP-1/PUMA pathway in non-small cell lung cancer cells: A new mechanism for an old compound. *Front. Pharmacol.* **2022**, *13*, 1045375. [CrossRef]
11. Yin, X.Y.; Fang, X.M. The effect of Platycodon grandiflorum and its historical change in the clinical application of Platycodonis radix. *Zhonghua Yi Shi Za Zhi* **2021**, *51*, 167–176.
12. Ji, M.-Y.; Bo, A.; Yang, M.; Xu, J.-F.; Jiang, L.-L.; Zhou, B.-C.; Li, M.-H. The Pharmacological Effects and Health Benefits of *Platycodon grandiflorus*—A Medicine Food Homology Species. *Foods* **2020**, *9*, 142. [CrossRef]
13. Sung, J.C.; Hee, K.G.; In, S.K.; Jae-Seoun, H.; Kwon-Seok, J.; Jun-Hyuck, Y.; Jin, K.Y. Root Rot of Balloon Flower (*Platycodon grandiflorum*) Caused by *Fusarium solani* and *Fusarium oxysporum*. *Plant Pathol. J.* **2013**, *29*, 440–445. [CrossRef]
14. Xiao, J.; Xu, X.; Wang, F.; Ma, J.; Liao, M.; Shi, Y.; Fang, Q.K.; Cao, H. Analysis of exposure to pesticide residues from Traditional Chinese Medicine. *J. Hazard. Mater.* **2019**, *365*, 857–867. [CrossRef]
15. Leong, F.; Hua, X.; Wang, M.; Chen, T.; Song, Y.; Tu, P.; Chen, X.J. Quality standard of traditional Chinese medicines: Comparison between European Pharmacopoeia and Chinese Pharmacopoeia and recent advances. *Chin. Med.* **2020**, *15*, 76. [CrossRef]
16. Bruzzoniti, M.C.; Checchini, L.; De Carlo, R.M.; Orlandini, S.; Rivoira, L.; Del Bubba, M. QuEChERS sample preparation for the determination of pesticides and other organic residues in environmental matrices: A critical review. *Anal. Bioanal. Chem.* **2014**, *406*, 4089–4116. [CrossRef]
17. Hu, L.; Li, M.; Sun, Y.; Zhao, H.; Chen, J.; Dai, X.; Kong, Z. Pesticide and myco-toxin residues in Astragalus: Transfer patterns, processing factors and risk assessment during Astragalus processing. *Food Chem.* **2025**, *481*, 144167. [CrossRef] [PubMed]
18. Huang, M.; Song, B.; Chen, Y.; Huang, P.; Yang, Y.; Lu, P. Untargeted screening of pesticide residues in coix seed by LC-MS/MS and dietary risk assessment. *Food Addit. Contam.* **2025**, *18*, 11–13. [CrossRef] [PubMed]
19. Xu, X.; Zhang, M.; Meng, X.; Chen, Y.; Leng, X.; Liang, S.; Zhao, D. Assessment of Pesticide Residues and Dietary Risks in Ginseng from Northeastern China. *Foods* **2025**, *14*, 1381. [CrossRef] [PubMed]
20. Zhang, R.; Chen, Y.; Zheng, P.; Dai, Y.; Li, S.S.; Jia, Y.Y.; Xie, R.; Wang, J.H. Determination of 101 pesticide residues in Platycodonis radix and extracts of Angelica sinensis by gel permeation chromatography-gas chromatography-ion trap mass spectrometry. *Chin. J. Chromatogr.* **2023**, *41*, 178–186. [CrossRef]
21. Yi, X.H.; Hua, Q.; Lu, Y.T. Determination of organophosphorus pesticide residues in the roots of Platycodon grandiflorum by solid-phase extraction and gas chromatog-raphy with flame photometric detection. *J. AOAC Int.* **2006**, *89*, 225–231. [CrossRef]
22. Tian, Y.; Zhang, J.Y.; Zheng, L.; Li, H.; Zhang, Y.; Tu, X. Determination of 34 Pesticide Residues in 24 Batches of Platycodonis Radix Using QuEChERS Combined With GC-MS/MS. *Pestic. Sci. Adm.* **2023**, *44*, 31–39.
23. Liu, J.; Tong, L.; Li, D.; Meng, W.; Sun, W.; Zhao, Y.; Yu, Z. Comparison of two extraction methods for the determination of 135 pesticides in Corydalis Rhizoma, Chuanxiong Rhizoma and Angelicae Sinensis Radix by liquid chromatography-triple quadrupole-mass spectrometry. Application to the roots and rhizomes of Chinese herbal medicines. *J. Chromatogr. B* **2016**, *1017–1018*, 233–240.
24. Liu, Q.; Zhang, H.; Qiao, X.; Dong, X.H.; Ren, P.C.; Wang, J.; Sun, R.Q.; Qi, Y.L.; Li, J.D.; Li, Z.; et al. Residue and dietary risk assessment of sixteen pesticides in sorghum by UHPLC-MS/MS. *J. Food Compos. Anal.* **2026**, *153*, 109053. [CrossRef]
25. Jensen, B.H.; Petersen, A.; Petersen, P.B.; Christensen, T.; Fagt, S.; Trolle, E.; Poulsen, M.E.; Hinge Andersen, J. Cumulative dietary risk assessment of pesticides in food for the Danish population for the period 2012–2017. *Food Chem. Toxicol.* **2022**, *168*, 113359. [CrossRef]
26. Zuo, T.T.; Zhang, L.; Wang, Y.; Nie, L.X.; Shen, M.R.; Liu, L.N.; Yu, J.D.; Jin, H.Y.; Wei, F.; Ma, S.C. Technical guidelines for risk assessment of heavy metals in traditional Chinese medicines. *Chin. Med.* **2023**, *18*, 69. [CrossRef] [PubMed]

https://doi.org/10.3390/foods15111835

27. Xiong, Q.; Wu, D.; Cheng, Y.X.; Hu, L.X.; Chen, Q.L.; Wu, H.Y.; Sun, Y.H.; Liu, Y.S.; Ying, G.G. Development and validation of a simultaneous method for the analysis of benzothiazoles and organic ultraviolet filters in various environmental matrices by GC-MS/MS. *Anal. Bioanal. Chem.* **2022**, *414*, 6541–6555. [CrossRef] [PubMed]
28. SANTE. *Analytical Quality Control and Method Validation Procedures for Pesticide Residues Analysis in Food and Feed*; SANTE/11312/2021; European Union: Brussels, Belgium, 2021.
29. Karaağaçlı, H.; Balkan, T.; Kızılarslan, M.; Kara, K. Pesticide residues in citrus fruits from Türkiye: Assessment of distribution, matrix effects, and health risks in lemon, mandarin, and orange. *J. Food Compos. Anal.* **2025**, *147*, 108091. [CrossRef]
30. Van Eeckhaut, A.; Lanckmans, K.; Sarre, S.; Smolders, I.; Michotte, Y. Validation of bioanalytical LC-MS/MS assays: Evaluation of matrix effects. *J. Chromatogr. B Anal. Technol. Biomed. Life Sci.* **2009**, *877*, 2198–2207. [CrossRef]
31. Wang, L.; Jia, H.; Liu, X.; Sun, Y.; Yang, M.; Hong, W.; Qi, H.; Li, Y.F. Historical contamination and ecological risk of organochlorine pesticides in sediment core in north-eastern Chinese river. *Ecotoxicol. Environ. Saf.* **2013**, *93*, 112–120. [CrossRef]
32. Guan, Y.; Wang, J.; Ni, H.; Zeng, E.Y. Organochlorine pesticides and polychlorinated biphenyls in riverine runoff of the Pearl River Delta, China: Assessment of mass loading, input source and environmental fate. *Environ. Pollut.* **2008**, *157*, 618–624. [CrossRef]
33. Zhao, L.; Liu, Q.S.J.; Jia, Y.N.; Lin, H.S.; Yu, Y.Y.; Chen, X.M.; Liu, Z.Q.; Li, W.X.; Fang, T.; Jiang, W.B.; et al. The Associations be-tween Organophosphate Pesticides (OPs) and Respiratory Disease, Diabetes Mellitus, and Cardiovascular Disease: A Review and Meta-Analysis of Observational Studies. *Toxics* **2023**, *11*, 741. [CrossRef] [PubMed]
34. Farkhondeh, T.; Mehrpour, O.; Forouzanfar, F.; Roshanravan, B.; Samarghandian, S. Oxidative stress and mitochondrial dysfunction in organophosphate pesticide-induced neurotoxicity and its amelioration: A review. *Environ. Sci. Pollut. Res.* **2020**, *27*, 24799–24814. [CrossRef]
35. Toniasso, S.d.C.C.; Brum, M.C.B.; Reis, F.L.D.; Sampaio, V.C.; Silva, L.P.d.; Pereira, R.M.; Bertoloto, J.C.F.; Klein, E.N.M.; Kassim, H.U.; Baldin, C.P.; et al. Association between exposure to pesticides and toxicity in humans: A review. *Eur. Rev. Med. Pharmacol. Sci.* **2025**, *29*, 12–22.
36. Rinsky, J.L.; Richardson, D.B.; Kreiss, K.; Nylander-French, L.; Freeman, L.E.B.; London, S.J.; Henneberger, P.K.; Hoppin, J.A. Animal production, insecticide use and self-reported symptoms and diagnoses of COPD, including chronic bronchitis, in the Agricultural Health Study. *Environ. Int.* **2019**, *127*, 764–772. [CrossRef]

https://doi.org/10.3390/foods15111835

separations

MDPI

Article

Effects of Hydrocarbons and Ionic Impurities on Foaming and Purification of UDS Desulfurization Solvent

Haiyang Wen [1,†], Qiyue Zhao [1,†], Yaolin Wang [1], Zhenwu Jiang [1], Yupeng Cui [1], Mengna Xu [1], Chuanlei Liu [1,*] and Hui Sun [1,2,3,*]

[1] School of Chemical Engineering, East China University of Science and Technology, Shanghai 200237, China; y30230223@mail.ecust.edu.cn (H.W.); y20220012@mail.ecust.edu.cn (Q.Z.); wangyaolin341623@163.com (Y.W.); y30230188@mail.ecust.edu.cn (Z.J.); y30230156@mail.ecust.edu.cn (Y.C.); xumengna2023@163.com (M.X.)
[2] International Joint Research Center of Green Energy Chemical Engineering, East China University of Science and Technology, Shanghai 200237, China
[3] Ministry Key Laboratory of Oil and Gas Fine Chemicals, School of Chemical Engineering and Technology, Xinjiang University, Urumqi 830046, China
* Correspondence: wllwliuchuanlei@163.com (C.L.); sunhui@ecust.edu.cn (H.S.)
† These authors contributed equally to this work.

Abstract

Severe foaming and a significant decrease in desulfurization performance were noted in a novel UDS solvent applied in a natural gas field in western Sichuan, China. The effects of hydrocarbon and ionic impurities on foaming behavior and the purification performance of candidate adsorbents were investigated. An extraction-gas chromatography method was established and validated for determining total hydrocarbons in amine solutions, enabling quantitative evaluation of hydrocarbon contamination. Controlled contamination experiments revealed that hydrocarbons had the strongest effect on foaming, while sulfate and chloride strongly promoted foam formation; organic acid anions showed only minor effects. Fixed-bed screening identified A-98FM anion-exchange resin as the most effective for anionic impurity removal and AC-02 activated carbon as the best candidate for hydrocarbon purification, with a cumulative adsorption capacity q_{0-12} of 14.86 mg/g over 12 h. Pore-structure and thermal-release analyses suggested that conventional pore descriptors alone could not fully explain the dynamic purification performance, while hydrocarbon-related loadings in spent AC-02 occupied accessible pore space and contributed to performance decay. Treatment of a field-aged UDS lean solvent further showed that reductions in target impurities were accompanied by lower foam height and shorter defoaming time. This work provides experimental support for impurity monitoring, foaming-risk identification, and adsorptive purification of UDS desulfurization solvent under flowback-contamination conditions.

Keywords: UDS desulfurization solvent; foaming behavior; hydrocarbon impurities; adsorption; activated carbon

1. Introduction

Against the backdrop of the clean and low-carbon transition and the rising share of natural gas in end-use energy, capacity expansion of acid gas removal (AGR) units has continued [1,2]. Among the available routes, including dry processes, wet processes, membrane separation, and biological approaches [2–4], regenerable-solvent-based wet desulfurization has become the mainstream option because of its large treatment capacity,

https://doi.org/10.3390/separations13050150

high process maturity, and broad applicability to feeds with different sulfur contents [1,5]. In this context, an efficient and stable desulfurization solvent is essential for meeting product-gas specifications and ensuring smooth operation [5–7]. To meet the demand for treating natural gas with high acid-gas content, a variety of blended desulfurization solvent systems, such as Sulfinol, UCARSOL, FLEXSORB, and the CT8 series, have been developed and applied both domestically and internationally. Compared with conventional single-solvent systems, blended desulfurization solvents generally offer advantages in desulfurization selectivity, purification efficiency, and adaptability to operating conditions, and have therefore been widely adopted in natural-gas purification processes.

Despite these advantages, impurity buildup and the resulting foaming of blended solvents remain major challenges during long-term operation. In practice, hydrocarbons entrained in the feed gas, external aqueous phases, and other contaminants may enter the desulfurization unit and gradually accumulate in the amine solution, together with solvent degradation products during repeated absorption-regeneration cycles. The presence of these impurities can alter the apparent properties of the solvent, such as density, viscosity, and surface tension [8–10], thereby increasing its tendency to foam and ultimately impairing the stable operation of the unit [6,11,12]. However, existing studies on blended amine systems have mainly focused on acid-gas absorption performance and solvent degradation, whereas foaming behavior under impurity accumulation has received much less attention [7,10,13–16]. In particular, the respective roles of hydrocarbon and ionic impurities in promoting foaming and stabilizing foam in blended desulfurization solvents have not yet been systematically clarified. As a result, field operations still lack a sufficiently clear basis for evaluating impurity-specific foaming risks and for identifying the key contaminants that should be preferentially controlled [9,17–20].

In parallel, a range of technologies has been proposed and applied by both industry and academia for purification and reclamation of degraded amine solvents, including distillation-based reclamation, side-stream ion exchange, electrodialysis, and bipolar membrane electrodialysis, adsorption, precipitation-crystallization, membrane separation, and solvent extraction [21–24]. Overall, existing purification and reclamation methods have become relatively mature in removing salts, heat-stable salts, and some degradation products, but they generally still suffer from high energy consumption and cost, risks of secondary pollution, as well as equipment fouling, membrane fouling, or scaling [24–31]. Adsorption technologies, represented by activated carbon, are commonly used to remove color bodies, degradation products, and some nonpolar organics under mild conditions with relatively simple flow sheets, and can be integrated into existing systems as a parallel side-loop purification circuit [30,32]. Nevertheless, for blended amine systems, systematic studies on impurity removal remain limited, particularly with respect to the removal of dissolved hydrocarbon impurities by activated carbon. The adsorption behavior, structural matching of adsorbents, service life, and practical applicability in blended desulfurization solvents cannot be directly inferred from conventional amine systems and still require targeted investigation [12,17,32]. Therefore, for impurity-contaminated blended desulfurization solvents, two closely related questions remain insufficiently answered: which impurities dominate a foaming risk, and which purification media are truly effective for their selective removal.

To address the need for removing acid gases and complex organosulfur compounds from natural gas, our team developed a proprietary UDS blended desulfurization solvent through industry-academia-research collaboration [17,33–37]. The formulation contains functional components such as a five-membered sulfur-containing heterocyclic compound (SUL) and a cyclic amine (CA). The added formulation components are used to enhance the removal of methyl mercaptan and COS. Compared with MDEA, UDS solvent shows

https://doi.org/10.3390/separations13050150

higher removal efficiencies for methyl mercaptan, COS, and total organosulfur, and has been applied in industrial natural-gas purification units. In one gas field in western Sichuan, improper upstream process control allowed part of the downhole flowback fluid to enter the main desulfurization unit together with the feed gas. The hydrocarbons and ionic impurities introduced by the flowback fluid subsequently entered the absorption-regeneration loop, contaminated the circulating amine solution, and led to operational instability accompanied by increased foaming risk. Under such field conditions, the UDS solvent provides a representative engineering case for examining how impurity accumulation affects foaming behavior and for evaluating practical purification strategies. However, for this system, systematic studies on impurity-specific foaming effects and corresponding impurity-control measures remain limited.

Against this background, this study focuses on the flowback-contamination scenario of UDS solvent in the target gas field, with emphasis on foaming risks induced by hydrocarbons and ionic impurities and on corresponding adsorptive purification measures. First, representative hydrocarbons and ionic impurities, selected to mimic the impurity composition of field-aged solvent, were introduced into fresh UDS solvent to construct controlled contamination systems. The impacts of impurity type and loading on the foaming behavior of the solvent were systematically examined. Second, considering the hydrocarbon composition of produced water from the gas field and the practical analytical requirements, an extraction-gas chromatography method applicable to the UDS solvent matrix was developed for quantitative determination of total hydrocarbons. Building on these results, commercial ion-exchange resins and activated carbons were evaluated under fixed-bed conditions to screen adsorbents suitable for removing anionic and hydrocarbon impurities from UDS solvent. Furthermore, pore-structure characterization and TG-FTIR-MS analysis were used to examine structural changes and thermally released species of spent AC-02, providing supporting evidence for discussing pore occupation and performance decay. In addition, a field-aged UDS lean solvent was treated with the selected adsorbents to examine their applicability in a real solvent matrix. This work provides an application-oriented experimental basis for impurity monitoring, foaming-risk identification, and adsorptive purification of UDS desulfurization solvent under flowback-contamination conditions.

2. Methods

2.1. Materials

Fresh UDS desulfurization solvent was provided by Shandong Jinlu Environmental Protection Technology Co., Ltd. (Jining, China), with the mass fraction of major active components exceeding 99.5%. The detailed formulation and component ratios of the solvent were not disclosed because of industrial confidentiality. In this study, the fresh solvent was diluted with deionized water to the industrial working concentration of 50 wt%. The basic properties and baseline foaming behavior of the 50 wt% fresh UDS working solution are provided in Table S1. Contaminated UDS solvent samples and produced water samples were supplied by a gas field in western Sichuan, China. Nitrogen (purity 99.999%) used for foaming tests was supplied by Shanghai Wetry Standard Gas Analysis Technology Co., Ltd. (Shanghai, China).

Formic acid, acetic acid, oxalic acid, hydrochloric acid, and sulfuric acid used to construct the representative anionic-impurity system were all of analytical grade and purchased from Shanghai Macklin Biochemical Co., Ltd. (Shanghai, China). Linear n-alkanes (n-C14 to n-C18), used as representative hydrocarbon impurities, were of analytical grade and purchased from the same supplier. n-Hexane (pesticide-residue grade), used as the extraction solvent, was supplied by Anpel Laboratory Technologies (Shanghai) Inc. (Shanghai, China). Anhydrous sodium sulfate (analytical grade), purchased from Shanghai

https://doi.org/10.3390/separations13050150

Macklin Biochemical Co., Ltd., was used for drying the extracts. All reagents were used as received without further purification.

The commercial activated carbons used in the purification experiments were purchased from Shanghai Deep Chemical Technology Co., Ltd. (Shanghai, China), including water-treatment carbons, solvent-recovery carbons, coconut-shell activated carbon for gold refining, cylindrical carbon for desulfurization and denitrification, activated carbon for pressure swing adsorption, and activated coke. These activated carbons were coded as AC-01 to AC-13. Their basic physicochemical properties, including particle-size range, morphology, bulk density, moisture, ash content, and iodine value, are summarized in Table S2. In addition, four commercial strong-base anion-exchange resins, namely XRK207, A-98FM, IRA-900, and A-654, were purchased from Cohesion (Beijing) Co., Ltd. (Beijing, China). Before use, the resins were pretreated by acid-base activation and then rinsed to neutrality.

2.2. *Experimental Procedures*

2.2.1. Extraction-GC Method for Hydrocarbon Determination in UDS Solvent

An extraction-gas chromatography (extraction-GC) method was developed for quantifying hydrocarbon impurities in the UDS solvent system. Produced water samples associated with flowback contamination were collected from the stripper outlet of a gas treatment station in western Sichuan, China, and were used to characterize the composition of hydrocarbon impurities. The samples were subjected to multiple liquid–liquid extractions with n-hexane, and the combined organic phases were concentrated by rotary evaporation to obtain a concentrated extract for compositional analysis. Based on the analytical results, representative hydrocarbon components were selected to prepare a standard oil. Referring to the general framework for petroleum hydrocarbon determination in water and considering the matrix characteristics of the UDS solvent [38], an n-hexane extraction-GC method applicable to the UDS solvent system was established. The quantitative reliability and applicable range of the method were subsequently evaluated through repeated measurements of calibration standards and spiked samples.

2.2.2. Preparation of Contaminated UDS Solvent Samples

To prepare model contaminated UDS solvent samples, a specified mass of fresh UDS solvent was placed in a container and spiked with either the standard oil or the selected anionic impurities at calculated dosages. For the hydrocarbon-contaminated samples, the standard oil was added to obtain a series of samples with different total hydrocarbon concentrations. For the anion-contaminated samples, formic acid, acetic acid, oxalic acid, hydrochloric acid, and sulfuric acid were separately added to the fresh solvent at the designed concentrations to construct systems containing only anionic impurities. In each case, the mixture was vigorously shaken and thoroughly mixed at room temperature to ensure homogenization. The resulting samples were then used for foaming behavior tests and, where applicable, for subsequent evaluation of activated-carbon adsorption performance.

2.2.3. Fixed-Bed Dynamic Adsorption Experiments

Adsorption experiments with activated carbons and strong-base anion-exchange resins were conducted in the same double-jacketed glass column, which had an inner diameter of 2.0 cm and a packed-bed height of 12.5 cm. The column temperature was maintained at 40 °C using an external thermostatic water bath. Before use, the activated carbons were dried in an oven at 105 °C, whereas the anion-exchange resins were pretreated as described in Section 2.1 and then packed into the column. During packing, the column was gently tapped to compact the bed and improve bed uniformity. Prior to operation, the packed bed was prewetted with the solvent to be treated, and visible bubbles were removed.

https://doi.org/10.3390/separations13050150

During the adsorbent-screening experiments, either hydrocarbon-contaminated UDS solvent or anion-containing UDS solvent was used as the feed. After prewetting, the feed solution was continuously introduced into the column at a flow rate of 5 mL/min. After several minutes of operation, when the residual liquid associated with prewetting had been essentially displaced and the flow had stabilized, the effluent was collected. The concentrations of total hydrocarbons or anionic impurities in the effluent were then determined, and the removal efficiency was calculated by comparing the effluent concentration with the initial feed concentration.

For dynamic-performance evaluation, representative activated carbon samples were tested under the same apparatus configuration and operating conditions. After prewetting and stabilization, the first effluent sample was collected, and the column was then operated continuously for 12 h. Subsequent effluent samples were collected at 1 h intervals. The total hydrocarbon concentration in the effluent at each sampling time, C_t, was determined, and the instantaneous removal efficiency relative to the initial feed concentration, C_0, was calculated as follows:

$$\eta_t = \frac{C_0 - C_t}{C_0} \times 100\% \tag{1}$$

The dynamic purification efficiency curves were obtained by plotting η_t as a function of time.

2.3. *Solvent Impurity Analysis*

2.3.1. Determination of Total Hydrocarbon in UDS Solvent by Extraction-GC

Hydrocarbon impurities in UDS solvent were quantified by an extraction-GC method. The representative standard oil mixture prepared in Section 2.2.1 was serially diluted with n-hexane to prepare a series of calibration standards with different total hydrocarbon concentrations. An external calibration curve was established by plotting the total hydrocarbon concentration against the corresponding total chromatographic peak area.

Sample pretreatment was performed as follows. Approximately 50 g of a hydrocarbon-contaminated UDS solvent sample was weighed into a 125 mL separatory funnel, followed by the addition of 10 g of n-hexane. The mixture was vigorously shaken and then allowed to stand for phase separation, after which the upper organic phase was collected. A second extraction was performed with an additional 10 g of n-hexane. The two organic extracts were then combined and dried over anhydrous sodium sulfate. If necessary, the extract was vacuum-concentrated at approximately 30 °C to about 1 mL to obtain the test extract. The masses of the extract before and after concentration were recorded to calculate the concentration factor, N. A blank control was prepared by replacing the sample with fresh UDS solvent and following the same procedure.

The extracts were analyzed on a gas chromatograph equipped with a flame ionization detector (FID) and an HP-5 nonpolar capillary column. The carrier-gas flow rate was 2.0 mL/min. The oven temperature program was 60 °C for 2 min, followed by heating at 15 °C/min to 280 °C and holding for 15 min. The inlet and detector temperatures were set at 280 and 300 °C, respectively. The injection volume was 0.2 μL. Splitless injection was used initially, and the mode was switched to split after 0.75 min with a split ratio of 30:1. Under these conditions, both the test extract and the blank extract were analyzed. The n-hexane solvent peak and the peaks present in the fresh-UDS blank extract were not included in the hydrocarbon peak-area integration. The remaining resolved peaks in the contaminated UDS extract were summed and used with the external calibration curve to calculate the total hydrocarbon content in the UDS solvent.

https://doi.org/10.3390/separations13050150

2.3.2. Determination of Anionic Impurities in UDS Solvent by Ion Chromatography (IC)

The concentrations of anionic species were determined by ion chromatography (IC). Analyses were performed on an ICS-1100 ion chromatograph equipped with a conductivity detector. An IonPac AS-11 anion-exchange column was used in combination with 20 mmol/L NaOH as the eluent and an ASRS-ULTRA anion suppressor. The detailed operating conditions are summarized in Table S3.

Before analysis, the samples were diluted with deionized water to bring their concentrations within the linear range of the instrument and filtered through a membrane filter to remove suspended particles. External calibration curves were established using a series of standard solutions containing the target anions. Qualitative identification was based on retention time, and quantitative analysis was performed using peak area to determine the mass concentration of each anionic impurity. The results were used to evaluate the removal performance of the anion-exchange resins toward anionic impurities.

2.4. Determination of Foaming Behavior

The foaming behavior of the UDS solvent was evaluated using a previously reported laboratory bubbling test for UDS solvents [17]. A transparent glass foaming tube (inner diameter, ca. 30 mm) fitted with a G3 sintered-glass frit at the bottom was placed in a constant-temperature water bath maintained at 40 °C. For each test, the UDS solvent was introduced into the foaming tube until the static liquid height reached 100 mm and then allowed to equilibrate. According to this bubbling-test protocol, high-purity N_2 was then introduced from the bottom at a constant flow rate of 250 mL/min. After the bubble distribution became stable, sparging was continued for 5 min. Immediately after the gas supply was stopped, the height difference between the upper foam boundary and the initial liquid level was recorded as the foam height. The foam was then continuously observed, and the time required for the foam to completely collapse and for a clear liquid surface to reappear was recorded as the defoaming time. Each sample was tested in triplicate, and the average value was used for subsequent analysis.

2.5. Characterization of Activated Carbon

The specific surface area and pore-structure parameters of the activated carbon samples were determined from N_2 adsorption–desorption isotherms measured at 77 K on a BSD-660 automated high-throughput gas adsorption and micropore analyzer (BSD Instrument Technology (Beijing) Co., Ltd., Beijing, China) using high-purity N_2 as the adsorptive. Prior to measurement, the samples were vacuum-degassed at 300 °C for 180 min. The specific surface area was calculated by the multipoint BET method. The micropore surface area and micropore volume were obtained by the t-plot method, and the external surface area was determined by t-plot extrapolation. The average pore diameter and mesopore volume were calculated by the BJH method based on the desorption branch, and the micropore size distribution was fitted using the NLDFT model. The isotherm types and pore-structure characteristics were identified according to IUPAC recommendations for subsequent comparison with hydrocarbon removal performance [39].

The surface adsorbates and thermal decomposition behavior of spent activated carbon were characterized by thermogravimetry coupled with Fourier transform infrared spectroscopy and mass spectrometry (TG-FTIR-MS). The measurements were carried out using a NETZSCH STA 449 F3 Jupiter simultaneous thermal analyzer coupled with a Bruker INVENIO S FTIR spectrometer and a NETZSCH QMS 403 D mass spectrometer [40]. The samples were heated from room temperature to 800 °C at a rate of 10 K/min under a high-purity nitrogen atmosphere. The TG signal recorded the mass change as a function of temperature, while FTIR and MS synchronously collected the infrared spectra and mass-to-charge ratio (m/z) signals of the evolved gases online. These data were used to identify representative adsorbed or decomposition products and their release temperature ranges after adsorption saturation, providing complementary evidence for the mechanistic analysis of hydrocarbon adsorption and activated-carbon deactivation.

3. Results and Discussion

3.1. Development and Validation of an Extraction-GC Method for Total Hydrocarbons Determination in UDS Solvent

3.1.1. Characterization of Hydrocarbon Impurities in Produced Water and Preparation of the Standard Oil

The concentrated extract obtained from field-produced gas-field water was first characterized by ^{1}H NMR, as shown in Figure S1a. The characteristic resonances at approximately 0.9 and 1.2 ppm were assigned to methyl ($-CH_3$) and methylene ($-CH_2-$) protons of hydrocarbons, respectively, indicating that the oil-phase impurities in the extract were predominantly hydrocarbon species. A weak signal near 5.0 ppm was attributed to residual water or trace hydroxyl-containing species. To further identify the individual components and their distribution, the concentrated extract was analyzed by GC-MS. Major peaks were identified by mass spectral library matching, and the relative abundance of each component was estimated from chromatographic peak-area integration. The GC-MS chromatogram is shown in Figure S1b, and the identified major compounds and their relative abundances are summarized in Table S4. The results indicate that the hydrocarbon impurities in the produced-water extract span a carbon-number range of C_{10}–C_{27}, with a more pronounced enrichment in the C_{14}–C_{18} range. In addition to n-alkanes, the extract also contained measurable amounts of branched and isoprenoid hydrocarbons, together with minor aromatic and naphthenic compounds.

Based on the above carbon-number distribution, and considering both compositional contribution and the availability of reference standards, n-alkanes in the C_{14}–C_{18} range were selected as representative components for preparation of the standard oil. This representative standard oil was intended for fit-for-purpose quantification of total hydrocarbons in the present UDS solvent system rather than for exact reproduction of the full hydrocarbon composition of the field samples. The mixing ratio was normalized according to the relative abundances of the corresponding n-alkanes in the produced-water samples, and the composition and mass fractions of the standard oil are listed in Table S5. The C_{14}–C_{18} n-alkanes were weighed into an amber bottle according to Table S5 and thoroughly mixed at room temperature until homogeneous. The resulting mixture was used as the standard oil for construction of the external calibration curve and preparation of model hydrocarbon-contaminated systems.

3.1.2. Development and Validation of the Extraction-GC Method

To quantify hydrocarbon impurities in UDS solvent, an extraction-gas chromatography method was developed using n-hexane extraction and GC-FID analysis. The method was adapted from extractable petroleum hydrocarbon analysis in water and adjusted for the UDS solvent matrix. A fresh-UDS blank was processed and analyzed under the same conditions as the contaminated samples.

The C14–C18 n-alkane mixture prepared in Section 3.1.1 was used as a surrogate calibration standard. Thus, the reported total hydrocarbon content represents standard-oil-equivalent concentration. This method was used for quantitative analysis of hydrocarbon impurities in UDS solvent samples.

The representative standard oil was serially diluted with n-hexane to prepare calibration standards at 200, 500, 1000, 3000, and 5000 mg/kg. The calibration curve was established using the summed peak areas of the standard oil-hydrocarbon peaks. The n-hexane solvent peak was not included in the integration. The GC-FID chromatogram of the standard oil and the calibration curve are shown in Figure S2. The linear regression coefficient was higher than 0.999, showing good linearity over the tested range.

The sample pretreatment procedure is described in Section 2.3.1. For each contaminated sample, the n-hexane solvent peak and the peaks present in the fresh-UDS blank extract were not included in the hydrocarbon peak-area integration. The remaining resolved peaks in the contaminated UDS extract were summed as A_x. Representative GC-FID chromatograms of the fresh-UDS blank extract and the 1000 mg/kg matrix-spiked UDS extract are shown in Figure S3, illustrating the matrix background and the hydrocarbon peaks used for integration. The total hydrocarbon content in the original UDS solvent, ω (mg/kg), was calculated as follows:

$$\omega = \frac{A_x - a}{b} \times \frac{m_1}{m} \times \frac{1}{N} \tag{2}$$

where A_x is the summed peak area used for total hydrocarbon calculation; a and b are the intercept and slope of the calibration curve; m is the sample mass (g); m_1 is the mass of the combined extract (g); and N is the concentration factor. When the extract was not concentrated, $N = 1$.

The precision and accuracy of the method were evaluated using UDS matrix-spiked samples. Fresh UDS solvent was spiked with the representative standard oil at 500, 1000, and 3000 mg/kg. Each level was analyzed in triplicate. The recovery, relative standard deviation (RSD), and relative error (RE) were calculated. The results are summarized in Table 1, and the replicate data are provided in Table S6.

Table 1. Validation results of the extraction-GC method for total hydrocarbon determination in UDS solvent.

Spiked Level (mg/kg)	Measured Mean (mg/kg)	Recovery (%)	RSD (%)	RE (%)	n
500	478.47	95.69	1.14	−4.31	3
1000	1044.38	104.44	1.68	4.44	3
3000	3033.43	101.11	1.25	1.11	3

The recoveries were 95.69–104.44%. The RSD values were 1.14–1.68%, and the RE values were −4.31–4.44%. These results indicate acceptable precision and accuracy for total hydrocarbon determination in the tested UDS solvent samples.

During method testing, the GC-FID signal was not stable enough for reliable integration when the standard oil concentration in the final n-hexane extract was below approximately 100 mg/kg. For a non-concentrated extract, this concentration corresponds to about 40 mg/kg in the original UDS solvent, based on an extract-to-sample mass ratio of approximately 20/50. When emulsification or turbidity occurred during extraction, centrifugation was used to improve phase separation.

3.2. *Effects of Impurities on the Foaming Behavior of UDS Solvent*

To evaluate the sensitivity of foaming behavior to typical impurities present in field amine solvents, fresh UDS solvent was used as the baseline system. Based on impurities commonly detected in field lean solvent, formate ($Formate^-$), acetate ($Acetate^-$), oxalate ($Oxalate^{2-}$), chloride (Cl^-), sulfate (SO_4^{2-}), and the representative standard oil were individually introduced to construct controlled contamination systems. Foaming behavior was evaluated using the bubbling test described above, with foam height (mm) and de-

https://doi.org/10.3390/separations13050150

foaming time (s) used as the comparative metrics. The corresponding trends are shown in Figure 1a,b.

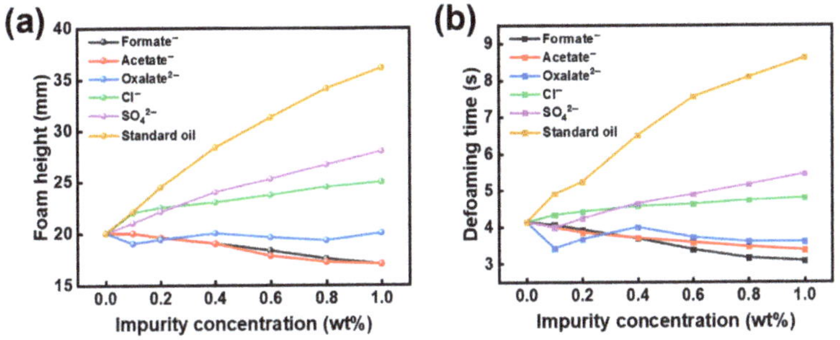

Figure 1. (**a**) Foam height and (**b**) defoaming time of UDS solvent versus impurity concentration.

Under identical test conditions, the fresh solvent exhibited a foam height of 20.0 mm and a defoaming time of 4.13 s. Within the concentration range of 0–1.0 wt%, the different impurity systems showed distinct effects on foaming tendency and foam stability. Overall, the organic acid anions (formate, acetate, and oxalate) had relatively minor effects, whereas the inorganic anions (Cl^- and SO_4^{2-}) and the representative standard oil showed varying degrees of foam promotion and stabilization.

Among the organic acid anions, formate and acetate exhibited similar trends. Over the investigated concentration range, both foam height and defoaming time decreased slightly with increasing concentration, indicating no evident tendency to promote foaming or stabilize foam. Oxalate showed an even weaker effect: the foam height remained close to that of the fresh solvent, and the defoaming time showed only minor fluctuations. Taken together, these three organic acid anions contributed only weakly to foaming in the present system.

In contrast, the two inorganic anions showed more pronounced effects on foaming behavior. With increasing concentration, chloride continuously increased both foam height and foam stability. At 0.1 wt%, the foam height increased to 22.0 mm and the defoaming time to 4.33 s; at 1.0 wt%, these values further increased to 25.0 mm and 4.80 s, respectively. Sulfate exhibited a stronger effect, particularly on defoaming time. At 0.4 wt%, the foam height and defoaming time had already increased to 24.0 mm and 4.64 s, respectively; at 1.0 wt%, they further increased to 28.0 mm and 5.45 s. These results indicate that sulfate exerts a stronger amplifying effect on both foam height and defoaming time than chloride and is therefore more likely to become a key risk factor under impurity accumulation in field operation. Consistently, Figure 1a,b show increasing trends with concentration, with a steeper rise for sulfate.

The representative standard oil showed the strongest effect on foaming behavior, causing marked foam promotion and stabilization with rapidly increasing intensity as concentration increased. At a standard-oil concentration of 0.4 wt%, the foam height and defoaming time increased to 28.4 mm and 6.51 s, respectively; at 1.0 wt%, they further increased to 36.1 mm and 8.61 s. These results indicate that hydrocarbon contamination is the dominant driver of both increased foam height and prolonged defoaming time in the present system. Comparing the magnitude of change induced by each impurity within the same concentration window further shows that hydrocarbon contamination contributes much more strongly to foaming risk than any single anionic impurity. Meanwhile, the accumulation of inorganic anions, especially sulfate and chloride, also significantly aggravates foaming and foam stability. Therefore, hydrocarbon contamination should be regarded as

https://doi.org/10.3390/separations13050150

the primary target for mitigation, while the accumulation of key inorganic anions should also be controlled to reduce their additional contributions to foaming risk.

These differences are related to the different effects of the impurities on the gas–liquid interface and foam-film stability. Formate, acetate, and oxalate have limited interfacial activity at the aqueous solution/vapor interface. Therefore, they only weakly affect foam-film drainage and rupture [41]. In contrast, Cl^- and SO_4^{2-} change the ionic strength and hydration environment of the liquid phase. These changes can modify the electrostatic and hydration conditions in foam films, slow film thinning, and delay bubble coalescence. The greater effect of SO_4^{2-} may be related to its divalent charge and stronger hydration, which can alter the liquid-film environment more effectively than Cl^- [42,43]. The standard oil acts as a hydrophobic organic phase in the UDS solvent. After mixing, dispersed hydrocarbon components can enter the gas–liquid interfacial region and make the interfacial layer less uniform. This hinders liquid drainage from foam films and delays bubble coalescence and rupture [44–46]. As a result, the standard oil produces the largest increase in both foam height and defoaming time.

3.3. *Adsorptive Removal Performance Toward Ionic and Hydrocarbon Impurities*

3.3.1. Removal of Anionic Impurities by Anion-Exchange Resins

The removal performance of four anion-exchange resins (XRK207, A-98FM, IRA-900, and A-654) and coconut-shell activated carbon toward typical anionic impurities in the solvent, namely formate, acetate, oxalate, chloride, and sulfate, was compared. The results are shown in Figure 2. The corresponding ion concentrations before and after adsorbent treatment are provided in Table S7.

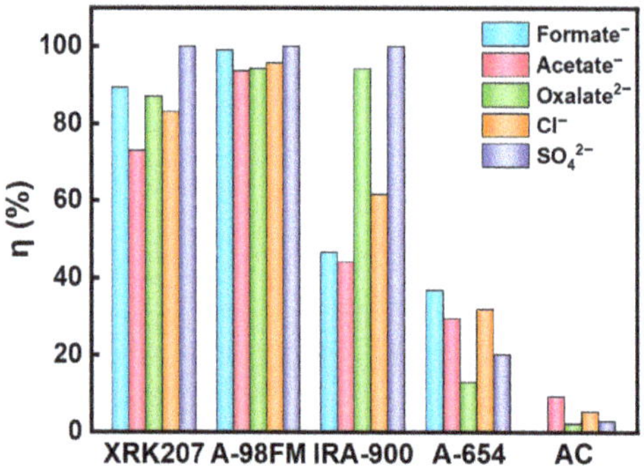

Figure 2. Removal efficiencies of typical anions by anion-exchange resins and coconut-shell activated carbon.

Overall, the anion-exchange resins showed substantially higher removal efficiencies for anionic impurities than coconut-shell activated carbon, although clear differences were also observed among the resins. Among the tested materials, A-98FM exhibited the best overall performance, with removal efficiencies of 99.00, 93.62, 94.22, and 95.61% for formate, acetate, oxalate, and chloride, respectively; the sulfate concentration decreased from 60.8 mg/L to below the detection limit (<0.02 mg/L). XRK207 also performed well, achieving removal efficiencies of 89.26, 72.80, 86.99, and 83.02% for formate, acetate, oxalate, and chloride, respectively, while likewise reducing sulfate to below the detection limit. IRA-900 showed relatively limited removal of formate and acetate, with efficiencies of 46.51 and 43.96%, respectively, while still maintaining a high removal efficiency for oxalate (94.18%). Its removal efficiency for chloride was 61.55%, and sulfate was again reduced to below the detection limit. In contrast, A-654 exhibited the poorest overall performance,

https://doi.org/10.3390/separations13050150

with removal efficiencies of 36.63, 29.38, 12.77, and 31.78% for formate, acetate, oxalate, and chloride, respectively, although sulfate could still be reduced to below the detection limit. These results indicate pronounced resin-dependent differences in removal performance and selectivity toward different anions under the present conditions. Overall, A-98FM showed the strongest comprehensive removal capability, followed by XRK207.

In contrast, the coconut-shell activated carbon showed no appreciable removal of the tested anionic impurities, as no significant decreases were observed in the concentrations of formate, acetate, oxalate, chloride, or sulfate after treatment. Taken together, these results indicate that effective control of ionic impurities in the present system relies primarily on anion-exchange resins.

3.3.2. Removal of Hydrocarbon Impurities by Activated Carbons

To evaluate the purification performance of different activated carbons toward hydrocarbon impurities in UDS solvent, comparative adsorption tests were conducted using 13 activated carbon samples. For these activated-carbon adsorption tests, the hydrocarbon-contaminated UDS feed was prepared with an initial total hydrocarbon content (C_0) of 200 mg/kg. The removal efficiencies calculated from the first 50 g effluent sample collected after prewetting had been displaced and the column flow had stabilized are shown in Figure 3. Because the particle size of AC-09 was too small, it caused an excessive pressure drop in the adsorption column and prevented stable solvent flow; therefore, AC-09 was excluded from the comparison. Overall, the purification performance varied markedly with carbon type and particle size, with removal efficiencies ranging from 8.92 to 74.64%.

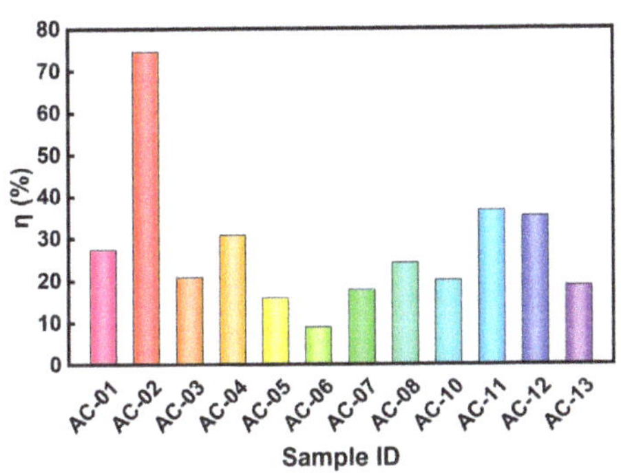

Figure 3. Removal efficiencies of total hydrocarbons by activated carbons.

Among the tested samples, the water-treatment activated carbons showed the best overall performance. AC-02 exhibited the highest removal efficiency, reaching 74.64%, whereas AC-01 from the same series but with a larger particle size showed a much lower removal efficiency of 27.35%, indicating a pronounced particle-size-related effect within this series. The solvent-recovery carbons (AC-03 and AC-04) showed intermediate performance, with removal efficiencies ranging from 20.78 to 30.72%. In the gold-refining series, AC-11 achieved a removal efficiency of 36.64%, higher than that of AC-10 (20.06%). The gas-purification carbons and AC-07 showed relatively limited performance, with most removal efficiencies below 36%; among them, AC-12 performed slightly better, with a removal efficiency of 35.25%. The cylindrical carbons used for desulfurization and denitrification were the least effective overall, especially AC-06, which showed a removal efficiency of only 8.92%.

Overall, under identical packing and operating conditions, samples with smaller particle sizes generally exhibited higher removal efficiencies, whereas larger granules and formed rod- or cylindrical-shaped carbons showed poorer purification performance. In

https://doi.org/10.3390/separations13050150

addition, activated carbons designed for different applications displayed substantially different compatibility with hydrocarbon removal from UDS solvent. The water-treatment series was better matched to the present system, whereas the desulfurization-denitrification carbons and some formed carbons were comparatively less suitable.

Taken together, the initial screening results provided the basis for selecting candidates for subsequent fixed-bed dynamic adsorption experiments. Considering both purification efficiency and sample representativeness, AC-02, AC-04, AC-07, AC-08, AC-11, AC-12, and AC-13 were selected for further continuous fixed-bed performance evaluation.

3.3.3. Dynamic Purification Performance of Selected Activated Carbons

To further evaluate the purification performance of the selected activated carbons toward hydrocarbon impurities in UDS solvent under continuous-flow fixed-bed conditions, dynamic adsorption experiments were conducted using AC-02, AC-04, AC-07, AC-08, AC-11, AC-12, and AC-13. The first 50 g effluent sample collected after prewetting had been displaced and the column flow had stabilized was defined as the 0 h data point, and subsequent 50 g effluent samples were collected at 1 h intervals.

The time-dependent removal efficiencies of the seven activated carbons are shown in Figure 4. Overall, the η_t–t profiles can be divided into a high-performance group (AC-02), a medium-performance group (AC-11 and AC-12), and a low-performance group (AC-04, AC-07, AC-08, and AC-13). These groups differed markedly in initial removal level, decay rate, and the extent of curve fluctuation. AC-02 maintained the highest removal efficiency throughout the experiment, with an initial η_t of 74.64% and a value of 38.06% at 12 h, indicating a relatively slow decay. AC-11 and AC-12 exhibited similar initial η_t values (36.64% and 35.25%, respectively), followed by a steady decline to 19.14% and 19.37% at 12 h; among them, AC-12 showed a slightly more gradual decrease in the middle and late stages. In contrast, the low-performance group showed generally low η_t values and more rapid decay over time. Specifically, AC-04 decreased rapidly from 30.72% to 2.69% at 12 h, indicating a rapid loss of dynamic purification capacity. AC-13 showed an approximately monotonic decrease from 18.77% to 5.79%. AC-07 and AC-08 exhibited some fluctuations at low removal levels, but their overall performance remained substantially lower than that of the medium-performance group.

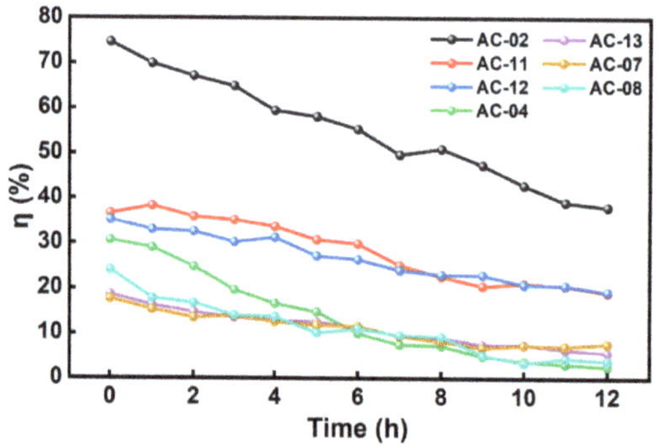

Figure 4. Dynamic purification efficiency curves of selected activated carbons during continuous fixed-bed purification of total hydrocarbons.

Although the dynamic purification efficiency curves in Figure 4 provide a direct comparison of different activated carbons, removal efficiency alone does not describe the mass-based adsorption performance. Therefore, the cumulative adsorption capacity over 0–12 h, q_{0-12} (mg/g), was calculated from the dynamic effluent concentration data. The

https://doi.org/10.3390/separations13050150

same hydrocarbon-contaminated UDS feed with C_0 = 200 mg/kg was used for all dynamic fixed-bed tests. For the hourly sampling data, $q_{0\text{–}12}$ was calculated by trapezoidal integration:

$$q_{0\text{–}12} = \frac{C_0 F_m}{m_{\text{ads}}} \sum_{i=0}^{n-1} \frac{\eta_i + \eta_{i+1}}{2} \Delta t \tag{3}$$

where C_0 is the initial total hydrocarbon content of the feed solution (mg/kg), F_m is the mass flow rate of the feed solution (kg/h), η_i is the removal efficiency (%) at the i-th sampling point, Δt is the sampling interval (h), and m_{ads} is the mass of activated carbon packed in the column (g). The calculation parameters are listed in Table S8.

To obtain comparable descriptors for the decay of dynamic purification efficiency, ηt was further fitted using a Weibull-type empirical model:

$$\eta_t = \eta_\infty + (\eta_0 - \eta_\infty) \exp\left[-\left(\frac{t}{\lambda}\right)^k\right] \tag{4}$$

where η_0 is the initial removal efficiency, η_∞ is the long-time limiting removal efficiency, and λ and k are the scale and shape parameters, respectively. Because the goal of this study is to characterize the decay of purification capability over operating time and to facilitate cross-sample comparison on the same basis, η_∞ = 0, and the model simplifies to:

$$\eta_t = \eta_0 \exp\left[-\left(\frac{t}{\lambda}\right)^k\right] \tag{5}$$

This Weibull-type equation was used as an empirical smoothing function rather than a mechanistic adsorption model. Based on the fitted parameters, characteristic times $t_{\eta=20\%}$ and $t_{\eta=10\%}$ were back-calculated from the fitted curves. These represent the time points when the removal efficiency decays to 20% and 10%, respectively, and were used to describe the decline in dynamic purification performance. Thus, $q_{0\text{–}12}$ was used to compare cumulative adsorption capacity, while $t_{\eta=20\%}$ and $t_{\eta=10\%}$ were used to compare the decay of dynamic purification efficiency. The corresponding results are summarized in Table 2.

Table 2. Dynamic purification performance and cumulative adsorption capacity of different activated carbon samples.

Sample	η_0 (%)	η_{12} (%)	$q_{0\text{–}12}$ (mg/g)	$t_{\eta=20\%}$ (h)	$t_{\eta=10\%}$ (h)	R^2
AC-02	74.64	38.06	14.86	22.82	33.90	0.988
AC-11	36.64	19.14	10.55	10.59	17.75	0.962
AC-12	35.25	19.37	8.68	11.23	23.62	0.978
AC-04	30.72	2.69	4.47	3.14	6.36	0.995
AC-13	18.77	5.79	4.14	—	6.66	0.985
AC-08	24.19	3.92	3.31	0.95	5.66	0.954
AC-07	17.75	7.78	2.91	—	6.59	0.950

Table 2 summarizes the dynamic purification performance and cumulative adsorption capacity of the activated carbon samples. AC-02 showed the best overall performance across multiple indicators, including η_0 (74.64%), η_{12} (38.06%), and $q_{0\text{–}12}$ (14.86 mg/g). Its threshold times, $t_{\eta=20\%}$ and $t_{\eta=10\%}$, reached 22.82 and 33.90 h, respectively, indicating the highest cumulative adsorption capacity and slower efficiency decay. AC-11 and AC-12 showed intermediate performance. Their $q_{0\text{–}12}$ values were 10.55 and 8.68 mg/g, respectively. AC-11 had a higher $q_{0\text{–}12}$, whereas AC-12 showed a longer $t_{\eta=10\%}$ (23.62 h), indicating that cumulative adsorption capacity and efficiency-decay behavior were not fully consistent. AC-04, AC-13, AC-08, and AC-07 showed lower dynamic purification

https://doi.org/10.3390/separations13050150

performance, with $q_{0\text{–}12}$ values of 4.47, 4.14, 3.31, and 2.91 mg/g, respectively. Their $t_{\eta=10\%}$ values were only 5.66–6.66 h, indicating rapid efficiency decay. The Weibull fits gave $R^2 > 0.950$ for all samples, supporting the use of the fitted curves to extract comparable decay-time parameters [47].

Overall, AC-02 exhibited the highest cumulative adsorption capacity and the most sustained dynamic purification performance. Therefore, it was selected for further characterization before and after adsorption.

3.4. Pore-Structure Characteristics and Their Relation to Dynamic Purification Performance

To investigate the structural basis for the differences in dynamic purification performance among the activated carbons during continuous fixed-bed operation, the pore-structure characteristics derived from N_2 adsorption–desorption measurements were compared with the dynamic performance metrics defined in Section 3.3.3. The pore descriptors considered in this analysis included BET specific surface area (S_{BET}), t-plot micropore volume (V_{micro}), total pore volume (V_{total}), external surface area (S_{ext}), and equivalent average pore diameter (D_{avg}). To further characterize mesopore-related transport features, the mesopore volume ($V_{meso,des}$) and modal pore diameter ($D_{mode,des}$) derived from the BJH desorption branch were also included as representative parameters [48]. The raw values of all descriptors are summarized in Table 3. For the construction of the heat map, the raw values of $q_{0\text{–}12}$ and the pore-structure descriptors were independently normalized to a 0–1 scale using min-max transformation, and the samples were arranged in descending order of $q_{0\text{–}12}$. Figure 5a presents the comparison between pore descriptors and dynamic purification performance. For consistency, the samples in Figure 5b–e are ordered in the same sequence.

Table 3. Summary of pore-structure parameters for different activated carbons.

Sample ID	S_{BET} (m^2/g)	V_{total} (cm^3/g)	V_{micro} (cm^3/g)	S_{ext} (m^2/g)	D_{avg} (nm)	$V_{meso,des}$ (cm^3/g)	$D_{mode,des}$ (nm)
AC-02	69.1	0.0485	0.021	18.9	2.78	0.0271	3.887
AC-11	1132.1	0.4701	0.4309	33.4	1.661	0.0682	1.901
AC-12	592.3	0.2738	0.2253	31.9	1.849	0.0746	3.935
AC-04	870.7	0.3569	0.3306	26.4	2.538	0.0478	1.869
AC-13	1032.5	0.4814	0.379	124.1	1.864	0.1773	1.867
AC-08	73.0	0.1036	0.013	41.9	6.624	0.091	3.863
AC-07	117.8	0.0525	0.0454	2.3	1.859	0.0091	1.859

As shown in Figure 5a, substantial differences in pore structure were observed among the activated carbons. Taking V_{micro} as an example, AC-11 exhibited the highest micropore volume (0.4309 cm^3/g), followed by AC-13 and AC-04 (0.3790 and 0.3306 cm^3/g, respectively). AC-12 showed an intermediate value (0.2253 cm^3/g), whereas AC-07, AC-02, and AC-08 exhibited relatively low V_{micro} values of 0.0454, 0.0210, and 0.0130 cm^3/g, respectively. S_{BET} generally followed a similar trend, exceeding 1000 m^2/g for AC-11 and AC-13 (1132.1 and 1032.5 m^2/g, respectively), while remaining only about 70 m^2/g for AC-02 and AC-08 (69.1 and 73.0 m^2/g, respectively). At the mesopore level, both BJH mesopore volume and modal pore diameter also varied among samples. For example, AC-13 showed a relatively high $V_{meso,des}$, whereas AC-12 and AC-08 exhibited comparatively larger $D_{mode,des}$ values. Overall, Figure 5a provides a comparative overview of the relative distributions of pore descriptors and dynamic purification performance across the selected samples.

https://doi.org/10.3390/separations13050150

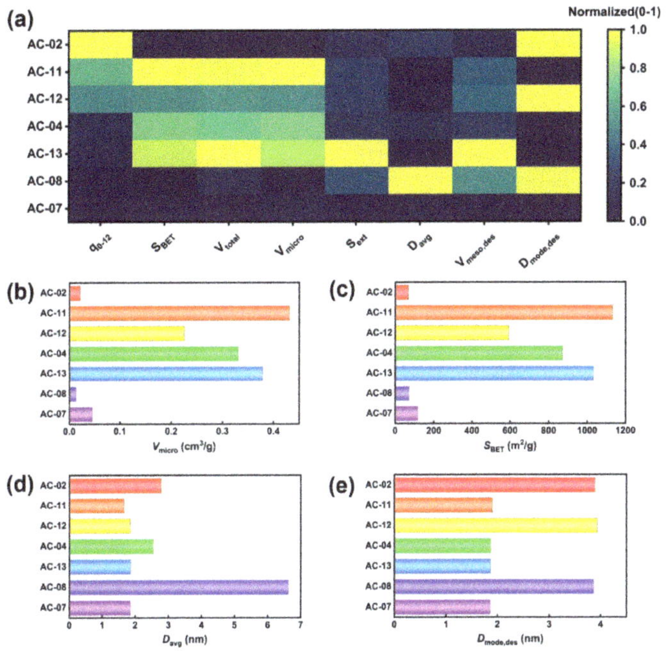

Figure 5. (**a**) Relationship between pore-structure descriptors and dynamic purification performance. (**b**) V_{micro}, (**c**) S_{BET}, (**d**) D_{avg} and (**e**) $D_{mode,des}$ of activated carbons ordered by q_{0-12}.

The comparison between pore structure and q_{0-12} shows a partial correspondence between micropore volume and cumulative adsorption capacity. AC-11 and AC-12 had relatively high q_{0-12} values and larger V_{micro}, whereas AC-08 and AC-07 showed both lower q_{0-12} values and lower micropore volumes. This trend suggests that micropores may provide part of the adsorption space for hydrocarbon impurities. However, the correspondence was not consistent across all samples. Some carbons with higher S_{BET}, V_{micro}, or $V_{meso,des}$ did not show correspondingly higher q_{0-12}. Thus, conventional pore-structure parameters are useful descriptors but are insufficient to fully account for fixed-bed hydrocarbon purification performance.

The differences among activated carbons should therefore be interpreted as the result of multiple factors. In fixed-bed operation, the effective use of adsorption sites may also be affected by pore accessibility, particle size, surface compatibility with the UDS solvent, and bed-level mass transfer. Accordingly, V_{micro} can be regarded as a relevant pore descriptor associated with hydrocarbon adsorption, but it should not be used alone to explain the dynamic purification performance.

3.5. Deactivation Behavior and Pore-Structure Evolution of Activated Carbon

3.5.1. Pore-Structure Evolution Before and After Fixed-Bed Adsorption

To examine the pore-structure evolution associated with the performance decay of AC-02 during continuous fixed-bed purification, the N_2 adsorption–desorption results of the sample before and after adsorption were compared. Changes in pore structure were characterized using S_{BET}, V_{micro}, V_{total}, S_{ext}, and D_{avg}. The corresponding results are summarized in Table 4, and the comparison of adsorption isotherms is shown in Figure 6.

https://doi.org/10.3390/separations13050150

Table 4. Changes in pore-structure parameters of AC-02 before and after adsorption.

Parameter	Unit	Fresh	Spent	Change (Spent vs. Fresh)
S_{BET}	m^2/g	69.1177	5.1016	−92.62%
V_{micro}	cm^3/g	0.021	0.0007	−96.67%
V_{total}	cm^3/g	0.0485	0.009	−81.44%
S_{ext}	m^2/g	18.8877	3.4796	−81.58%
D_{avg}	nm	2.8068	7.0566	151.41%
V_{micro}/V_{total}	—	0.433	0.0778	—

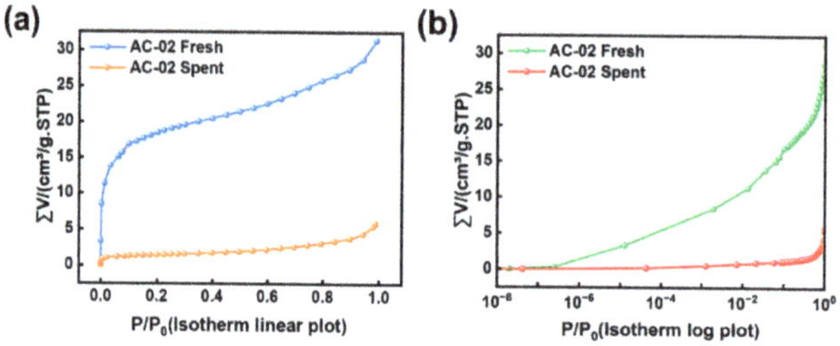

Figure 6. N_2 adsorption–desorption isotherms of AC-02 before and after fixed-bed adsorption. N_2 adsorption–desorption isotherms of AC-02 before and after fixed-bed adsorption: (**a**) linear plot; (**b**) log plot.

Before adsorption, AC-02 exhibited an S_{BET} of 69.12 m^2/g, a V_{micro} of 0.0210 cm^3/g, a V_{total} of 0.0485 cm^3/g, a S_{ext} of 18.89 m^2/g, and a D_{avg} of 2.81 nm. After fixed-bed adsorption, S_{BET}, V_{micro}, V_{total}, and S_{ext} decreased to 5.10 m^2/g, 0.0007 cm^3/g, 0.0090 cm^3/g, and 3.48 m^2/g, respectively, corresponding to decreases of 92.62, 96.67, 81.44, and 81.58%, whereas D_{avg} increased to 7.06 nm. These results indicate that the pore volume and surface area accessible for adsorption were substantially reduced after fixed-bed operation, with the most pronounced loss occurring in the micropore-related contribution. In addition, V_{micro}/V_{total} decreased from 0.433 to 0.078, indicating that part of the originally accessible micropore space became occupied or inaccessible after adsorption.

As shown in Figure 6, fresh AC-02 exhibited a pronounced rapid uptake in the low-pressure region ($P/P_0 < 10^{-4}$), which is characteristic of micropore filling. For example, the adsorption uptake had already reached 3.37 cm^3/g at $P/P_0 \approx 1.25 \times 10^{-5}$. The isotherm further turned upward at the high-pressure end ($P/P_0 \to 1$), reaching 31.44 cm^3/g, indicating that mesopores and/or macropores also contributed to adsorption. In contrast, after fixed-bed adsorption, the isotherm of AC-02 shifted markedly downward over the entire pressure range. The rapid uptake in the low-pressure region nearly disappeared, and the adsorption uptake at the medium-to-high-pressure end increased only to about 5.86 cm^3/g at $P/P_0 \approx 0.99$. This result indicates that the pore volume and accessible internal surface area, especially the micropore-related contribution, were greatly diminished after adsorption.

Taken together, the pore-structure parameters and adsorption isotherms show a substantial reduction in accessible pore volume and internal surface area after adsorption. The loss of low-pressure uptake further indicates that the micropore-related contribution was strongly reduced. This structural evolution is consistent with the decline in dynamic purification efficiency during fixed-bed operation and supports pore occupation as an important factor in the performance decay of AC-02.

https://doi.org/10.3390/separations13050150

3.5.2. Thermal-Release and Pyrolysis Evidence of Loaded Species on Spent Carbon

To further examine the species associated with the performance decay of AC-02 after fixed-bed operation and to elucidate the compositional characteristics of the organic loadings accumulated within the pore channels, the spent AC-02 sample was analyzed by thermogravimetry coupled with Fourier transform infrared spectroscopy and mass spectrometry (TG-FTIR-MS). Under programmed heating in a nitrogen atmosphere, this technique simultaneously records sample mass change and monitors the infrared bands and mass-spectral ion signals of the evolved gases, thereby revealing the stage-dependent characteristics of thermally released species from spent activated carbon over different temperature ranges, including both desorption and thermal cracking products. In this section, the main mass-loss intervals and the temperature ranges of the DTG rate peaks were first identified from the TG and DTG curves. The FTIR bands and MS channels were then correlated with these temperature intervals to analyze the evolved products in each stage.

As shown in Figure 7a, the spent AC-02 sample exhibited only a small mass loss in the low-temperature range of 40–150 °C, with no pronounced overall change. When the temperature increased to around 150 °C, continuous mass loss became evident and could be divided into three intervals: 150–350 °C, with a mass loss of about 4.87%; 350–550 °C, with a mass loss of about 9.26%, representing the major contribution over the entire run; and 550–800 °C, with a further gradual mass loss of about 6.38%, indicating that thermal release continued even in the high-temperature range. The corresponding DTG curve displayed a pronounced mass-loss-rate peak in the intermediate-temperature region, with the main peak centered at about 440 °C, indicating that this stage corresponded to the most concentrated release rate. In the high-temperature region, a weaker secondary rate peak was observed at about 570 °C. These stage-wise mass-loss features and DTG peak temperatures provide the basis for assigning the FTIR and MS signals to specific temperature intervals.

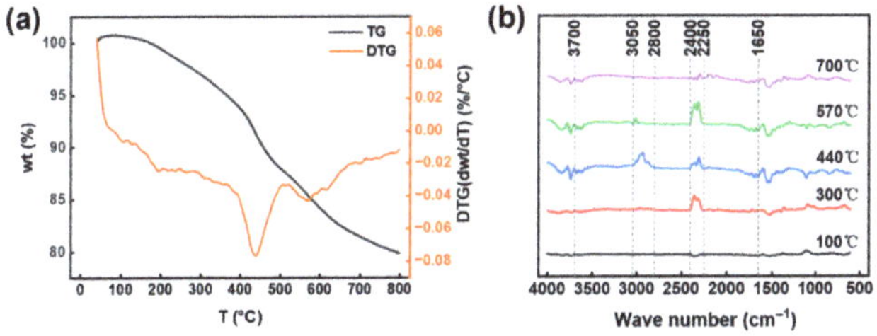

Figure 7. (**a**) TG and DTG curves of spent AC-02. (**b**) FTIR spectra of evolved gases at selected temperatures during TG-FTIR analysis of spent AC-02.

Accordingly, five representative temperature points, namely 100, 300, 440, 570, and 700 °C, were selected from the onset of mass loss, the vicinity of the main DTG peak, and the region near the secondary high-temperature peak. The corresponding gas-phase FTIR spectra obtained from the TG-FTIR measurement were extracted for comparison. The spectra were baseline-corrected and vertically offset to facilitate visualization of the relative changes in product bands across different temperature ranges (Figure 7b).

At 100 °C, the gas-phase FTIR spectrum was essentially featureless, with no obvious characteristic absorption bands, suggesting that only negligible amounts of volatile products were released at the initial heating stage. As the temperature increased to 300 °C, the water-related bands were still present but became weaker overall, whereas the characteristic CO_2 absorption band at 2400–2250 cm^{-1} began to increase significantly, indicating that the

https://doi.org/10.3390/separations13050150

thermal evolution of oxygen-containing structures had gradually started to contribute to gas-phase release.

When the temperature was further raised to 440 °C, the hydrocarbon C-H stretching bands in the 3050–2800 cm^{-1} region and the CH_3/CH_2 deformation bands at 1450–1470 cm^{-1} became much more pronounced and showed their strongest response in this temperature range. This observation suggests that this interval corresponds to the concentrated release of adsorbed hydrocarbon impurities and related organic residues through volatilization and/or cracking. Upon further heating, the characteristic CO_2 absorption band became more intense at 570 °C. In this intermediate-to-high-temperature range, the enhancement of CO_2 can be attributed to the removal and cracking of oxygen-containing structures in the adsorbed organic residues, possibly together with the thermal evolution of oxygen-containing functional groups on the activated-carbon surface. At 700 °C, the overall spectral bands weakened markedly and entered a tailing stage, with only weak signals remaining.

No NH_3 or amine-related characteristic absorption bands showing significant enhancement with increasing temperature were observed in the present spectra, indicating that volatile or cracking products associated with amines made only a limited contribution under the present test conditions and at the selected temperature points. Therefore, no clear evidence of substantial retention of active solvent components was observed in the spent activated carbon. Overall, the sample exhibited more pronounced alkyl-related spectral responses in the intermediate-temperature region, whereas the CO_2 characteristic band increased in the intermediate-to-high-temperature region and then entered a decaying tail at high temperature. These results indicate a stage-wise thermal-release behavior of the spent activated carbon, with intensified hydrocarbon-related release in the intermediate-temperature region, followed by enhanced release or transformation of oxygen-containing structures at higher temperatures.

To further identify the volatile and cracking products released from the spent sample during heating, MS ion-current signals of several representative m/z channels were tracked, and the curves were normalized by peak intensity to compare the peak profiles and the corresponding temperature ranges of different channels (Figure 8). In the low-m/z region, channels such as m/z = 2, 16, 18, and 44 served as reference signals for small molecules. Among them, m/z = 44 maintained a relatively pronounced response in the intermediate-to-high-temperature region, consistent with the enhancement trend of the CO_2 band observed in TG-FTIR, indicating the release of oxygen-containing small molecules during this stage. Meanwhile, hydrocarbon-related fragment channels showed more concentrated peak profiles in the intermediate-temperature region: the ion currents of m/z = 41, 42, 55, 57, and 71 increased synchronously in the range of about 400–500 °C and exhibited closely aligned main peaks around 440 °C (Figure 8b). These m/z channels are typical alkyl fragment-ion indicators commonly observed during thermal cracking of aliphatic hydrocarbons, and they are consistent with the enhanced alkyl C-H stretching and CH_3/CH_2 deformation bands observed by TG-FTIR in the same temperature range. Taken together, the major organic loadings released from the spent sample in the intermediate-temperature stage are more characteristic of aliphatic hydrocarbons and their related cracking fragments, which is consistent with the purification target of externally introduced hydrocarbon impurities during fixed-bed operation.

In addition to hydrocarbon fragments, several channels displayed weak but distinguishable synchronous peaks during the same thermal-release stage, which can be used to screen for the co-release of low-abundance solvent-related species (Figure 8c). Among them, m/z = 58 and 115 may be regarded as screening channels for relatively polar organic fragments, while m/z = 64 and 91 may serve as screening indicators for sulfur-containing species and aromatic fragments. Their responses appearing near 440 °C, coincident with the main release stage, suggest that, in addition to hydrocarbons, the system may also contain small amounts of polar solvent-derived components or derivative fragments released simultaneously, possibly associated with polar residues and degradation products from the solvent system.

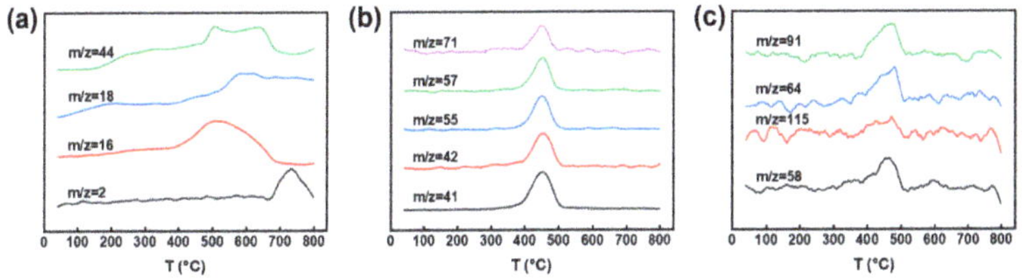

Figure 8. Temperature-resolved MS ion-current profiles of selected m/z channels during TG-MS analysis of spent AC-02: (**a**) m/z = 2, 16, 18, and 44; (**b**) m/z = 41, 42, 55, 57, and 71; (**c**) m/z = 58, 64, 91, and 115.

However, because the overall intensities of these channels were much lower than those of the hydrocarbon fragment channels, these non-hydrocarbon signals were assigned to minor co-released species under the present test conditions. The TG-MS results, therefore, indicate that hydrocarbon-related fragments were the main organic signals detected from spent AC-02 during thermal release. Combined with the pore-structure loss discussed in Section 3.5.1, these results support that hydrocarbon-related loadings occupied accessible pore space and were an important factor associated with the decay of dynamic purification performance of AC-02. At the same time, the weak amine-related signals suggest that substantial adsorption of active solvent components was not evident under the present conditions.

3.6. Practical Applicability in Field-Aged UDS Lean Solvent

To further examine the applicability of the selected adsorbents in a real solvent matrix, a field-aged UDS lean solvent was treated with AC-02 and A-98FM, respectively. The results are summarized in Table 5. The original field-aged lean solvent contained 424.03 mg/kg total hydrocarbons and 4288.8 mg/L total anionic impurities. Its foam height and defoaming time were 230 mm and 45.63 s, respectively.

Table 5. Treatment of a field-aged UDS lean solvent by selected adsorbents.

Sample	Total Hydrocarbons (mg/kg)	Total Anionic Impurities (mg/L)	Foam Height (mm)	Defoaming Time (s)
Field-aged lean solvent	424.03	4288.8	230	45.63
After AC-02 treatment	235.83	—	185	32.45
After A-98FM treatment	—	165.07 *	135	26.76

* Detailed concentrations of individual anions are provided in Table S9.

After AC-02 treatment, the total hydrocarbon content decreased to 235.83 mg/kg. The foam height decreased to 185 mm, and the defoaming time decreased to 32.45 s. After A-98FM treatment, the total anionic impurity concentration decreased to 165.07 mg/L. The foam height and defoaming time decreased to 135 mm and 26.76 s. These results indicate that AC-02 and A-98FM showed their respective removal effects in the field-aged UDS lean solvent. The reductions in hydrocarbon and ionic impurities were also accompanied by lower foam height and shorter defoaming time.

https://doi.org/10.3390/separations13050150

In field operation, the resin and activated carbon can be managed according to their respective target impurities. The anion-exchange resin is periodically regenerated with NaOH solution to restore its exchange capacity, whereas activated carbon used for hydrocarbon removal is usually replaced at regular intervals. The regeneration or replacement interval can be adjusted according to the changes in target impurity concentrations before and after treatment, together with the required solvent purification level. Spent carbon and spent alkaline regenerant should be handled according to the waste-management requirements of the desulfurization unit.

4. Conclusions

In this study, an extraction-GC method suitable for the UDS solvent matrix was established for the quantitative determination of total hydrocarbons, and the effects of representative impurities on the foaming behavior of UDS solvent were evaluated under controlled contamination conditions. The results demonstrated that hydrocarbon impurities showed the strongest effect on foaming behavior, while SO_4^{2-} and Cl^- also promoted foam formation and prolonged defoaming time, whereas organic acid anions showed relatively limited effects. For impurity control, A-98FM exhibited the best overall performance for anionic impurity removal, and AC-02 showed the best continuous fixed-bed performance for hydrocarbon purification. Pore-structure and deactivation analyses further suggested that hydrocarbon-related loadings occupied accessible pore space, especially micropore-related regions, and were an important factor associated with the decline in purification performance. Treatment of a field-aged UDS lean solvent further showed that AC-02 reduced total hydrocarbon content, whereas A-98FM reduced total anionic impurity concentration; both treatments were accompanied by lower foam height and shorter defoaming time. The present study provides experimental support for impurity monitoring, foaming-risk identification, and adsorptive purification of UDS desulfurization solvent under flowback-contamination conditions.

Supplementary Materials: The following supporting information can be downloaded at: https://www.mdpi.com/article/10.3390/separations13050150/s1, Figure S1: (a) ^{1}H NMR spectrum of the concentrated extract from produced water, (b) GC-MS of the concentrated extract from produced water; Figure S2: (a) GC chromatogram of the standard oil, (b) external calibration curve for total hydrocarbon quantification; Figure S3: Representative GC-FID chromatograms of (a) fresh-UDS blank extract and (b) 1000 mg/kg matrix-spiked UDS extract; Table S1: Basic properties of the 50 wt% fresh UDS working solution at 40 °C; Table S2: Basic properties of activated carbon samples; Table S3: Detailed ion chromatography conditions; Table S4: Composition analysis of major hydrocarbon impurities in produced water; Table S5: Composition of the standard oil; Table S6: Replicate determination results and method validation metrics; Table S7: Ion concentrations of UDS solvent before and after adsorbent treatment; Table S8: Packed masses of activated carbons used for q_{0-12} calculation; Table S9: Concentrations of individual anions in field-aged UDS lean solvent before and after A-98FM treatment.

Author Contributions: Conceptualization, H.S.; methodology, H.W.; software, H.W. and Q.Z.; validation, C.L. and H.S.; formal analysis, H.W., Q.Z., Y.W., Z.J. and Y.C.; investigation, H.W., Z.J. and M.X.; resources, H.S.; data curation, H.W., Q.Z., Z.J. and M.X.; writing—original draft preparation, H.W. and Q.Z.; writing—review and editing, H.W., Q.Z., Y.W., Z.J., Y.C., M.X., C.L. and H.S.; visualization, H.W. and Q.Z.; supervision, C.L. and H.S.; project administration, H.S.; funding acquisition, H.S. All authors have read and agreed to the published version of the manuscript.

Funding: This research was funded by the National Natural Science Foundation of China (Grant numbers 22178109 and 21878097) and the Natural Science Foundation of Shanghai (Grant number 21ZR1417700).

Data Availability Statement: The original contributions presented in this study are included in the article/Supplementary Materials. Further inquiries can be directed to the corresponding authors.

Conflicts of Interest: The authors declare no conflicts of interest.

References

1. Zahid, U.; Sakheta, A.; Lee, C.-J. Techno-Economic Analysis of Acid Gas Removal from Associated and Non-Associated Sour Gas Using Amine Blend. *Int. J. Greenh. Gas Control* **2020**, *98*, 103078. [CrossRef]
2. Peters, L.; Hussain, A.; Follmann, M.; Melin, T.; Hägg, M.-B. CO_2 Removal from Natural Gas by Employing Amine Absorption and Membrane Technology—A Technical and Economical Analysis. *Chem. Eng. J.* **2011**, *172*, 952–960. [CrossRef]
3. George, G.; Bhoria, N.; AlHallaq, S.; Abdala, A.; Mittal, V. Polymer Membranes for Acid Gas Removal from Natural Gas. *Sep. Purif. Technol.* **2016**, *158*, 333–356. [CrossRef]
4. Haider, J.; Saeed, S.; Qyyum, M.A.; Kazmi, B.; Ahmad, R.; Muhammad, A.; Lee, M. Simultaneous Capture of Acid Gases from Natural Gas Adopting Ionic Liquids: Challenges, Recent Developments, and Prospects. *Renew. Sustain. Energy Rev.* **2020**, *123*, 109771. [CrossRef]
5. Jassim, M.S. Sensitivity Analyses and Optimization of a Gas Sweetening Plant for Hydrogen Sulfide and Carbon Dioxide Capture Using Methyldiethanolamine Solutions. *J. Nat. Gas Sci. Eng.* **2016**, *36*, 175–183. [CrossRef]
6. Verma, N.; Verma, A. Amine System Problems Arising from Heat Stable Salts and Solutions to Improve System Performance. *Fuel Process. Technol.* **2009**, *90*, 483–489. [CrossRef]
7. Pal, P.; AbuKashabeh, A.; Al-Asheh, S.; Banat, F. Role of Aqueous Methyldiethanolamine (MDEA) as Solvent in Natural Gas Sweetening Unit and Process Contaminants with Probable Reaction Pathway. *J. Nat. Gas Sci. Eng.* **2015**, *24*, 124–131. [CrossRef]
8. Alhseinat, E.; Pal, P.; Ganesan, A.; Banat, F. Effect of MDEA Degradation Products on Foaming Behavior and Physical Properties of Aqueous MDEA Solutions. *Int. J. Greenh. Gas Control* **2015**, *37*, 280–286. [CrossRef]
9. Alhseinat, E.; Pal, P.; Keewan, M.; Banat, F. Foaming Study Combined with Physical Characterization of Aqueous MDEA Gas Sweetening Solutions. *J. Nat. Gas Sci. Eng.* **2014**, *17*, 49–57. [CrossRef]
10. Abukashabeh, A.S.; Alhseinat, E.; Al-Asheh, S.; Banat, F. Thermophysical Properties of Fresh and Lean Thermally Degraded N-Methyldiethanolamine. *J. Nat. Gas Sci. Eng.* **2014**, *19*, 317–323. [CrossRef]
11. Shahouni, R.; Abbasi, M.; Akrami, M. A Comprehensive Review on Control and Mitigating the Foam Formation in Amine-Based Solvents. *Chem. Eng. J. Adv.* **2024**, *19*, 100618. [CrossRef]
12. Chen, Q.; Peng, X.; Xue, J.; Li, J.; Pan, C. Identification and Quantification of Organic Contaminants and Evaluation of Their Effects on Amine Foaming in the Natural Gas Sweetening Industry. *ACS Omega* **2022**, *7*, 46421–46427. [CrossRef]
13. Pal, P.; AbuKashabeh, A.; Al-Asheh, S.; Banat, F. Accumulation of Heat Stable Salts and Degraded Products during Thermal Degradation of Aqueous Methyldiethanolamine (MDEA) Using Microwave Digester and High Pressure Reactor. *J. Nat. Gas Sci. Eng.* **2014**, *21*, 1043–1047. [CrossRef]
14. Pal, P.; Banat, F. Comparison of Thermal Degradation between Fresh and Industrial Aqueous Methyldiethanolamine with Continuous Injection of H_2S/CO_2 in High Pressure Reactor. *J. Nat. Gas Sci. Eng.* **2016**, *29*, 479–487. [CrossRef]
15. Gouedard, C.; Picq, D.; Launay, F.; Carrette, P.-L. Amine Degradation in CO_2 Capture. I. A Review. *Int. J. Greenh. Gas Control* **2012**, *10*, 244–270. [CrossRef]
16. Rochelle, G.T. Thermal Degradation of Amines for CO_2 Capture. *Curr. Opin. Chem. Eng.* **2012**, *1*, 183–190. [CrossRef]
17. Ke, Y.; Shen, B.; Sun, H.; Liu, J.; Xu, X. Study on Foaming of Formulated Solvent UDS and Improving Foaming Control in Acid Natural Gas Sweetening Process. *J. Nat. Gas Sci. Eng.* **2016**, *28*, 271–279. [CrossRef]
18. Alhseinat, E.; Keewan, M.; Banat, F. Impact of Dissolved and Undissolved Organics on Foaming of Industrial Amine. *Int. J. Greenh. Gas Control* **2017**, *60*, 156–161. [CrossRef]
19. Alhseinat, E.; Amr, M.; Jumah, R.; Banat, F. Removal of MDEA Foam Creators Using Foam Fractionation: Parametric Study Coupled with Foam Characterization. *J. Nat. Gas Sci. Eng.* **2015**, *26*, 502–509. [CrossRef]
20. Edalatpour, A.; Abbasi, M.; Riahi, S.; Sanaye Mozaffari Sabet, N.; Tavakoli, O. Investigation of Foaming Tendency of Aqueous Mixture of MDEA+IPAE for Carbon Dioxide Absorption. *J. CO_2 Util.* **2022**, *62*, 102079. [CrossRef]
21. Dumée, L.; Scholes, C.; Stevens, G.; Kentish, S. Purification of Aqueous Amine Solvents Used in Post Combustion CO_2 Capture: A Review. *Int. J. Greenh. Gas Control* **2012**, *10*, 443–455. [CrossRef]
22. Wang, T.; Hovland, J.; Jens, K.J. Amine Reclaiming Technologies in Post-Combustion Carbon Dioxide Capture. *J. Environ. Sci.* **2015**, *27*, 276–289. [CrossRef] [PubMed]
23. Garg, B.; Verheyen, T.V.; Pearson, P.; Feron, P.; Cousins, A. A Technology Review for Regeneration of Sulfur Rich Amine Systems. *Int. J. Greenh. Gas Control* **2018**, *75*, 243–253. [CrossRef]
24. Tavan, Y.; Moradi, M.; Rostami, A.; Azizpour, H. Theoretical and Industrial Aspects of Amine Reclaiming Unit to Separate Heat Stable Salts. *Sep. Purif. Technol.* **2020**, *237*, 116314. [CrossRef]

https://doi.org/10.3390/separations13050150

25. Pal, P.; Banat, F.; AlShoaibi, A. Adsorptive Removal of Heat Stable Salt Anions from Industrial Lean Amine Solvent Using Anion Exchange Resins from Gas Sweetening Unit. *J. Nat. Gas Sci. Eng.* **2013**, *15*, 14–21. [CrossRef]
26. Meng, H.; Zhang, S.; Li, C.; Li, L. Removal of Heat Stable Salts from Aqueous Solutions of N-Methyldiethanolamine Using a Specially Designed Three-Compartment Configuration Electrodialyzer. *J. Membr. Sci.* **2008**, *322*, 436–440. [CrossRef]
27. Zhou, Z.; Lin, Y.; Jin, Y.; Guan, K.; Matsuyama, H.; Yu, J. Removal of Heat-Stable Salts from Lean Amine Solution Using Bipolar Membrane Electrodialysis. *J. Membr. Sci.* **2022**, *645*, 120213. [CrossRef]
28. Wang, Y.; Fang, M.; Huang, Y.; Wang, T.; Li, S.; Lu, X.; Xie, Y. Removal of Heat Stable Salts from Degraded Amine Solvent by "BMED+ED" Two-Stage Electrodialysis Unit. *Int. J. Greenh. Gas Control* **2023**, *127*, 103939. [CrossRef]
29. Ghorbani, A.; Bayati, B.; Poerio, T.; Argurio, P.; Kikhavani, T.; Namdari, M.; Ferreira, L.M. Application of NF Polymeric Membranes for Removal of Multicomponent Heat-Stable Salts (HSS) Ions from Methyl Diethanolamine (MDEA) Solutions. *Molecules* **2020**, *25*, 4911. [CrossRef]
30. Edathil, A.A.; Alhseinat, E.; Banat, F. Removal of Heat Stable Salts from Industrial Lean Methyldiethanolamine Using Magnetic Alginate/Iron Oxide Hydrogel Composite. *Int. J. Greenh. Gas Control* **2019**, *83*, 117–127. [CrossRef]
31. Chen, F.; Chi, Y.; Zhang, M.; Yang, K.; Fu, C. Removal of Heat Stable Salts from N-Methyldiethanolamine Wastewater Using Electrodialysis: A Pilot-Scale Study. *Desalin. Water Treat.* **2020**, *196*, 48–57. [CrossRef]
32. Aleghafouri, A.; Hasanzadeh, N.; Mahdyarfar, M.; SeifKordi, A.; Mahdavi, S.M.; Zoghi, A.T. Experimental and Theoretical Study on BTEX Removal from Aqueous Solution of Diethanolamine Using Activated Carbon Adsorption. *J. Nat. Gas Sci. Eng.* **2015**, *22*, 618–624. [CrossRef]
33. Zhang, J.H.; Shen, B.X.; Sun, H.; Liu, J.C. A Study on the Desulfurization Performance of Solvent UDS for Purifying High Sour Natural Gas. *Pet. Sci. Technol.* **2011**, *29*, 48–58. [CrossRef]
34. Zhang, J.H.; Shen, B.X.; Liu, J.C.; Sun, H.; Hu, W.J. Absorption Selectivities of Solvents for Organosulfurs in High Sour Natural Gas. *Energy Sources Part A* **2014**, *36*, 822–829. [CrossRef]
35. Zhang, F.; Shen, B.; Sun, H.; Liu, J.; Liu, L. Rational Formulation Design and Commercial Application of a New Hybrid Solvent for Selectively Removing H_2 S and Organosulfurs from Sour Natural Gas. *Energy Fuels* **2016**, *30*, 12–19. [CrossRef]
36. Liu, L.; Shen, B.; Sun, H.; Xu, S. Kinetics Study of CO_2 Absorption into Solutions of UDS. *Pet. Sci. Technol.* **2012**, *30*, 2563–2570. [CrossRef]
37. Zhang, F.; Shen, B.-X.; Sun, H.; Liu, J.-C. Removal of Organosulfurs from Liquefied Petroleum Gas in a Fiber Film Contactor Using a New Formulated Solvent. *Fuel Process. Technol.* **2015**, *140*, 76–81. [CrossRef]
38. Adeniji, A.O.; Okoh, O.O.; Okoh, A.I. Analytical Methods for the Determination of the Distribution of Total Petroleum Hydrocarbons in the Water and Sediment of Aquatic Systems: A Review. *J. Chem.* **2017**, *2017*, 5178937. [CrossRef]
39. Thommes, M.; Kaneko, K.; Neimark, A.V.; Olivier, J.P.; Rodriguez-Reinoso, F.; Rouquerol, J.; Sing, K.S.W. Physisorption of Gases, with Special Reference to the Evaluation of Surface Area and Pore Size Distribution (IUPAC Technical Report). *Pure Appl. Chem.* **2015**, *87*, 1051–1069. [CrossRef]
40. Dittmann, D.; Eisentraut, P.; Goedecke, C.; Wiesner, Y.; Jekel, M.; Ruhl, A.S.; Braun, U. Specific Adsorption Sites and Conditions Derived by Thermal Decomposition of Activated Carbons and Adsorbed Carbamazepine. *Sci. Rep.* **2020**, *10*, 6695. [CrossRef]
41. Minofar, B.; Jungwirth, P.; Das, M.R.; Kunz, W.; Mahiuddin, S. Propensity of Formate, Acetate, Benzoate, and Phenolate for the Aqueous Solution/Vapor Interface: Surface Tension Measurements and Molecular Dynamics Simulations. *J. Phys. Chem. C* **2007**, *111*, 8242–8247. [CrossRef]
42. Sett, S.; Karakashev, S.I.; Smoukov, S.K.; Yarin, A.L. Ion-Specific Effects in Foams. *Adv. Colloid Interface Sci.* **2015**, *225*, 98–113. [CrossRef] [PubMed]
43. Schelero, N.; von Klitzing, R. Ion Specific Effects in Foam Films. *Curr. Opin. Colloid Interface Sci.* **2015**, *20*, 124–129. [CrossRef]
44. Salonen, A.; Lhermerout, R.; Rio, E.; Langevin, D.; Saint-Jalmes, A. Dual Gas and Oil Dispersions in Water: Production and Stability of Foamulsion. *Soft Matter* **2012**, *8*, 699–706. [CrossRef]
45. Khormali, A. Increasing the Efficiency of Demulsification Treatment in Petroleum Industry Using a Multicomponent Demulsifier Package. *Pet. Chem.* **2023**, *63*, 539–552. [CrossRef]
46. Khormali, A. Using a New Mixture of Reagents for Effective Inhibition of Corrosion and Salt Precipitation in the Petroleum Industry. *J. Chem. Pet. Eng.* **2021**, *55*, 257–276. [CrossRef]
47. Chu, K.H. Fixed Bed Adsorption of Chromium and the Weibull Function. *J. Hazard. Mater. Lett.* **2021**, *2*, 100022. [CrossRef]
48. Zhang, Z.; Jiang, C.; Li, D.; Lei, Y.; Yao, H.; Zhou, G.; Wang, K.; Rao, Y.; Liu, W.; Xu, C.; et al. Micro-Mesoporous Activated Carbon Simultaneously Possessing Large Surface Area and Ultra-High Pore Volume for Efficiently Adsorbing Various VOCs. *Carbon* **2020**, *170*, 567–579. [CrossRef]

https://doi.org/10.3390/separations13050150

molecules

MDPI

Article

Caffeine Biotransformation in Zebrafish Larvae: Integrated LC-MS/MS Quantification and FTIR Spatial Mapping

Adrian Bartoszek *, Anna Kozub-Pędrak, Alicja Wielgosz, Agata Sumara, Kacper Przykaza, Grzegorz Kalisz and Anna Stachniuk

Department of Bioanalytics, Medical University of Lublin, Kazimierza Jaczewskiego St. 8b, 20090 Lublin, Poland; 51112@umlub.edu.pl (A.K.-P.); alicja.wielgosz@umlub.edu.pl (A.W.); agata.sumara@umlub.edu.pl (A.S.); kacper.przykaza@umlub.edu.pl (K.P.); anna.stachniuk@umlub.edu.pl (A.S.)

* Correspondence: adrian.bartoszek@umlub.edu.pl

Abstract

Caffeine (CAF) is one of the most widely consumed psychoactive substances worldwide. It is primarily metabolized to paraxanthine (PAR), theobromine (THR), and theophylline (THY). While CAF metabolism has been extensively characterized in humans and rodent models, corresponding data in zebrafish (*Danio rerio*) larvae remain limited. This study provides a comprehensive characterization of CAF biotransformation in zebrafish larvae using validated LC-QQQ-MS for quantitative metabolite profiling and FTIR imaging for spatially resolved tissue distribution mapping. Zebrafish larvae at 4 days post fertilization (dpf) were exposed to CAF at concentrations of 15, 25, and 50 mg/L for 18 h. The LC–MS/MS analysis demonstrated dose-dependent conversion of CAF to PAR (up to 4.54%), THR (up to 1.32%), and THY (up to 0.27%). The PAR/CAF and THR/CAF ratios increased as CAF concentration rose, while the THY/CAF ratio declined. In parallel, FTIR imaging was applied to visualize the spatial distribution of CAF and selected metabolites in larval tissue sections, confirming the presence of CAF, PAR, and THR and revealing their localization within the tissue. However, THY was not detected by this method. Metabolite localization differed across larval tissues, indicating compartmentalized metabolic processes. This study provides the first comprehensive characterization of CAF metabolism in zebrafish larvae, revealing low but detectable CYP450-mediated metabolic activity prior to full hepatic maturation. The findings support the utility of zebrafish larvae as a model for developmental pharmacokinetic studies.

Keywords: caffeine; caffeine metabolism; zebrafish larvae; mass spectrometry; FTIR

1. Introduction

Caffeine (CAF) remains one of the most extensively ingested stimulants worldwide. This compound falls within the category of compounds known as purine alkaloids [1]. CAF is present in beverages commonly consumed daily, such as coffee and tea [2]. Various botanical sources contain CAF within their seeds, fruits, and foliage. Notable examples include cacao beans, found in chocolate; yerba mate leaves, used for herbal tea; and guarana berries incorporated into various beverages and supplements [3]. For centuries, coffee and tea have been integral to cultural customs and social interactions. Moreover, individuals utilize coffee to enhance alertness and productivity in their daily activities [3]. In North America, approximately 90% of the population consumes beverages containing CAF [4]. Furthermore, van Dam et al. (2020) highlight that although it exhibits numerous health benefits, current evidence does not justify recommending CAF or coffee consumption for disease prevention [3]. However, they suggest that moderate intake of coffee or tea can be incorporated into the lifestyle of adults without specific health conditions, excluding pregnant or lactating individuals [3]. On the contrary, CAF is associated with adverse effects related to excessive CAF consumption. These effects include anxiety, restlessness, nervousness, dysphoria, insomnia, excitement, psychomotor agitation, and rambling speech. It is important to note that these symptoms are not limited to these and can vary from person to person [3,5]. Moreover, it is noteworthy that CAF undergoes hepatic metabolism primarily via demethylation, a process facilitated predominantly by the cytochrome P450 enzyme family, notably CYP1A2, accounting for approximately 95% of the conversion. This enzymatic process leads to the production of CAF's primary metabolites, including PAR, THR, and THY, which undergo subsequent metabolism to uric acid and are ultimately excreted via urine [6]. Caffeine and its principal derivatives exert various effects on physiological functions. CAF itself enhances the basal metabolic rate and acts as a mild stimulant for the central nervous system, heart, and smooth muscles. Moreover, PAR, the most abundant and biologically active metabolite, acts as a central nervous system stimulant with comparable activity to CAF but tends to have lower toxicity and fewer anxiety-inducing effects. THR also exerts stimulating effects, although they are generally milder than CAF. Lastly, THY is a widely used bronchodilator with a limited therapeutic range in the bloodstream [7].

Zebrafish (*Danio rerio*) has become an established in vivo model in biomedical research, largely due to its genetic similarity to humans (~70%) and broad applicability across different experimental settings [8]. This model offers several advantages, including low maintenance requirements, rapid development, and fewer ethical constraints compared to mammalian systems. In accordance with the principles of the 3Rs (Replacement, Reduction, and Refinement), the use of zebrafish larvae allows for a reduction in animal usage while maintaining the relevance of biological observations [9]. At early developmental stages (4–5 days post fertilization (dpf)), zebrafish larvae already exhibit functional physiological processes, including absorption and metabolic activity, despite incomplete hepatic maturation. In addition, their small size and optical transparency, combined with the possibility of high-throughput experimental design, support their application in pharmacokinetic and toxicological studies [10].

As of now, no investigations have been conducted on the primary three CAF metabolites in zebrafish embryos, larvae, or adults. In fish studies, tested compounds are usually dissolved in the medium. The investigated dose is often described as the concentration in the medium rather than in larvae. The relationship between the external concentration of the exogenous compound in the medium and its internal exposure is crucial for accurately interpreting the observed results, as it is the internal concentration that drives pharmacological and toxicological effects. Ignoring this critical issue has been reported to result in unfavorable outcomes in drug research [11]. Therefore, research on CAF should involve quantifying drugs and their metabolites to interpret the data obtained accurately. Furthermore, assessing the internal concentration based on the compound's physicochemical properties has proven to be quite challenging [12]. Another significant challenge is the small size of the larvae, necessitating the use of highly sensitive methods. It should be emphasized that compounds' absorption and elimination depend on the larvae's age [13]. In zebrafish larvae, xenobiotics present in the exposure medium are primarily absorbed via passive diffusion through the skin and gills. Due to their small size and high surface-to-volume ratio, uptake of small molecules is generally efficient, although internal concentrations may differ from nominal exposure levels [13].

In recent years, the integration of multiple analytical techniques has become increasingly common in biochemical and metabolic research, providing a more comprehensive

https://doi.org/10.3390/molecules31071095

understanding of complex biological systems. Liquid chromatography–mass spectrometry (LC-MS) and Fourier-transform infrared spectroscopy (FTIR) are two complementary methods that have proven invaluable in such studies. LC-MS combines the separation power of LC with the molecular specificity of MS, enabling the simultaneous detection and quantification of multiple compounds with high sensitivity. This makes it a gold-standard method for metabolite profiling and pharmacokinetic studies in complex biological matrices [14]. FTIR spectroscopy, on the other hand, offers a rapid, non-invasive approach to identifying and analyzing functional groups within molecules based on their vibrational properties. When combined with imaging, FTIR allows for the spatial mapping of molecular distributions in tissues, offering valuable insights into structural and functional dynamics [15].

The aim of this study was to investigate CAF biotransformation in zebrafish larvae, with particular focus on its primary metabolites PAR, THR, and THY (Figure 1). Quantitative analysis was performed using LC-MS/MS, while FTIR spectroscopy was applied to confirm the presence of the compounds and provide spatial information on their distribution within larval tissues.

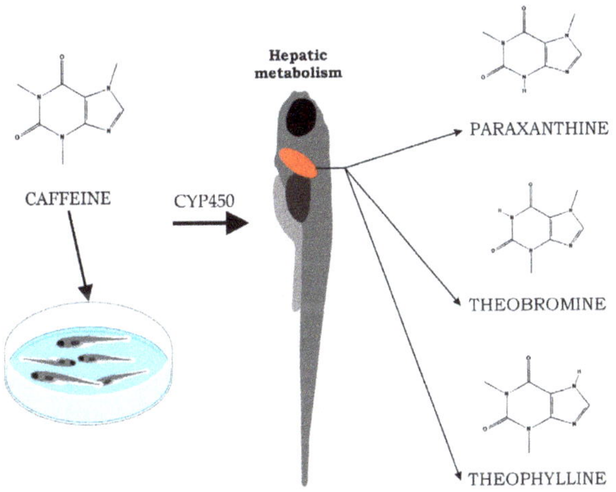

Figure 1. Schematic representation of the in vivo pathways of CAF metabolism in zebrafish larvae leading to the formation of the major metabolites PAR, THR and THY. Schematic drawings were obtained from Scidraw.io.

2. Results

2.1. *Quantification of CAF, PAR, THR and THY Levels in the Zebrafish Larvae*

The zebrafish larvae were treated with three different concentrations of CAF: 15 mg/L, 25 mg/L, and 50 mg/L. Firstly, the contents of CAF and its three metabolites were measured (Figure 2). CAF concentrations differed among all investigated solutions, measuring 504.7, 692.2 and 1444.7 nmol/L, respectively. The contents of each metabolite, namely PAR, THR, and THY, were significantly higher in the group administered 50 mg/L CAF compared to the groups receiving lower concentrations of 25 and 15 mg/L (65.3 vs. 18.4 and 12.3 nmol/L; 19 vs. 4.9 and 3.2 nmol/L; 2.8 vs. 1.6 and 1.4 nmol/L, respectively). However, there was no significant difference in metabolite concentration between the group administered 15 and 25 mg/L CAF (0.692 vs. 0.505 nmol/L; 0.018 vs. 0.012 nmol/L; 0.005 vs. 0.003 nmol/L, respectively) (Figure 2). Importantly, the absence of significant differences in metabolite concentrations between 15 and 25 mg/L groups, when contrasted with the marked increase at 50 mg/L, suggests a non-linear activation of CAF biotransformation pathways in zebrafish larvae.

https://doi.org/10.3390/molecules31071095

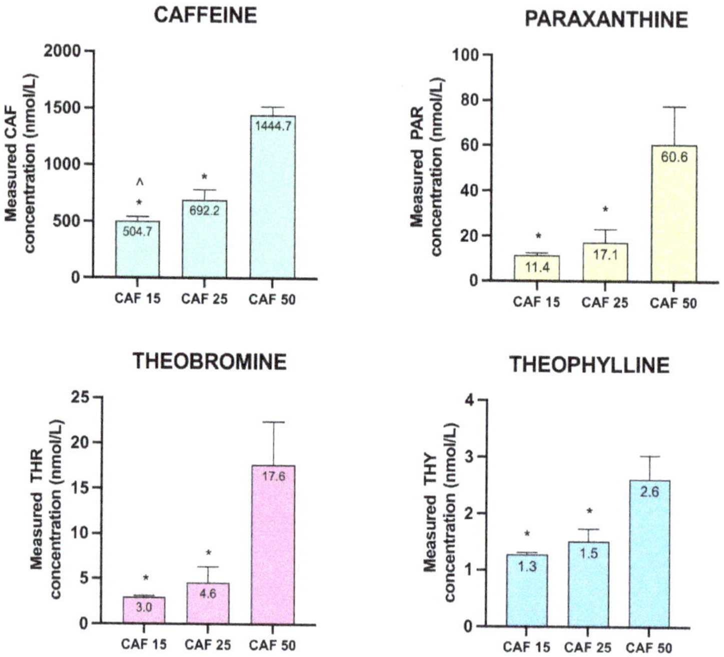

Figure 2. Contents of CAF, PAR, THR, and THY for larvae treated with CAF. CAF concentration applied at 15, 25, and 50 mg/L. * $p < 0.05$, as compared to CAF 50 mg/L group, ^ $p < 0.05$, as compared to CAF 25 mg/L group.

The LC-QQQ-MS method demonstrated high sensitivity and satisfactory precision for all four analytes. The limits of detection (LOD) and quantification (LOQ) were: CAF 5.358 and 17.859 nmol/L; PAR 0.430 and 1.434 nmol/L; THR 1.118 and 3.725 nmol/L; THY 0.389 and 1.298 nmol/L, respectively (Table 1). Precision, expressed as relative standard deviation (RSD, %), did not exceed 9.01% at the lowest and 4.26% at the highest calibration concentration, confirming the method's reproducibility across the working range. The low LOQ values for PAR, THR, and THY were critical for the reliable quantification of metabolites present at trace levels, especially THY, whose concentrations (1.4–2.8 nmol/L) were close to but consistently above the LOQ threshold.

Table 1. Multiple reaction monitoring parameters for caffeine and its primary metabolites, obtained using LC-QQQ-MS in positive electrospray ionization mode.

Analyte	Formula	RT (min)	RT Window (min)	Precursor Ion *m/z*	Product Ion *m/z*	Frag. (V)	CE (V)	LOD (nmol/L)	LOQ (nmol/L)
Caffeine	$C_8H_{10}N_4O_2$	5.29	0.5	195.1	110	136	25	5.358	17.859
					138	136	21		
					83	136	29		
					69.1	136	33		
					56.1	136	37		
Paraxanthine	$C_7H_8N_4O_2$	3.33	0.5	181.1	124	118	22	0.430	1.434
					96	118	26		
					69.1	118	38		
					67	118	42		
					55.1	118	34		

https://doi.org/10.3390/molecules31071095

Table 1. *Cont.*

Analyte	Formula	RT (min)	RT Window (min)	Precursor Ion *m/z*	Product Ion *m/z*	Frag. (V)	CE (V)	LOD (nmol/L)	LOQ (nmol/L)
Theobromine	$C_7H_8N_4O_2$	2.28	0.5	181.1	138	121	18	1.118	3.725
					110.1	121	22		
					83.1	121	30		
					67.1	121	38		
Theophylline	$C_7H_8N_4O_2$	3.68	0.5	181.1	124	103	22	0.389	1.298
					94	103	26		
					69.1	103	34		

LOD—limit of detection; LOQ—limit of quantification; RT—retention time; Frag—fragmentor; CE—collision energy.

2.2. Metabolites-to-CAF Ratio

Based on these results, we assessed the ratio of CAF metabolites to CAF (Figure 3).

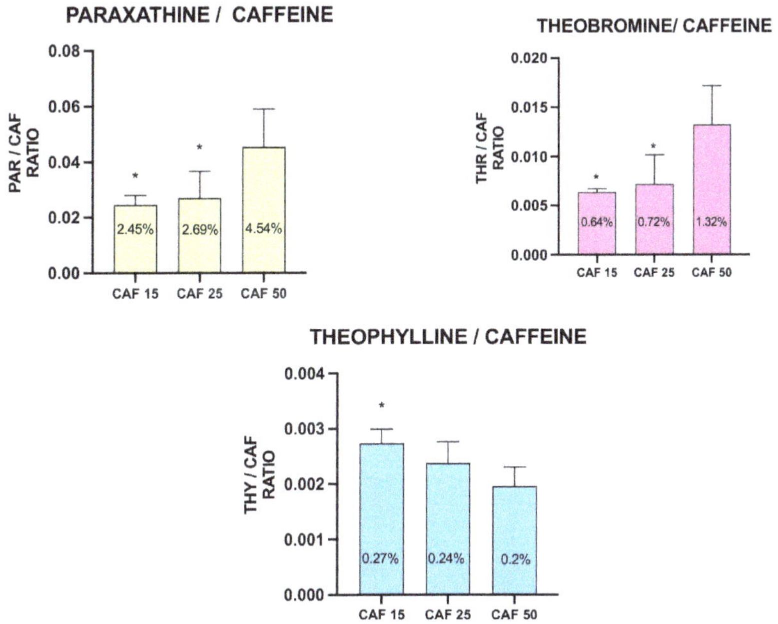

Figure 3. PAR/CAF, THR/CAF, and THY/CAF ratios for the larvae treated with CAF. CAF concentration applied: 15, 25, and 50 mg/L, * $p < 0.05$.

Similarly to PAR alone, the PAR/CAF ratio was significantly higher in the group incubated with 50 mg/L CAF compared to lower CAF doses and no difference between these lower concentrations was observed. The same relationship was observed for the THR/CAF ratio, but the ratio was lower compared to PAR. The opposite effect was observed for THY as the THY/CAF ratio decreased with higher CAF doses, and the difference was observed only between the 15 and 50 mg/L CAF groups (Figure 3). This divergent behavior indicates that THY formation may be driven by a distinct metabolic branch with limited inducibility or inhibition at higher levels of substrate.

2.3. CAF Metabolites Assessment

Then, we summarized the data to get a full insight into the CAF metabolism (Figure 4).

https://doi.org/10.3390/molecules31071095

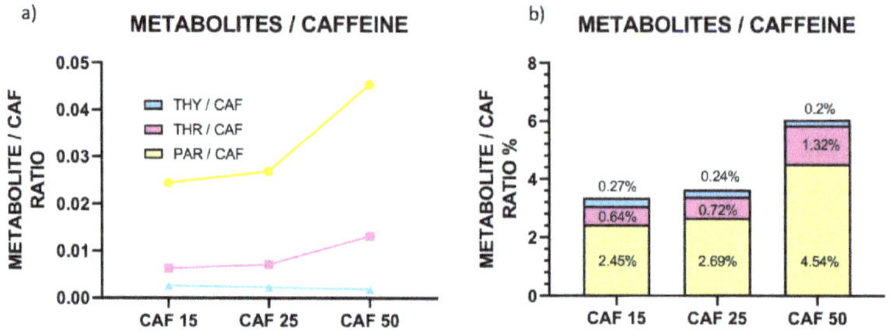

Figure 4. (**a**) raw values of PAR/CAF, THR/CAF, and THY/CAF ratios for the larvae treated with CAF (**b**) % of PAR/CAF, THR/CAF, and THY/CAF ratios. CAF concentration applied: 15, 25, and 50 mg/L.

As presented in Figure 4, the greater the CAF dosage, the more the drug is metabolized to PAR and THR. The most pronounced increase is observed for PAR. Inversely, the greater the CAF concentration, the less of the drug is transformed to THY (Figure 4).

Additionally, FTIR measurements were performed on metabolite standards of CAF, PAR, THR and THY. In parallel, control larvae spectra were recorded and averaged (background spectrum). Based on the evaluation of analytes and background spectrum, bands at 1072 and 1674 cm^{-1} were selected for CAF, 1120 and 1275 cm^{-1} for PAR, and 1266 and 1308 cm^{-1} for THR [16,17]. It was not possible to indicate suitable bands for THY. The mean values of absorbance for CAF, PAR, and THR (calculated as the sum of two bands) for the control and CAF 50 mg/L groups are presented in Table 2. Both CAF and its two major breakdown products were detected in the larvae's bodies; a statistically significant difference in band intensities for CAF, PAR, and THR was observed between groups (Figure 5).

Table 2. Mean absorbance intensity of peaks assigned to CAF, PAR, and THR, in 5 dpf zebrafish larvae ± SD, *** $p \leq 0.0001$.

Treatment	CAF	PAR	THR
CTR	0.25 ± 0.18 ***	0.11 ± 0.09 ***	0.12 ± 0.09 ***
CAF 50	0.65 ± 0.40 ***	0.36 ± 0.25 ***	0.41 ± 0.29 ***

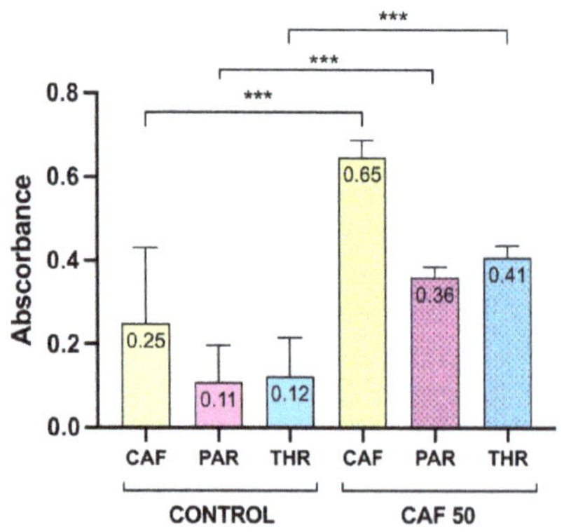

Figure 5. Comparison of CAF, PAR, and THR absorbance in control and CAF 50 groups; bands at 1072 and 1674 cm^{-1} were selected for CAF; 1120 and 1275 cm^{-1} for PAR and 1308 cm^{-1} for THR; *** $p \leq 0.0001$.

https://doi.org/10.3390/molecules31071095

FTIR imaging was employed as a complementary approach to LC-MS/MS to provide spatial information on the distribution of CAF and its metabolites within intact zebrafish larval tissue sections. Spatial distribution data were collected as a novel approach in zebrafish studies, as presented by Kalisz et al. [18]. Notably, FTIR analysis provided spatially resolved information revealing heterogeneous patterns of CAF and its metabolites in larval tissues. This observation suggested localized metabolism or metabolite accumulation, in contrast to the low and uniform signal observed in control larvae.

Following the spectra analysis, imaging of the head area was performed in control and CAF 50 mg/L groups, with the visualization of the abovementioned selected bands. To mitigate spectral overlap, the averaged FTIR spectrum of control larvae was used as a biological background reference and subtracted prior to map visualization, thereby removing the endogenous tissue contribution common to both groups. Presented in Figure 6, the maps show that in the control group, absorbance for all compounds is lower compared to treated larvae.

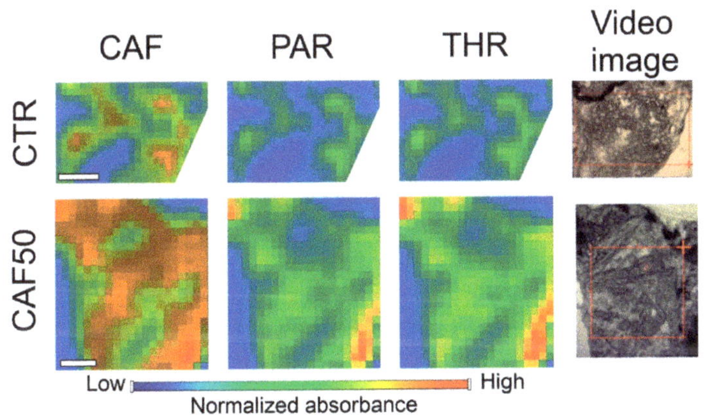

Figure 6. Spatial distribution of CAF and its metabolites (PAR and THR) in horizontal cross-sections of the head region mapped (red rectangles) of *Danio rerio* 5 dpf larvae. FTIR imaging in CTR and CAF50 samples, the white bar is 100 μm.

The absorbance patterns are not uniform across the treated samples, which points to localized or compartmentalized effects within the tissue (Figure 6).

3. Discussion

Zebrafish larvae are increasingly used in pharmacological and toxicological research due to their genetic similarity to humans and well-characterized developmental biology [19,20]. Nevertheless, the fate of CAF and its primary metabolites in this model has remained largely unexplored. Understanding CAF biotransformation in this organism is particularly relevant given that internal exposure determines the pharmacological and toxicological outcome. The present study addresses this gap by combining LC-QQQ-MS quantification with spatially resolved FTIR imaging to provide a systematic and tissue-level characterization of CAF biotransformation at the early larval stage.

Previous research indicates that the development of hepatic primordium in zebrafish starts at 28 h post fertilization (hpf), followed by hepatic outgrowth between 60 and 72 hpf. By 120 hpf, liver function, including CYP metabolism, is nearly fully established. Notably, evidence suggests that CYP metabolic functions, particularly CYP1A2, resemble those of human CYP isoforms even before complete liver development, as confirmed by both mRNA expression and metabolite analysis in larvae aged 24–120 hpf [21]. The metabolism of CAF shows considerable variability among individuals and is influenced by multiple factors, including age, gender, genetic predisposition, and environmental conditions [22]. The PAR to CAF ratio is a recognized biomarker reflecting CYP1A2 activity, with elevated ratios correlating with accelerated CAF metabolism and increased enzymatic activity of CYP1A2. Evaluating this ratio has the potential to predict individual responses to CAF and assess the propensity for adverse reactions associated with its consumption [23].

So far, only one study assessed the PAR/CAF ratio in zebrafish larvae [21]. Nawaji et al. found that only about 5% of CAF (10 mg/L) at 2 dpf and 5 dpf was transformed to PAR, suggesting that zebrafish embryos and larvae possess metabolic activity similar to that of human CYP1A2, although at a relatively low level [21]. In our study, the activity was 2.45–4.54%, depending on the dose. Our investigation revealed no statistically significant variance in the PAR/CAF ratio between larvae subjected to CAF concentrations of 15 and 25 mg/L. However, larvae treated with 50 mg/L CAF exhibited a notable elevation in the PAR/CAF ratio relative to the lower dosage cohorts. This dose-dependent increase indicates inducible or concentration-dependent activation of CYP-driven enzymatic pathways in zebrafish larvae. Nevertheless, the PAR/CAF ratio can be useful as a functional marker of metabolic activity rather than a fixed constant related to specific species.

Basnet et al. employed zebrafish embryos as a model system to investigate the overall toxicity and cardiovascular impacts of eight methylxanthines, including aminophylline, CAF, diprophylline, doxofylline, etophylline, 3-isobutyl-1-methylxanthine (IBMX), pentoxifylline, and THY [19]. They reported that CAF, IBMX, pentoxifylline, and THY induced pronounced embryotoxic and teratogenic effects. In contrast, aminophylline, doxofylline, and etophylline elicited such effects only at higher concentrations, whereas diprophylline displayed minimal (<10%) developmental toxicity. Structural abnormalities in the heart were induced by most of these drugs in 20–40% of the injected embryos at the highest dose, resulting in fatal outcomes within 120 hpf. Additionally, all drugs elicited a transient increase in heart rate at 48 hpf, which normalized by 96 hpf. This observed functional impact of methylxanthines parallels findings from studies conducted in humans and other vertebrates. Unfortunately, the referenced study did not delve into CAF metabolism; however, it did note the influence of the metabolite on the zebrafish organism [19]. In humans, CAF is reported to be transformed into PAR, THR and THY at rates of 84%, 12% and 4%, respectively [24]. In rats, CAF is demethylated to its three major metabolites in approximately equal quantities [25,26]. Our study presents that CAF in the zebrafish model is transformed into PAR, THR and THY at rates up to 4.54%, 1.32% and 0.27%, respectively. Zhao et al. conducted a study to perform a comparative transcriptome analysis of hepatocytes isolated from developing zebrafish embryos at various developmental stages (60, 72, and 96 hpf) [27]. This analysis unveiled notable alterations in gene expression profiles, elucidating pivotal biological processes and molecular pathways implicated in the proliferation and functional maturation of hepatocytes throughout liver development. These discoveries provide crucial insights into the intricate dynamics governing hepatocyte differentiation and maturation, thereby enriching our comprehension of liver development in zebrafish. In the present publication, we suggest that CAF metabolism may differ in the zebrafish model from that in the human body, which may also be important in the context of the mentioned molecular pathways. It is well known that CAF interacts with molecular cell pathways [28], which may be potentially significant for further use of the *Danio rerio* model in studies of the pathophysiology of diseases.

In this study, we also targeted CAF and metabolites using FTIR spectroscopy and spectral mapping. This technique is sensitive to changes in the secondary structure of proteins, as added chemical compounds may introduce it [29,30]. FTIR spectroscopy is a powerful tool for analyzing CAF and its metabolites. Recording their fingerprint vibrational modes

https://doi.org/10.3390/molecules31071095

enables both qualitative and semi-quantitative assessments without requiring extensive sample preparation. It provides a label-free approach in studies of molecular structures or metabolic pathways in cell biology [31]. Particularly, Ashengroph evaluated products of CAF biotransformation in an aqueous environment with *Salinvibrio costicola* [32]. Despite the fact that all metabolites were confirmed by LC-MS, only separated and purified THR was confirmed by FTIR. The important role of FTIR detection of THR and CAF was also postulated in the determination of food content, contamination [33], or crystallization [34]. The analysis of PAR and THY was less reported, possibly due to problems with low concentration and partially overlapping spectral features in metabolism studies. THY is also studied with FTIR as an active pharmaceutical ingredient [35]. Spectral overlap in the fingerprint region represents a general analytical challenge in biological FTIR imaging; proteins (Amide III, 1230–1300 cm^{-1}), nucleic acids (νas/νs PO_2^-, 1080–1240 cm^{-1}), carbohydrates (C–O–C stretching, 1030–1080 cm^{-1}), and lipids (C–O stretching, 1050–1100 cm^{-1}) all absorb in the same fingerprint region used for methylxanthine detection, which likely contributes to the limited reporting of PAR and THY in tissue-imaging studies. While FTIR has been successfully used in pharmaceutical and microbial systems, its application in higher model organisms, such as zebrafish larvae, remains underexplored. Spatially resolved data coupling zebrafish larvae and FTIR imaging is a novel approach, as only one report on the distribution of amides and lipids was reported in the available literature [18]. Our study addresses this gap by providing information on localized CAF biotransformation with visualization in larvae. This approach complements traditional methods such as liquid chromatography–tandem mass spectrometry (LC-QQQ-MS) and offers insights into molecular changes in models. Future studies could focus on improving the sensitivity for detecting low-abundance metabolites and the application of spatially resolved FTIR imaging in model organisms that could further validate its utility in metabolic studies, as a part of the "spectralomic" approach [31].

Comprehending CAF metabolism and the involvement of cytochrome P-450 is crucial for anticipating individual reactions to CAF and identifying variables that may impact CAF metabolism. This knowledge can inform strategies for the prudent and efficient utilization of CAF-containing products, particularly in individuals concurrently taking medications that interact with CAF. Furthermore, research indicates that a higher CAF metabolite ratio may be correlated with a reduced risk of specific health ailments, such as Parkinson's disease, whereas a lower ratio may be associated with heightened susceptibility to cardiovascular complications.

The present study focused on the three primary CAF metabolites known from mammalian systems (PAR, THR, and THY). Due to the targeted nature of the LC-MS/MS method applied, the analysis was limited to predefined compounds. However, the formation of additional minor metabolites in zebrafish larvae cannot be excluded and would require future studies for comprehensive characterization.

4. Materials and Methods

4.1. Animals

Wild-type zebrafish (*Danio rerio*) of the AB strain (Experimental Medicine Centre, Medical University of Lublin, Poland) were housed under controlled conditions at temperatures between 26 and 28.5 °C. Water pH was maintained in the range of 6.9 to 7.5, conductivity values were kept between 550 and 700 μS/cm, and a 14/10 h light-to-dark photoperiod was applied. Following fertilization, embryos were kept in E3 embryo medium and incubated under standard light/dark conditions using an IN 110 incubator (Memmert GmbH, Buechenbach, Germany). Larvae at 4 days post-fertilization (dpf) were selected for use in all experimental procedures. Upon completion of the experiments, larvae were euth-

https://doi.org/10.3390/molecules31071095

anized by immersion in ice-cold water at a ratio of 5 parts ice to 1 part water (0 to 4 °C) for 20 min. Subsequently, specimens were frozen at −22 °C to ensure cessation of brain activity. All procedures were carried out in full compliance with the National Institute of Health Guidelines for the Care and Use of Laboratory Animals. The European Community Council Directive 2010/63/EU of 22 September 2010 was also followed. As the study involved larvae at a developmental stage not exceeding 5 dpf, approval from the Local Ethical Commission was not required.

4.2. Chemicals

CAF, PAR, THR, and THY were obtained from Sigma-Aldrich (Saint Louis, MO, USA) at pharmaceutical secondary standards with a purity > 99% (LC-MS grade). Stock solutions were prepared by dissolving each compound in deionized water. The solutions were subsequently diluted in E3 embryo medium to reach the desired final concentrations. The E3 medium was composed of 17.4 μM NaCl, 0.21 μM KCl, 0.12 μM $MgSO_4$ and 0.18 μM $Ca(NO_3)_2$, with pH adjusted to a range of 7.1 to 7.3.

4.3. Treatments for the Assessment of CAF and Its Metabolites Concentrations

A total of five larvae per well were placed in a 48-well plate containing 500 μL of either E3 embryo medium or CAF at concentrations of 15, 25, and 50 mg/L. Preincubation was carried out at 28 °C for 18 h. The applied CAF concentrations and exposure duration were established based on the available literature and previous research conducted in a home laboratory using zebrafish larvae as a model organism [36–39].

4.4. Sample Preparation

Larvae were rinsed three times with E3 embryo medium and subsequently transferred to tubes at a density of 20 larvae per sample with 5 samples per experimental group. Tissue homogenization was performed by sonication (4 cycles of 5 s each) in 150 μL of 100 mM NH_4HCO_3. The homogenate was then kept on ice for 15 min and centrifuged at 15,000 *g* for 15 min at 4 °C. A volume of 100 μL of the resulting supernatant was combined with a methanol and ethanol solution (1:1) to achieve a final ratio of 1:3. The mixture was vortexed for 30 s and incubated at 20 °C for 15 min. A subsequent centrifugation step was performed at 15,000 *g* for 10 min at 4 °C. Finally, 250 μL of the obtained supernatant was transferred to a chromatography vial.

4.5. LC-QQQ-MS Analysis

Chromatographic separation was carried out with the Agilent 1290 Infinity II LC system coupled to the Agilent 6470 Triple Quad tandem mass spectrometer equipped with an electrospray ionization source, Jet Stream Technology (Agilent Technologies, Santa Clara, CA, USA). The chromatographic separations were conducted using an RRHD Zorbax Eclipse Plus C18 column (Agilent Technologies; 2.1 × 100 mm; 1.8 μm) at a flow rate of 0.3 mL/min. The mobile phases were 0.1% formic acid in water (A) and 0.1% formic acid in methanol (B). The run time was 7 min, with the gradient program 0–4 min, 13–18% B; 4.01–6 min, 60–75% B; 6.01–7 min, 95% B and final column conditioning of 3 min. The injection volume was set to 2 μL, and the column temperature was held at 40 °C. The mass spectrometer was operated in positive electrospray ionization mode (ESI+) with the following settings: ion source gas, N_2; ion source gas temperature, 300 °C; ion source gas flow rate, 12 L/min; nebulizer pressure, 40 psi; sheath gas temperature, 350 °C; sheath gas flow rate, 12 L/min; and capillary voltage, 4000 V. The needle was washed after each injection with a mixture of methanol and water (80:20, *v*/*v*). Full acquisition details and optimized MS parameters are provided in Table 1.

https://doi.org/10.3390/molecules31071095

Data acquisition was carried out using Agilent Mass Hunter Data Acquisition software (version B.09.00). Subsequent data processing was performed with Agilent Mass Hunter Qualitative Analysis software (version B.10.02) (Agilent Technologies, Santa Clara, CA, USA). The amount of CAF and its three primary metabolites, PAR, THR, and THY, in control larvae was determined from the corresponding matrix-matched calibration curve. The linear ranges of calibration curves were estimated using the coefficient of determination ($R^2 \geq 0.99$). LOD and LOQ values were derived from signal-to-noise ratios of 3:1 and 10:1, respectively. Calculations were based on the most intense MRM transition recorded at the lowest concentration level of the calibration sample. Concentration results were reported as mean values across five injections per group.

4.6. Fourier Transform Infrared Spectroscopy

Fourier transform infrared spectroscopy (FTIR) measurements were carried out in the reflection mode using an FTIR spectrometer (Nicolet 8700, Thermo, Waltham, MA, USA) attached to an infrared microscope (Nicolet Continuum, Thermo, Waltham, MA, USA) with an MCT-A detector. The microscope was equipped with a $\times 15$ IR objective. Spectra within the 4000–650 cm^{-1} spectral range were measured with 120 scans at 8 cm^{-1} spectral resolution using Happ–Genzel apodization and two levels of zero filling. Area maps were acquired in automated point-mapping mode with a stage step size of 20 $\times$ 20 μm and an aperture of 25 μm, yielding a spatial resolution of 20 μm per pixel. Background spectra were collected from a region free of tissue using 120 scans. Frozen larvae from control and CAF 50 mg/L were subjected to horizontal sectioning with Leica 19650UV (Leica Microsystems, Wutzlar, Germany) to 10 μm slices embedded in gelatin from porcine skin (Sigma-Aldrich, St. Louis, MO, USA) and placed on reflective, aluminum-coated slides. The assessed physical characteristics of the samples (size and thickness) were standardized for the investigations. Initially, the FTIR spectra for CAF, PAR, THR, and THY standards (Sigma Aldrich) were recorded concurrently with 100 spectra randomly collected from control and CAF 50 mg/L larvae whole bodies. Finally, in imaging mode, the head area measurements were carried out step-by-step, with a 20 μm step in the x/y plane.

FTIR spectral data were analyzed using the OMNIC 8 software (8.2.0.387 Thermo Fisher Scientific, Waltham, MA, USA) and Orange (ver. 3.36.2, University of Ljubljana, Ljubljana, Slovenia) software. The data were preprocessed with algorithmic baseline correction with a rubber band algorithm, Gaussian smoothing (SD: 5) and vector normalization. The absorbance intensity was used for statistical evaluation of larvae spectral data.

4.7. Statistical Analysis

All statistical analyses were carried out in GraphPad Prism 8 (GraphPad Software, San Diego, CA, USA). Depending on the experimental design, one-way or two-way ANOVA was applied. One-way ANOVA was followed by Tukey's post hoc test. A p-value < 0.05 was considered statistically significant. Results are expressed as mean $\pm$ standard deviation (SD). Larvae were randomly assigned to experimental groups prior to the start of the experiment.

5. Limitations

This study provides novel insights into CAF biotransformation in zebrafish larvae; however, several limitations should be acknowledged. First, the applied LC-MS/MS approach was based on targeted analysis, focusing on CAF and its three primary metabolites (PAR, THR and THY) known from mammalian systems. Consequently, the presence of additional minor or alternative metabolites cannot be excluded and would require untargeted metabolomics approaches for comprehensive characterization. FTIR imaging

https://doi.org/10.3390/molecules31071095

was carefully performed using two bands per analyte and the mean control spectrum as background to reduce the risk of false attribution; however, complete spectral separation was not achieved. It was not possible to identify analytically suitable bands for THY, which was therefore excluded from FTIR mapping. Collectively, FTIR-derived band intensities should be interpreted as semiquantitative indicators of spatial distribution, complementary to the LC-MS/MS quantification data. The study was conducted at a single developmental stage (4–5 dpf), which represents a period of ongoing physiological maturation. Therefore, the observed metabolic activity may not fully reflect the metabolic capacity of later developmental stages or adult organisms.

6. Conclusions

We conducted a comprehensive investigation into CAF metabolism in zebrafish larvae, focusing on its three primary metabolites: PAR, THY, and THR. To characterize this biotransformation quantitatively, a validated LC-QQQ-MS method was developed and applied for the simultaneous determination of CAF and its primary metabolites, complemented by FTIR imaging for spatially resolved metabolite mapping in larval tissues. The quantitative method achieved LOD values of 5.358, 0.430, 1.118, and 0.389 nmol/L and LOQ values of 17.859, 1.434, 3.725, and 1.298 nmol/L for CAF, PAR, THR, and THY, respectively, with precision (RSD) not exceeding 9.01%. Zebrafish larvae exposed to CAF at 15, 25, and 50 mg/L revealed dose-dependent biotransformation: CAF was converted to PAR at up to 4.54%, to THR at up to 1.32%, and to THY at up to 0.27% of the administered dose. Measured internal CAF concentrations reached 504.7, 692.2, and 1444.7 nmol/L, while PAR, THR, and THY peaked at 65.3, 19.0, and 2.8 nmol/L, respectively, at the highest exposure level. Complementary FTIR imaging confirmed the presence of CAF, PAR, and THR in larval tissues and revealed their heterogeneous spatial distribution, with statistically significant differences in absorbance between control and CAF 50 mg/L groups. To our knowledge, this study is the first to assess CAF metabolites in zebrafish larvae, offering a novel analytical approach applicable to various zebrafish-related research endeavors. Our findings underscore the presence of CYP450-mediated demethylation activity in zebrafish larvae before full liver development, highlighting the significance of early-stage zebrafish studies in pharmacological research. Further investigations involving older larvae and adult fish are warranted to elucidate the complete spectrum of CAF metabolism in zebrafish. The integration of quantitative LC-MS/MS methods with spatially resolved FTIR imaging described in this study contributes to the development of the zebrafish model by extending its application towards more mechanistically informed approaches for potential xenobiotic investigation of metabolism and tissue-specific biochemical responses.

Author Contributions: Conceptualization, A.B.; Formal analysis, A.B.; Investigation, A.B., A.S. (Agata Sumara), A.K.-P., G.K., A.W. and A.S. (Anna Stachniuk); Methodology, A.B.; Supervision, A.S. (Anna Stachniuk); Visualization: A.B. and G.K.; Writing—original draft, A.B.; Writing—review and editing, K.P., A.S. (Anna Stachniuk) and G.K. All authors have read and agreed to the published version of the manuscript.

Funding: This research was funded by the Ministry of Science (Poland), project No. 0088/DIA/2020/49 granted to AB.

Institutional Review Board Statement: Not applicable.

Informed Consent Statement: Not applicable.

Data Availability Statement: The raw data supporting the conclusions of this article will be made available by the authors, without undue reservation.

Acknowledgments: All animal studies were conducted at the Center for Experimental Medicine of the Medical University of Lublin. The authors express their gratitude for the opportunity to conduct experiments.

Conflicts of Interest: The authors declare no conflicts of interest.

Abbreviations

The following abbreviations are used in this manuscript:

ANOVA	Analysis of variance
ARRIVE	Animal Research: Reporting of In Vivo Experiments
ATR-FTIR	Attenuated Total Reflectance Fourier-Transform Infrared Spectroscopy
CAF	Caffeine
CE	Collision energy
CTR	Control group
CYP	Cytochrome P450
CYP1A2	Cytochrome P450 1A2
dMRM	Dynamic Multiple Reaction Monitoring
dpf	Days post-fertilization
E3	Embryo medium (E3 medium)
ESI+	Electrospray ionization, positive mode
FTIR	Fourier-transform infrared spectroscopy
hpf	Hours post-fertilization
LC	Liquid chromatography
LC-MS	Liquid chromatography–mass spectrometry
LC-MS/MS	Liquid chromatography–tandem mass spectrometry
LC-QQQ-MS	Liquid chromatography–triple quadrupole mass spectrometry
LOD	Limit of detection
LOQ	Limit of quantification
MCT-A	Mercury Cadmium Telluride Detector (Type A)
MRM	Multiple reaction monitoring
PAR	Paraxanthine
RSD	Relative standard deviation
RT	Retention time
SD	Standard deviation
THR	Theobromine
THY	Theophylline
UV	Ultraviolet

References

1. Fredholm, B.B.; Bättig, K.; Holmén, J.; Nehlig, A.; Zvartau, E.E. Actions of Caffeine in the Brain with Special Reference to Factors That Contribute to Its Widespread Use. *Pharmacol. Rev.* **1999**, *51*, 83–133. [CrossRef]
2. Fredholm, B.B. Adenosine, Adenosine Receptors and the Actions of Caffeine. *Pharmacol. Toxicol.* **1995**, *76*, 93–101. [CrossRef]
3. van Dam, R.M.; Hu, F.B.; Willett, W.C. Coffee, Caffeine, and Health. *N. Engl. J. Med.* **2020**, *383*, 369–378. [CrossRef]
4. Fulgoni, V.L.; Keast, D.R.; Lieberman, H.R. Trends in Intake and Sources of Caffeine in the Diets of US Adults: 2001-2010. *Am. J. Clin. Nutr.* **2015**, *101*, 1081–1087. [CrossRef]
5. Lara, D.R. Caffeine, Mental Health, and Psychiatric Disorders. *J. Alzheimer's Dis.* **2010**, *20*, S239–S248. [CrossRef] [PubMed]
6. Chen, F.; Hu, Z.Y.; Parker, R.B.; Laizure, S.C. Measurement of Caffeine and Its Three Primary Metabolites in Human Plasma by HPLC-ESI-MS/MS and Clinical Application. *Biomed. Chromatogr.* **2017**, *31*, e3900. [CrossRef] [PubMed]
7. Choi, E.J.; Bae, S.H.; Park, J.B.; Kwon, M.J.; Jang, S.M.; Zheng, Y.F.; Lee, Y.S.; Lee, S.-J.; Bae, S.K. Simultaneous Quantification of Caffeine and Its Three Primary Metabolites in Rat Plasma by Liquid Chromatography-Tandem Mass Spectrometry. *Food Chem.* **2013**, *141*, 2735–2742. [CrossRef] [PubMed]

https://doi.org/10.3390/molecules31071095

8. Howe, K.; Clark, M.D.; Torroja, C.F.; Torrance, J.; Berthelot, C.; Muffato, M.; Collins, J.E.; Humphray, S.; McLaren, K.; Matthews, L.; et al. The Zebrafish Reference Genome Sequence and Its Relationship to the Human Genome. *Nature* **2013**, *496*, 498–503. [CrossRef] [PubMed]
9. Russell, W.M.S.; Burch, R.L. The Principles of Humane Experimental Technique. *Med. J. Aust.* **1960**, *1*, 500. [CrossRef]
10. Kantae, V.; Krekels, E.H.J.; Ordas, A.; González, O.; Van Wijk, R.C.; Harms, A.C.; Racz, P.I.; Van Der Graaf, P.H.; Spaink, H.P.; Hankemeier, T. Pharmacokinetic Modeling of Paracetamol Uptake and Clearance in Zebrafish Larvae: Expanding the Allometric Scale in Vertebrates with Five Orders of Magnitude. *Zebrafish* **2016**, *13*, 504–510. [CrossRef]
11. Morgan, P.; Van Der Graaf, P.H.; Arrowsmith, J.; Feltner, D.E.; Drummond, K.S.; Wegner, C.D.; Street, S.D.A. Can the Flow of Medicines Be Improved? Fundamental Pharmacokinetic and Pharmacological Principles toward Improving Phase II Survival. *Drug Discov. Today* **2012**, *17*, 419–424. [CrossRef] [PubMed]
12. Geier, M.C.; James Minick, D.; Truong, L.; Tilton, S.; Pande, P.; Anderson, K.A.; Teeguardan, J.; Tanguay, R.L. Systematic Developmental Neurotoxicity Assessment of a Representative PAH Superfund Mixture Using Zebrafish. *Toxicol. Appl. Pharmacol.* **2018**, *354*, 115–125. [CrossRef] [PubMed]
13. van Wijk, R.C.; Krekels, E.H.J.; Kantae, V.; Harms, A.C.; Hankemeier, T.; van der Graaf, P.H.; Spaink, H.P. Impact of Post-Hatching Maturation on the Pharmacokinetics of Paracetamol in Zebrafish Larvae. *Sci. Rep.* **2019**, *9*, 2149. [CrossRef]
14. Klupczynska, A.; Misiura, M.; Miltyk, W.; Oscilowska, I.; Palka, J.; Kokot, Z.J.; Matysiak, J. Development of an LC-MS Targeted Metabolomics Methodology to Study Proline Metabolism in Mammalian Cell Cultures. *Molecules* **2020**, *25*, 4639. [CrossRef] [PubMed]
15. Benavente, L.; Coetsier, C.; Venault, A.; Chang, Y.; Causserand, C.; Bacchin, P.; Aimar, P. FTIR Mapping as a Simple and Powerful Approach to Study Membrane Coating and Fouling. *J. Memb. Sci.* **2016**, *520*, 477–489. [CrossRef]
16. Gunasekaran, S.; Sankari, G.; Ponnusamy, S. Vibrational Spectral Investigation on Xanthine and Its Derivatives—Theophylline, Caffeine and Theobromine. *Spectrochim. Acta A Mol. Biomol. Spectrosc.* **2005**, *61*, 117–127. [CrossRef] [PubMed]
17. González-González, J.S.; Zúñiga-Lemus, O.; Hernández-Galindo, M.D.C. Hydrated Solid Forms of Theophylline and Caffeine Obtained by Mechanochemistry. *IOSR J. Pharm.* **2017**, *7*, 28–30.
18. Kalisz, G.; Budzynska, B.; Sroka-Bartnicka, A. The Optimization of Sample Preparation on Zebrafish Larvae in Vibrational Spectroscopy Imaging. *Spectrochim. Acta A Mol. Biomol. Spectrosc.* **2025**, *326*, 125288. [CrossRef]
19. Basnet, R.M.; Guarienti, M.; Memo, M. Zebrafish Embryo as an In Vivo Model for Behavioral and Pharmacological Characterization of Methylxanthine Drugs. *Int. J. Mol. Sci.* **2017**, *18*, 596. [CrossRef]
20. Ali, S.; van Mil, H.G.J.; Richardson, M.K. Large-Scale Assessment of the Zebrafish Embryo as a Possible Predictive Model in Toxicity Testing. *PLoS ONE* **2011**, *6*, e21076. [CrossRef]
21. Nawaji, T.; Yamashita, N.; Umeda, H.; Zhang, S.; Mizoguchi, N.; Seki, M.; Kitazawa, T.; Teraoka, H. Cytochrome P450 Expression and Chemical Metabolic Activity before Full Liver Development in Zebrafish. *Pharmaceuticals* **2020**, *13*, 456. [CrossRef]
22. Nehlig, A. Interindividual Differences in Caffeine Metabolism and Factors Driving Caffeine Consumption. *Pharmacol. Rev.* **2018**, *70*, 384–411. [CrossRef]
23. Lajin, B.; Schweighofer, N.; Goessler, W.; Obermayer-Pietsch, B. The Determination of the Paraxanthine/Caffeine Ratio as a Metabolic Biomarker for CYP1A2 Activity in Various Human Matrices by UHPLC-ESIMS/MS. *Talanta* **2021**, *234*, 122658. [CrossRef]
24. Cornelis, M.C.; Kacprowski, T.; Menni, C.; Gustafsson, S.; Pivin, E.; Adamski, J.; Artati, A.; Eap, C.B.; Ehret, G.; Friedrich, N.; et al. Genome-Wide Association Study of Caffeine Metabolites Provides New Insights to Caffeine Metabolism and Dietary Caffeine-Consumption Behavior. *Hum. Mol. Genet.* **2016**, *25*, 5472–5482. [CrossRef]
25. Orrú, M.; Guitart, X.; Karcz-Kubicha, M.; Solinas, M.; Justinova, Z.; Barodia, S.K.; Zanoveli, J.; Cortes, A.; Lluis, C.; Casado, V.; et al. Psychostimulant Pharmacological Profile of Paraxanthine, the Main Metabolite of Caffeine in Humans. *Neuropharmacology* **2013**, *67*, 476–484. [CrossRef] [PubMed]
26. Berthou, F.; Guillois, B.; Riche, C.; Dreano, Y.; Jacqz-Aigrain, E.; Beaune, P.H. Interspecies Variations in Caffeine Metabolism Related to Cytochrome P4501a Enzymes. *Xenobiotica* **1992**, *22*, 671–680. [CrossRef]
27. Zhao, Y.; Li, X.; Song, G.; Li, Q.; Yan, H.; Cui, Z. Comparative Transcriptome Analysis Provides Novel Molecular Events for the Differentiation and Maturation of Hepatocytes during the Liver Development of Zebrafish. *Biomedicines* **2022**, *10*, 2264. [CrossRef]
28. Chian, S.; Jiang, Z.; Jiang, L.; Wang, K.; Fan, Y.; Liao, T.; Chen, W.; Yao, W. Caffeine-Induced Neurotoxicity Mediated by Nrf2 Pathway in PC12 Cells and Zebrafish Larvae. *J. Appl. Toxicol.* **2022**, *42*, 629–637. [CrossRef]
29. Shi, M.; Huang, L.Y.; Nie, N.; Ye, J.H.; Zheng, X.Q.; Lu, J.L.; Liang, Y.R. Binding of Tea Catechins to Rice Bran Protein Isolate: Interaction and Protective Effect during In Vitro Digestion. *Food Res. Int.* **2017**, *93*, 1–7. [CrossRef] [PubMed]
30. Li, Z.; Li, Z.; Ma, H.; Fu, S.; Liu, G.; Hao, C.; Liu, Y. Molecular Insight into Binding Behavior of Caffeine with Lactoferrin: Spectroscopic, Molecular Docking, and Simulation Study. *J. Dairy Sci.* **2023**, *106*, 8249–8261. [CrossRef] [PubMed]
31. Byrne, H.J. Spectralomics—Towards a Holistic Adaptation of Label Free Spectroscopy. *Vib. Spectrosc.* **2024**, *132*, 103671. [CrossRef]

https://doi.org/10.3390/molecules31071095

32. Ashengroph, M. Salinivibrio Costicola GL6, a Novel Isolated Strain for Biotransformation of Caffeine to Theobromine Under Hypersaline Conditions. *Curr. Microbiol.* **2017**, *74*, 34–41. [CrossRef]
33. Abdel Salam, J.; Saleh, A.A.; El Nenaiey, T.T.; Yang, H.; Shoeib, T.; El-Sayed, M.M.H. Mono- and Multicomponent Biosorption of Caffeine and Salicylic Acid onto Processed Cape Gooseberry Husk Agri-Food Waste. *ACS Omega* **2023**, *8*, 20697–20707. [CrossRef]
34. Gołdyn, M.; Komasa, A.; Pawlaczyk, M.; Lewandowska, A.; Adamska, E.B. Salts of Purine Alkaloids Caffeine and Theobromine with 2,6-Dihydroxybenzoic Acid as Coformer: Structural, Theoretical, Thermal and Spectroscopic Studies. *Acta Crystallogr. C Struct. Chem.* **2021**, *77*, 713–724. [CrossRef]
35. Tatsumi, Y.; Shimoyama, Y.; Kazarian, S.G. Analysis of the Dissolution Behavior of Theophylline and Its Cocrystal Using ATR-FTIR Spectroscopic Imaging. *Mol. Pharm.* **2024**, *21*, 3233–3239. [CrossRef]
36. Bartoszek, A.; Fornal, E. Evaluation of Caffeine Modulation of Topiramate Effect on Locomotor Activity of Zebrafish Larvae in Pentylenetetrazol-Induced Seizure Model. *PLoS ONE* **2025**, *20*, e0317241. [CrossRef] [PubMed]
37. Bartoszek, A.; Paszkowska, E.; Kozub-Pędrak, A.; Sumara, A.; Fornal, E. Unveiling the Influence of Caffeine on Topiramate: Metabolomic Marker Analysis Using Liquid Chromatography-Mass Spectrometry. *Front. Mol. Biosci.* **2025**, *12*, 1549993. [CrossRef]
38. Bartoszek, A.; Trzpil, A.; Kozub, A.; Fornal, E. Optimization of the Zebrafish Larvae Pentylenetetrazol-Induced Seizure Model for the Study of Caffeine and Topiramate Interactions. *Int. J. Mol. Sci.* **2023**, *24*, 12723. [CrossRef] [PubMed]
39. Bartoszek, A.; Sumara, A.; Kozub-Pędrak, A.; Trzpil, A.; Stachniuk, A.; Fornal, E. Caffeine Decreases Topiramate Levels in Zebrafish Larvae in a Pentylenetetrazol-Induced Seizure Model. *Int. J. Mol. Sci.* **2024**, *25*, 3309. [CrossRef]

https://doi.org/10.3390/molecules31071095

separations

MDPI

Article

Comprehensive Discovery and Characterization of Chemical Constituents in Huangqintang Decoction Using Off-Line Two-Dimensional Liquid Chromatography and High-Resolution Mass Spectrometry

Yan Fang, Yi Nan, Xijie Tian, Junyu Zhang, Xiaojuan Chen, Juan Song, Haizhen Liang * and Baiping Ma *

Academy of Military Medical Sciences, Beijing 100850, China
* Correspondence: lhzhcm@163.com (H.L.); mabaiping@sina.com (B.M.)

Abstract

Traditional Chinese prescriptions are characterized by complex chemical constituents and wide variations in constituent content, which pose a substantial challenge to their comprehensive characterization. As a classic traditional Chinese prescription known for its heat-clearing and detoxifying properties, Huangqintang Decoction (HQD) is composed of Scutellariae Radix, Paeoniae Radix Rubra, Glycyrrhizae Radix et Rhizoma, and Jujubae Fructus. In this study, we developed an off-line two-dimensional liquid chromatography that addressed the limitations of traditional analysis of unfractionated extracts, such as restricted peak capacity, which often obscured trace components. By coupling with ultra-high performance liquid chromatography-quadrupole time-of-flight mass spectrometry (UHPLC-Q-TOF/MS), this study successfully performed rapid identification or characterization of the complete chemical profile of HQD. Notably, beyond high-throughput identification, this approach leveraged characteristic fragment ions and reversed-phase chromatographic behaviors to differentiate some isomers of flavonoid glycosides and triterpenoid saponins, demonstrating its depth in structural identification. Flavonoid glycoside isomers were distinguished by diagnostic neutral losses, while flavanones and chalcones were characterized by retro-Diels–Alder (RDA) and β-rearrangement, respectively. Isomers of triterpenoid saponins were inferred from aglycone-specific pathways alongside RDA cleavages. Ultimately, a total of 192 compounds were identified, including 88 flavonoids, 80 triterpenoids, 7 monoterpene glycosides, 3 fatty acid amides, 3 phenylethanoid glycosides, 4 coumarins, 3 saccharides, 1 organic acid, and 3 others. This study demonstrated that the off-line two-dimensional liquid chromatography analysis strategy significantly enhanced chromatographic resolution and expanded the coverage of trace components. It presented an effective strategy for comprehensive compound identification in complex traditional Chinese medicine prescriptions.

Keywords: Huangqintang decoction; chemical constituents; isomer; off-line two-dimensional liquid chromatography analysis strategy; UHPLC-Q-TOF/MS

1. Introduction

Traditional Chinese prescriptions (TCPs) are a concentrated manifestation of the "holistic view" and "syndrome differentiation and treatment" concepts in traditional Chinese medicine (TCM). Their core lies in achieving coordinated integration and regulation through multiple components, targets, and pathways by adhering to the formulation principles of "Jun (emperor)—Chen (minister)—Zuo (adjuvant)—Shi (courier)," thereby offering therapeutic advantages that single compounds do not possess [1,2]. However, it is precisely this complex design, grounded in holistic effects, that presents fundamental challenges to modern scientific research. The crux lies in how to systematically characterize a mixture sample characterized by complex chemical constituents and significant disparities in the content of its various components. The challenges in elucidating the material basis of TCPs significantly hinder the exploration of their pharmacological mechanisms and the improvement of quality control standards [3,4].

Huangqintang decoction (HQD) is a classic TCP for the treatment of heat-induced diarrhea and dysentery. It is first documented in the "Treatise on Cold Damage Disorders" (*Shanghan Lun* in Chinese) [5–7]. The prescription comprises Scutellariae Radix (*Scutellaria baicalensis* Georgi, Huang-Qin in Chinese, SR), Paeoniae Radix Rubra (*Paeonia lactiflora* Pall., Chi-Shao, PRR), Glycyrrhizae Radix et Rhizoma (*Glycyrrhiza uralensis* Fisch., *Glycyrrhiza inflata* Bat., *Glycyrrhiza glabra* L., Gan-Cao, GRR), and Jujubae Fructus (*Ziziphus jujuba* Mill., Da-Zao, JF) at a ratio of 3:2:2:2 [8–23]. According to the principle of formulation, SR serves as the emperor herb, PRR acts as the minister herb, GRR and JF function as the adjuvant and courier herbs. These four herbs work together to exert therapeutic effects. However, due to its complex components and significant differences in the content of various constituents, systematic research on the elucidation of the full chemical spectrum of HQD remains insufficient. Therefore, it is urgent to establish rapid and efficient methods for the characterization and identification of chemical components. This will not only help elucidate the material basis of their pharmacological effects but also hold significant value for deeply understanding the "holistic view" of TCPs.

Charged aerosol detection operates by detecting the charge of aerosol particles, where the charge amount is proportional to the analyte mass. As the detection signal exhibits a nonlinear response over a wide concentration range, the power function value algorithm is employed to transform this nonlinear response into a linear calibration curve, thereby enabling more accurate quantitative analysis [24,25]. For the problem of different ultraviolet absorption of components in TCPs, different detector signals such as ultraviolet (UV) and CAD can be used to trigger automated fraction collection, enabling precise isolation of specific components for further analysis [26]. The first-dimensional liquid chromatography (^{1}D LC) system integrates the HPLC platform with a vanquish fraction collector (VFC) and automated workflow. Through multi-detectors real-time monitoring and signal-triggered automated fraction collection, it enables complete process control from method development and sample analysis to target fraction collection. At its core, the system operates via an automated workflow of "real-time monitoring, intelligent triggering, and precise collection." The integrated LC-automated fraction collector and workflow enables accurate determination and optimization of the delay volume, ensuring synchronization between detection and collection, fundamental to high purity and recovery. To safeguard sample integrity, the system precisely controls fraction collection needle movement and employs various washing protocols to minimize cross-contamination, while the closed storage module with temperature and light control ensures fraction stability [27]. Thus, the ^{1}D LC system can achieve the purpose of targeted removal of the main components in HQD and maintaining the integrity of trace components, laying the foundation for subsequent comprehensive component characterization and identification.

Ultra-high performance liquid chromatography coupled-quadrupole time-of-flight mass spectrometry (UHPLC-Q-TOF/MS) is characterized by high resolution, sensitivity, mass accuracy, and a broad scanning range [28,29]. By leveraging the accurate mass measurements and characteristic tandem mass spectrometry (MS/MS) information ob-

https://doi.org/10.3390/separations13040110

tained through MS/MS analysis, in conjunction with database searches, rapid structural elucidation of chemical constituents in TCPs can be achieved. In existing research, UHPLC-Q-TOF/MS technology has been applied to the chemical characterization of HQD. For instance, Sun et al. [30] identified 68 chemical constituents and 35 blood-absorbed components in HQD using UHPLC-Q-TOF/MS. Similarly, Tang et al. [31] tentatively identified 139 chemical constituents in HQD via UHPLC-Q-TOF/MS and revealed differences in metabolic profiles within an ulcerative colitis model. However, all these valuable studies are based on traditional analytical strategies, and their identification depth remains limited by strong interference from the complex matrix, manifested as co-elution of high-abundance components in chromatography and severe ion suppression effects in mass spectrometry. Consequently, although the major constituents of HQD have been partially characterized, its trace components, which are typically present at low abundance but may possess high biological activity, remain largely unexplored. These trace components may play a critical role in the synergistic therapeutic effects of TCPs, whose efficacy often arises from the combined action of multiple constituents rather than a single compound. Therefore, a comprehensive chemical profiling covering both major and trace constituents is essential for establishing a complete chemical basis for understanding the pharmacological mechanisms of HQD. To address this, the present study developed an innovative off-line ^{1}D LC separation and second-dimension (^{2}D) UHPLC-QTOF/MS analysis combined strategy. The core advantage of this off-line ^{2}D separation and analysis strategy lies in the targeted removal of main components, thereby achieving a significant enhancement in the identification rate of trace components. This approach enables the construction of a comprehensive chemical database for HQD and provides a powerful tool for investigating its pharmacodynamic material basis and multi-component synergy.

2. Results

2.1. *Enrichment of Trace Components*

In this work, it was observed through CAD analysis that the HQD samples had a complex composition comprising multiple main components, which posed a challenge for comprehensive constituent analysis. By comparing the retention times and fragmentation characteristics with reference standards, seven main constituents in HQD were characterized, namely albiflorin, paeoniflorin, baicalin, wogonoside, wogonin, baicalein, and glycyrrhizic acid. Simply amplifying the response signals of the UHPLC-CAD chromatogram or increasing the sample injection volume did not effectively resolve the issue of the overall chromatographic peak abundance being affected by the prominence of the main components in HQD (Figure S1). Therefore, the ^{1}D LC system was utilized to remove the main constituents and enrich the trace components in the sample, thereby achieving the characterization and identification of all constituents in HQD. Firstly, the chromatographic methods were systematically optimized to ensure suitable baseline separation with adequate resolution and stable peak shapes for multi-component detection. After fixing the column temperature at 35 °C and the injection volume at 1 μL, different chromatographic columns were evaluated under identical gradient elution conditions. Based on a comprehensive assessment of chromatographic peak separation (Figure S2A), the Agilent ZORBAX SB-C18 column was selected for subsequent analysis. The elution conditions were further optimized, as shown in Figure S2B. The optimized conditions were specified in Section 4.3. As shown in Figure S2(Bc), in the HPLC-CAD/UV system, seven main components in HQD could be preliminarily characterized, and these main components could achieve good separation. Based on the optimized conditions, the specific fraction collection strategy employed in this study was as follows: The start and end points of the target chromatographic peaks were identified by UV and CAD detectors, which generated trigger signals

https://doi.org/10.3390/separations13040110

to drive the fraction collection valve to switch within milliseconds. Within the specific time window defined by the trigger signal (RTs 8.10~8.75 min, 9.35~9.95 min, 24.20~25.50 min, 31.75~32.75 min, 37.5~39.00 min, 44.50~46.50 min, and 65.00~66.00 min), the chromatographic flow path was altered to direct the fraction containing the high-abundance dominant constituents to a specific sample vial, thereby achieving the targeted removal of the main components. Outside this time window, the system maintained the default flow path, continuously and completely collecting all remaining fractions enriched with trace components. These collected fractions were subsequently subjected to ^{2}D UHPLC-Q-TOF/MS analysis. As shown in Figure 1A, the seven peaks with the highest peak areas were removed via the system, effectively enriching the remaining constituents. Consequently, the chromatographic peak richness was significantly enhanced, as demonstrated in Figure 1B.

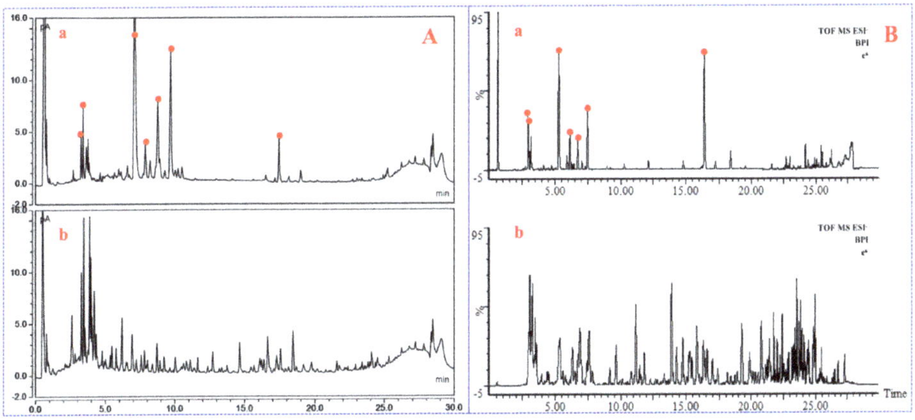

Figure 1. (**A**) UHPLC-CAD chromatograms of HQD before (**a**) and after (**b**) ^{1}D LC system processing. (**B**) UHPLC-Q-TOF/MS chromatograms of HQD before (**a**) and after (**b**) ^{1}D LC system processing. The seven main constituents are indicated by red dots.

2.2. *Workflow of Chemical Constituents Identification in HQD*

Based on the optimized conditions, the chemical constituents of HQD were comprehensively identified and analyzed using ^{2}D UHPLC-Q-TOF/MS analysis. A high-sensitivity data acquisition mode was employed to maximize MS/MS triggering for trace components. The base peak ion (BPI) chromatograms of HQD in both negative and positive ionization modes are presented in Figure 2A,B. Figures 3 and 4 present a summary diagram of the structural types of various constituents identified in HQD. According to the BPI profiles, compounds in HQD exhibited stronger intensity and greater abundance in negative ion mode compared to positive ion mode. To facilitate chemical identification, an in-house database was established, compiling information—including compound name, molecular formula, exact mass, and fragmentation patterns—of 1022 known constituents that have been reported for the four medicinal herbs of HQD. By integrating RT, accurate mass, MS/MS fragmentation behavior, and comparison with reference standards where available, a total of 192 compounds were identified or tentatively characterized. These comprised 88 flavonoids, 80 triterpenoids, 7 monoterpene glycosides, 3 fatty acid amides, 3 phenylethanoid glycosides, 4 coumarins, 3 saccharides, 1 organic acid, and 3 others. Among these 192 compounds, 11 were identified at Metabolomics Standards Initiative (MSI) Level 1, 146 were annotated at MSI Level 2.1, and 15 were tentatively characterized at MSI Level 2.2. Detailed information on these compounds was presented in the Supplementary Information Table S1.

https://doi.org/10.3390/separations13040110

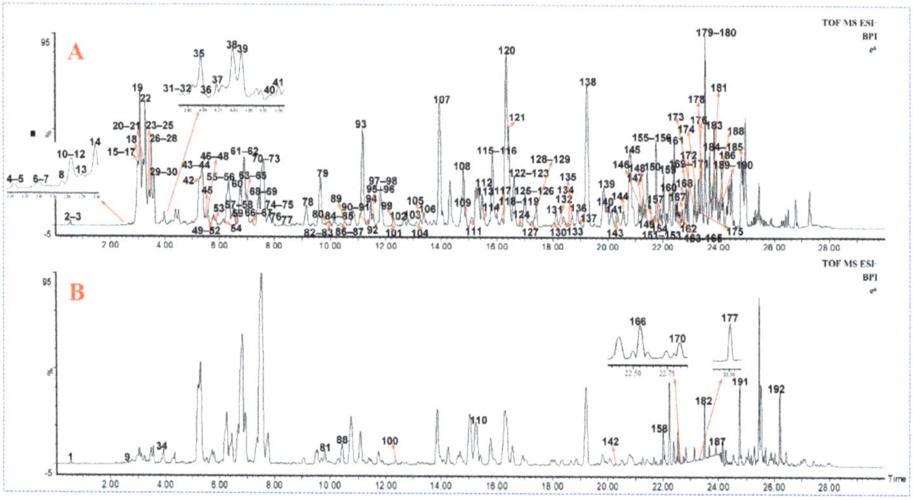

Figure 2. The qualitative analysis and comparisons on the chemical constituents of HQD by UHPLC-Q-TOF/MS. Base peak ion (BPI) chromatograms and the identification results in negative ion modes (**A**) and positive ion modes (**B**).

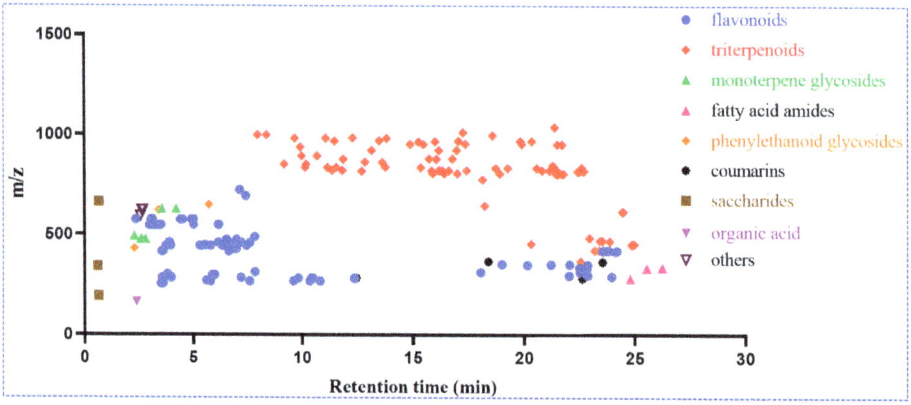

Figure 3. UHPLC-Q-TOF/MS-based qualitative analysis and comparison of the chemical constituents in HQD, summarizing the structural types of the identified compounds and presenting a 2D scatter plot (m/z vs. RT) of all 192 compounds.

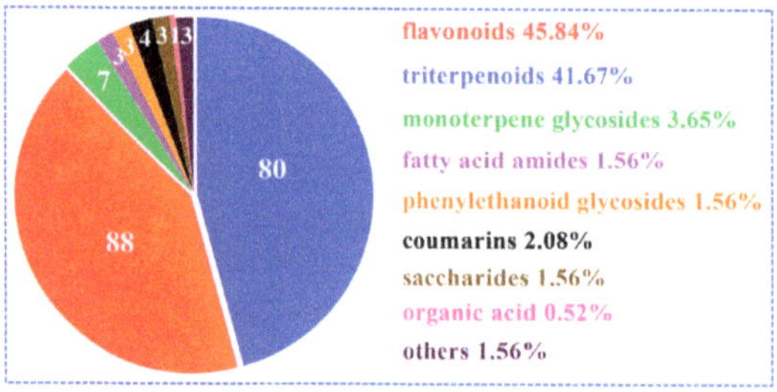

Figure 4. UHPLC-Q-TOF/MS-based qualitative analysis and comparison of the chemical constituents in HQD, summarizing the structural types of the identified compounds and presenting a pie chart illustrating the proportion of different component types.

2.2.1. Characterization of Flavonoid Constituents in HQD

HQD was found to contain a wide variety of natural constituents. Flavonoids were identified as the primary active components in HQD [32]. Previous studies have reported that flavonoids derived from HQD possessed antibacterial, anti-inflammatory, and hepatoprotective activities, which constituted the fundamental basis for the pharmacological activities of HQD [33,34]. Flavonoids predominantly existed in glycosylated forms. These compounds are combined with sugar molecules through glycosidic bonds via oxygen atoms (O) or carbon atoms (C), thereby forming different glycoside structures. This glycosylation not only improved their solubility and bioavailability but also modulated their pharmacological activity [35]. A total of 88 flavonoids were chemically identified or tentatively characterized in HQD, most of which originated from SR and GRR.

Taking compound **24** as an example, the compound **24** with an RT of 3.53 min presented a quasi-molecular ion peak at 417.1197 $[M-H]^-$ that was observed in negative ion mode. Therefore, the molecular formula was inferred to be $C_{21}H_{22}O_9$. The fragment ion at m/z 255.0662 $[M-H-C_6H_{10}O_6]^-$ was generated by neutral loss of glucose residue with 162 Da from the parent ion. Then, RDA cleavage occurred, generating fragment ions with m/z 135.0085 $[M-H-C_6H_{10}O_6-C_8H_8O]^-$ and m/z 119.0500 $[M-H-C_6H_{10}O_6-C_7H_4O_3]^-$ (Figure 5A). Compared with our in-house database and reference standard, the compound was chemically defined as liquiritin [30]. As shown in Figure 5B, the quasi-molecular ion was observed at m/z 445.0781 $[M-H]^-$, and the elemental composition was calculated as $C_{21}H_{18}O_{11}$, which further led to a series of product ions at m/z 269.0453 $[M-H-C_6H_8O_6]^-$, 241.0503 $[M-H-C_6H_8O_6-CO]^-$ and 223.0401 $[M-H-C_6H_8O_6-CO-H_2O]^-$, resulting from the loss of one molecule of dehydrated glucuronic acid, CO, and H_2O group in MS/MS spectrum. Comparison with the literature, in-house databases, and reference material identified compound **43** as baicalin [30].

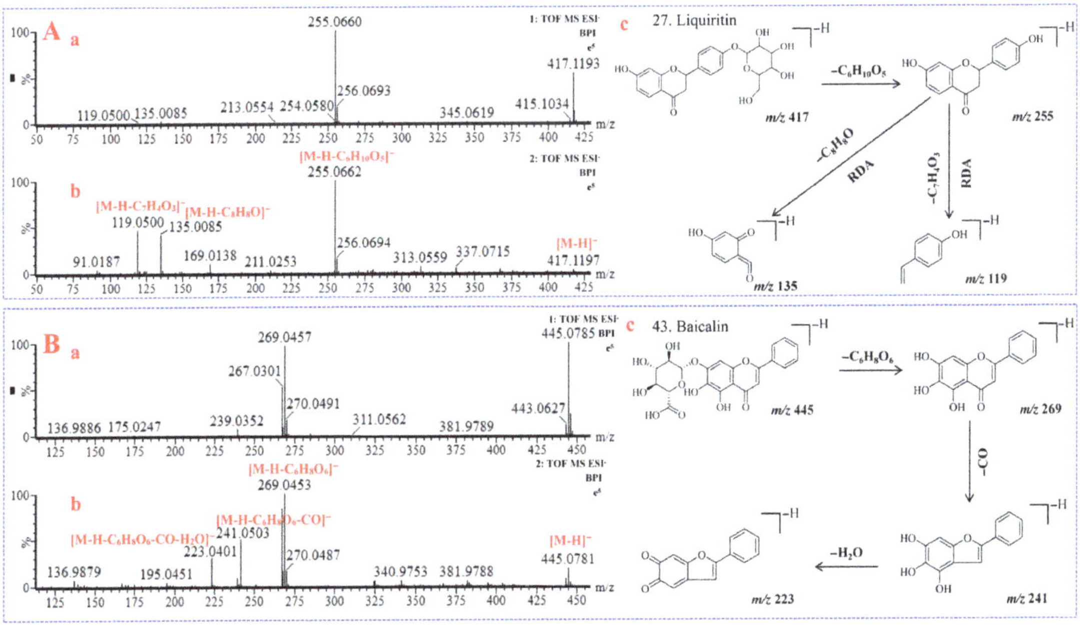

Figure 5. The MS1 (**a**), MS2 spectra (**b**) and characteristic fragmentation pathways (**c**) of liquiritin (**A**) and baicalin (**B**).

https://doi.org/10.3390/separations13040110

In addition, there were many isomers among the flavonoids in HQD. Based on the precise mass spectrometry fragment ions and the reversed-phase chromatography retention behavior, some isomers of flavonoid glycosides were identified and distinguished. Based on the analysis of the quasi-molecular ion peak, it could be calculated that the elemental composition of compounds **43**, **44**, **57** and **61** was all $C_{21}H_{18}O_{11}$. In the negative ion detection mode, the typical fragmentation patterns of flavonoid 7-*O*-glucuronides and 8-*O*-glucuronides are as follows: The 7-*O*-linked isomers undergo heterolytic cleavage of the glycosidic bond, resulting in the neutral loss of a glucuronic acid unit (176 Da) and the generation of an aglycone ion (m/z 269). In contrast, the 8-*O*-linked isomers undergo characteristic rearrangement fragmentation, preferentially losing a dehydrated glucuronic acid residue (148 Da) to form a key intermediate ion (m/z 297) [36,37]. Compound **44** presented characteristic dehydroxylation ion m/z 401.0542 and aglycone ion m/z 269.0452. Its key RDA cleavage fragment m/z 151.0554 clearly pointed to the apigenin aglycone, and combined with the typical 7-*O*-glycosidic bond cleavage pattern, it was identified as apigenin-7-*O*-glucuronide (Figure 6A). Compound **57** (RT 6.33 min) mainly generated the aglycone ion m/z 269.0450, accompanied by the characteristic fragment $[C_7H_3O_2]^-$ (m/z 119.2386) of the norwogonin B ring, which conformed to the 7-*O*-connection feature; therefore, it was putatively annotated to norwogonin-7-*O*-glucuronide (Figure 6B). Compound **61** (at 6.77 min) exhibited the characteristic rearrangement cleavage pathway at the 8-*O* position (producing m/z 297.0748 and its secondary fragment m/z 145.0282), and according to its longest retention time, it was putatively annotated to norwogonin-8-*O*-glucuronide (Figure 6C).

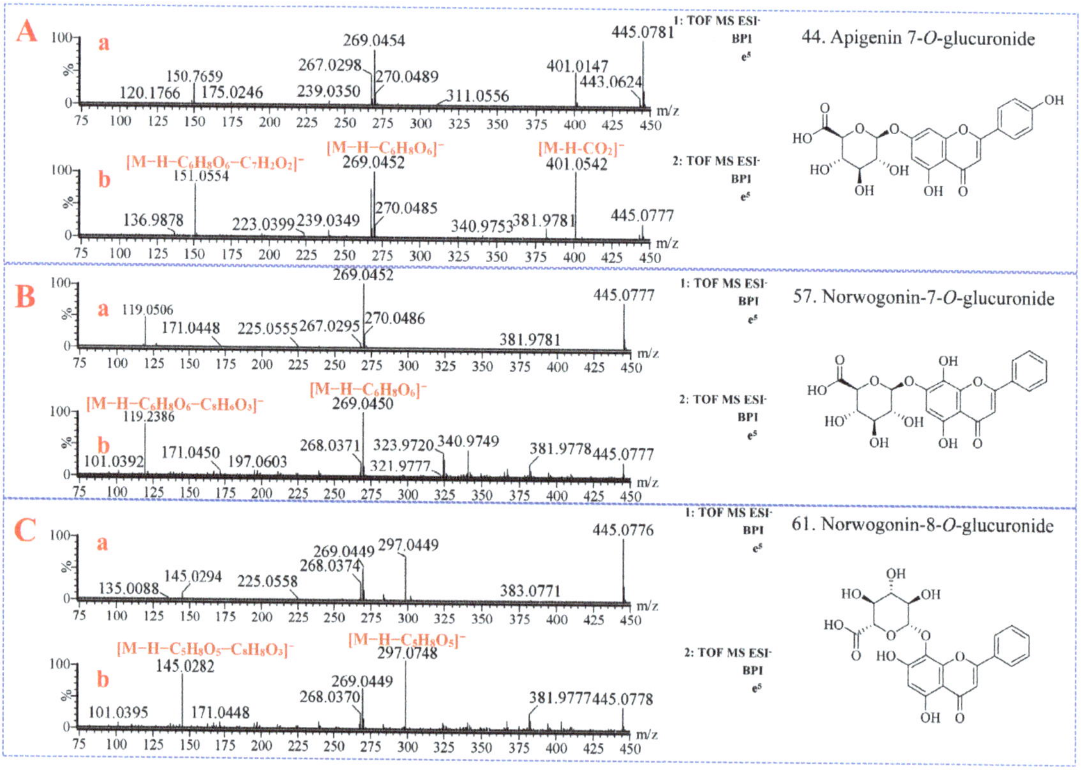

Figure 6. The MS1 (**a**) and MS2 spectra (**b**) of three isomers. (**A**) For apigenin 7-*O*-glucuronide, (**B**) for norwogonin-7-*O*-glucuronide and (**C**) for norwogonin-8-*O*-glucuronide.

https://doi.org/10.3390/separations13040110

Based on the quasi-molecular ion peaks, the elemental composition of compounds **21**, **24**, **42**, and **54** was inferred to be $C_{26}H_{30}O_{13}$. In negative ion mode, a series of deglycosylation fragments was observed for compounds **21** and **24**, including 417 $[M-H-132]^-$, 255 $[M-H-132-162]^-$, and 135 $[M-H-132-162-119]^-$. This indicated the successive loss of a pentose unit (132 Da), a hexose unit (162 Da), and a neutral $C_7H_3O_2$ fragment. This fragmentation pattern was consistent with the characteristic retro-Diels–Alder (RDA) cleavage of flavanone glycosides, which generates the diagnostic fragment ion m/z 135 $[C_8H_7O_2]^-$. Additionally, extensive phytochemical studies on *Glycyrrhiza* plants have shown that the main flavanone glycosides are apiosylglucosides [38,39]. Therefore, it was deduced that the aglycones of compounds **21** (Figure 7A) and **24** (Figure 7B) were the liquiritigenin and that the attached sugar moieties were apiose and glucose. In contrast, apart from similar deglycosylation fragments, compounds **42** (Figure 7C) and **54** (Figure 7D) exhibited a prominent fragment at m/z 147 $[C_9H_7O_2]^-$ in their MS/MS spectra, which is a characteristic ion generated from chalcone aglycones via β-cleavage and rearrangement of the enone skeleton. Combined with chromatographic behavior analysis, chalcone compounds, due to their larger conjugated system and stronger hydrophobicity, typically showed significantly longer retention times in reversed-phase chromatography compared to flavanones. It was inferred based on these differences that the aglycones of compounds **42** and **54** were the isoliquiritigenin. Furthermore, while both compounds **21** and **24** were identified as diglycosides of liquiritigenin, differences were observed in the abundance of their fragment ions derived from the aglycone ion (m/z 255). The significantly higher abundance of the m/z 135 fragment in compound **24** suggested that its A-ring structure remained intact and unsubstituted, facilitating efficient RDA cleavage. This observation aligned with the structural feature of the sugar chain being attached at the 4′-*O*-position on the B-ring. Additionally, when the sugar chain is linked to the B-ring, the hydrophobic part of the aglycone is more exposed, leading to slightly stronger chromatographic retention. Therefore, compound **24** was tentatively identified as liquiritigenin-4′-*O*-apiosylglucoside, and compound **21** as liquiritigenin-7-*O*-apiosylglucoside. Both compounds **42** and **54** were diglycosides of isoliquiritigenin, with the key distinction lying in the relative abundance of the characteristic fragment m/z 147. The notably higher abundance of this fragment in compound **42** was consistent with the scenario where the 2′-hydroxyl group on the A-ring of isoliquiritigenin remains free. This free hydroxyl group can promote an intramolecular cyclization rearrangement via a neighboring group effect, making the formation of m/z 147 the dominant fragmentation pathway. Based on this characteristic, the compound was putatively annotated as licraside, as reported in the literature [30]. In contrast, the relatively lower abundance of m/z 147 in compound **54** suggested that its 2′-position was likely glycosylated, forming 2′-*O*-β-D-apiosyl-(1→2)-β-D-glucopyranosyl-isoliquiritigenin. Searches in SciFinder, PubChem, and related databases yielded no prior reports of this structure, leading to the tentative identification that compound **54** is a novel compound. Further structural validation by NMR would be required to confirm the novelty of compound **54**. The approach of utilizing characteristic differences in mass spectrometry and chromatographic behavior enables the systematic characterization and identification of positional isomers, encompassing both functional group isomers and glycosidic linkage isomers.

https://doi.org/10.3390/separations13040110

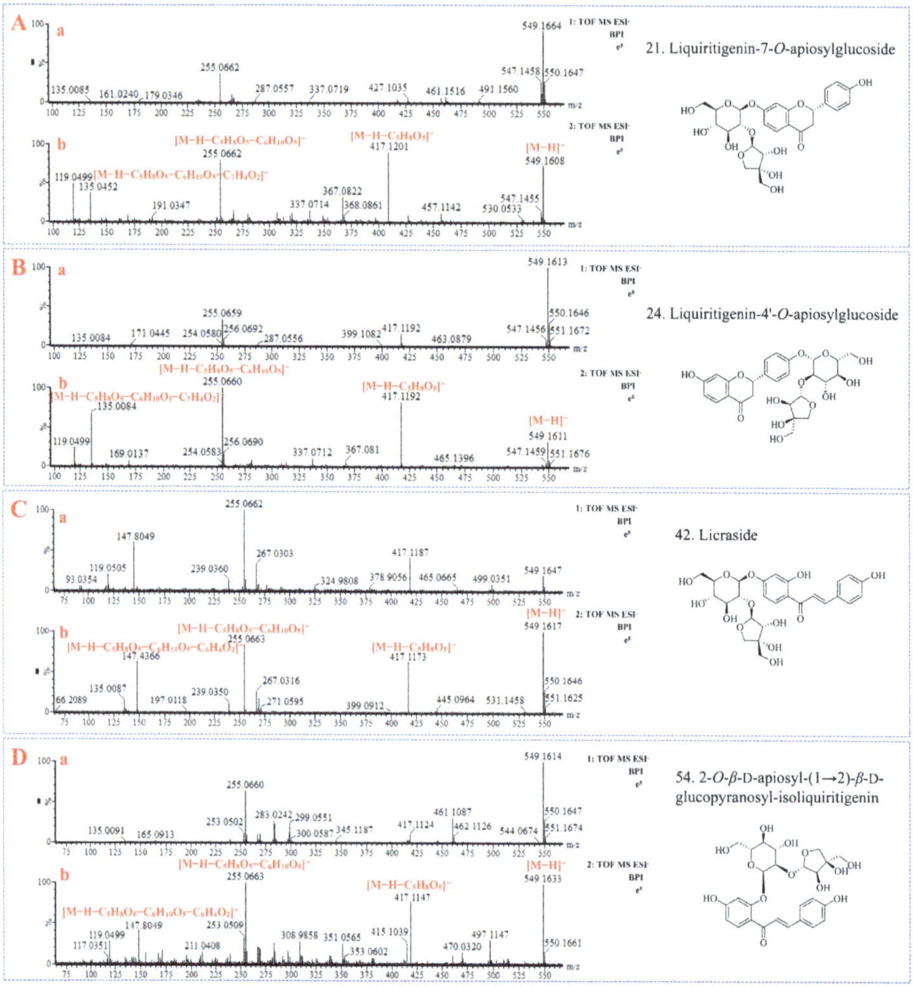

Figure 7. The MS1 (**a**) and MS2 spectra (**b**) of four isomers. (**A**) For liquiritigenin-7-*O*-apiosylglucoside, (**B**) for liquiritigenin-4′-*O*-apiosylglucoside, (**C**) for licraside and (**D**) for 2-*O*-*β*-D-apiosyl-(1→2)-β-D-glucopyranosyl-isoliquiritigenin.

2.2.2. Characterization of Triterpenoid Constituents in HQD

Triterpenoids have gathered significant attention due to their remarkable pharmacological activities. Studies reported that triterpenoids from HQD have anti-inflammatory and hepatoprotective activities [5,40]. In HQD, this class of compounds was predominantly characterized from GRR and JF, with a total of 80 triterpene derivatives identified, primarily consisting of oleanane-type and lupane-type categories. The fragmentation patterns of triterpenoid saponins usually involve the rupture of glycosidic bonds, leading to the formation of various aglycone ions. Licorice saponin A3 and glycyrrhizic acid were used to illustrate the identification process. A precursor ion of $[M-H]^-$ was observed at m/z 983.4520, with a retention time of 9.65 min. In negative ion mode, collision-induced dissociation was performed, resulting in a series of secondary fragment ions, including m/z 821.3993 $[M-H-C_6H_{10}O_5]^-$, m/z 645.3651 $[M-H-C_6H_{10}O_5-C_6H_8O_6]^-$, and m/z 469.3324 $[M-H-C_6H_{10}O_5-C_6H_8O_6-C_6H_8O_6]^-$. Based on the mass spectrometric data, it is suggested that the precursor ion $[M-H]^-$ first generated the fragment ion at

https://doi.org/10.3390/separations13040110

m/z 821.3993 by the loss of one molecule of glucose. Subsequently, this fragment ion further lost one molecule of glucuronic acid, forming the fragment ion at m/z 645.3651. Finally, the fragment ion at m/z 645.3651 lost another molecule of glucuronic acid, resulting in the fragment ion at m/z 469.3324. Comparing these data with existing literature and databases, compound **76** was identified as licorice saponin A3 [41]. The MS/MS mass spectrum of licorice saponin A3 is shown in Figure 8A. The precursor ion $[M-H]^-$, observed at m/z 821.3987 at 16.33 min, underwent fragmentation to yield the ion at m/z 351.0634, resulting from the neutral loss of glycyrrhetinic acid from the parent ion. Subsequently, the fragmented ion at m/z 193.0355 was generated through the neutral loss of dehydrated glucuronic acid from the ion at m/z 351.0634. Compared with our in-house database and reference standard, the compound was chemically defined as glycyrrhizic acid [30]. The MS/MS mass spectrum of glycyrrhizic acid is shown in Figure 8B.

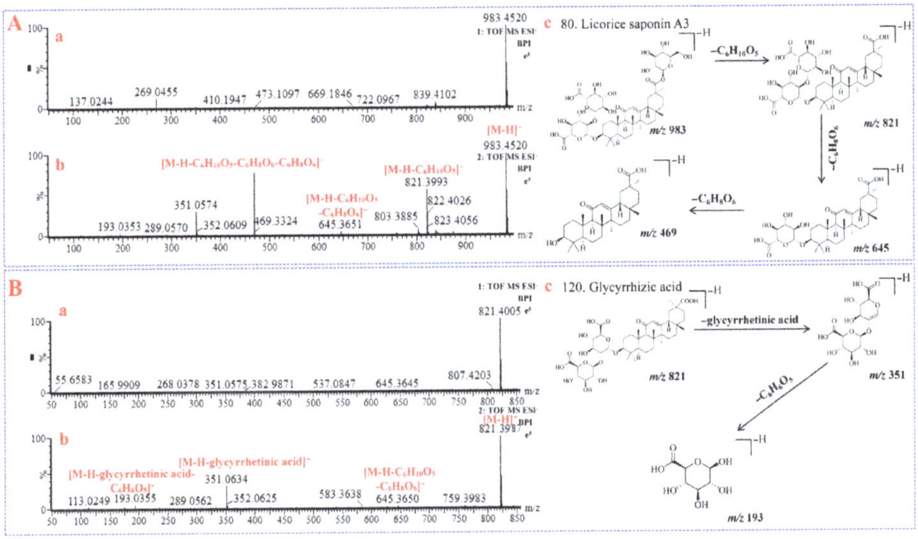

Figure 8. The MS1 (**a**), MS2 spectra (**b**) and characteristic fragmentation pathways (**c**) of licorice saponin A3 (**A**) and glycyrrhizic acid (**B**).

The data indicated that the molecular formulas of compounds **114, 120, 122, 167,** and **168** were all deduced to be $C_{42}H_{62}O_{16}$. Among them, compounds **114** and **122** generated characteristic fragment ions (m/z 351 and 193) in the MS/MS mass spectrum that were identical to those of glycyrrhizic acid, leading to the inference that they are isomers of glycyrrhizic acid. In contrast, compounds **167** and **168** exhibited distinct fragmentation pathways. The mass spectrum of compound **167** (Figure 9A) was dominated by a decarboxylation loss, yielding a prominent $[M-H-CO_2]^-$ ion (m/z 777.4104), while compound **168** (Figure 9B) mainly showed the rearrangement loss of $[M-H-CH_2O]^-$ (m/z 791.3825) of the hydroxymethyl group. Further analysis revealed that their respective aglycone ions, generated after deglycosylation, continued these trends in secondary fragmentation: the aglycone from compound **167** underwent further CO_2 loss, while that from compound **168** underwent further CH_2O loss. Based on these fragmentation signals, compound **167** was tentatively identified as Licorice saponin H2, which contains a carbonyl group at the C-11 position of its aglycone core, while compound **168** was putatively annotated as Licorice saponin K2, which contains a hydroxymethyl group at the C-4 position [41].

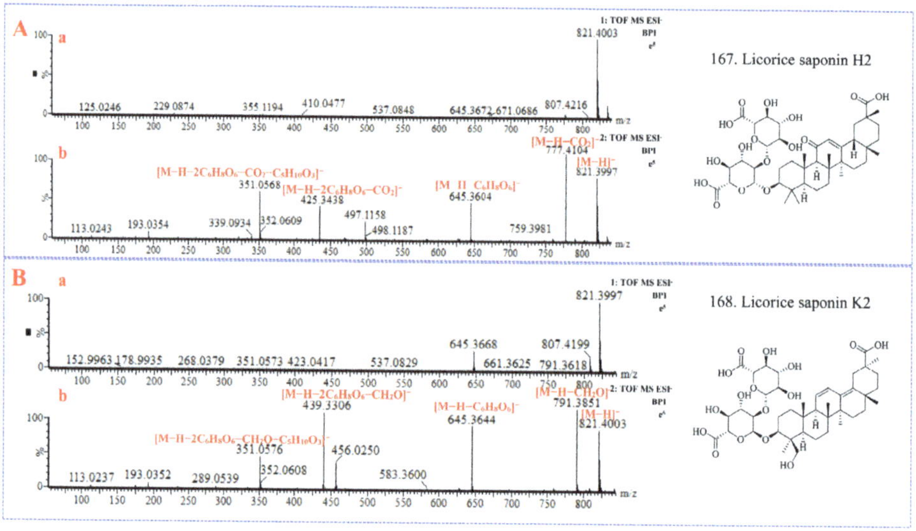

Figure 9. The MS1 (**a**) and MS2 spectra (**b**) of the two isomers. (**A**) For licorice saponin H2, and (**B**) for licorice saponin K2.

2.2.3. Characterization of Monoterpene Glycoside Constituents in HQD

Monoterpene glycosides are the most important type of active components in the medicinal herb PL of HQD. Pharmacological studies have shown that PL has multiple pharmacological effects, including anti-atherosclerosis, anti-diabetes, and anti-tumor activities [42]. Paeoniflorin and galloylpaeoniflorin are representative monoterpene glycosides found in PL, which are used to illustrate the identification process.

The precursor ion $[M-H]^-$ with an m/z of 479.1561 was detected at 2.62 min, yielding fragment ions at m/z 449.1455, 327.1084, and 165.0487. The product ion m/z 449.1455 was determined to have the molecular formula $C_{22}H_{26}O_{10}$, indicating a loss of 30 Da compared to the precursor ion, which suggested the elimination of a molecular CHOH from the glucose moiety. Subsequent fragmentation involved the loss of a benzoyl group, resulting in the formation of the ion at m/z 327.1084. The characteristic product ion at m/z 165.0487 was inferred to be $[M-H-CH_2O-\text{Benzoyl}-C_6H_{10}O_5]^-$, and this fragment originated from the pinane basic framework intrinsic to the monoterpene glycosides (Figure 10A). It showed similar fragmentation pathways with reference standard paeoniflorin [43]. Therefore, we defined the compound as paeoniflorin. The precursor ion peak of the compound with a RT of 3.53 min is m/z 631.1666 $[M-H]^-$. Based on the elemental composition, its possible molecular formula is inferred to be $C_{30}H_{32}O_{15}$. Furthermore, the fragment ions are generated from the parent ion through the sequential loss of neutral fragments: m/z 509.1303 $[M-H-C_7H_6O_2]^-$, 313.0561 $[M-H-C_7H_6O_2-C_{10}H_{12}O_4]^-$, 255.0662 $[M-H-C_7H_6O_2-C_{10}H_{12}O_4-C_2H_2O_2]^-$, 169.0137 $[M-H-C_7H_6O_2-C_{10}H_{12}O_4-C_2H_2O_2-C_3H_2O_3]^-$ were observed. The fragment ions are assigned as follows: the ion at m/z 509.1303 $[M-H-C_7H_6O_2]^-$ is generated from the parent ion by the loss of a benzoic acid. Subsequently, the further loss of the pinane nucleus yields the fragment ion at m/z 313.0561 $[M-H-C_7H_6O-C_{10}H_{12}O_4]^-$. The galloyl glucose residues then undergo cleavage and rearrangement, losing a molecule of $C_2H_2O_2$ to produce the fragment at m/z 225.0662 $[M-H-C_7H_6O_2-C_{10}H_{12}O_4-C_2H_2O_2]^-$. Finally, the loss of a molecule of $C_3H_2O_3$ produces the characteristic gallic acid fragment ion at m/z 169.0137 $[M-H-C_7H_6O_2-C_{10}H_{12}O_4-C_2H_2O_2-C_3H_2O_3]^-$. Based on its characteristic secondary fragmentation data and mass spectral behavior, compound **26** was identified

https://doi.org/10.3390/separations13040110

as galloylpaeoniflorin [43]. The fragmentation pathway in negative ion mode is shown in Figure 10B.

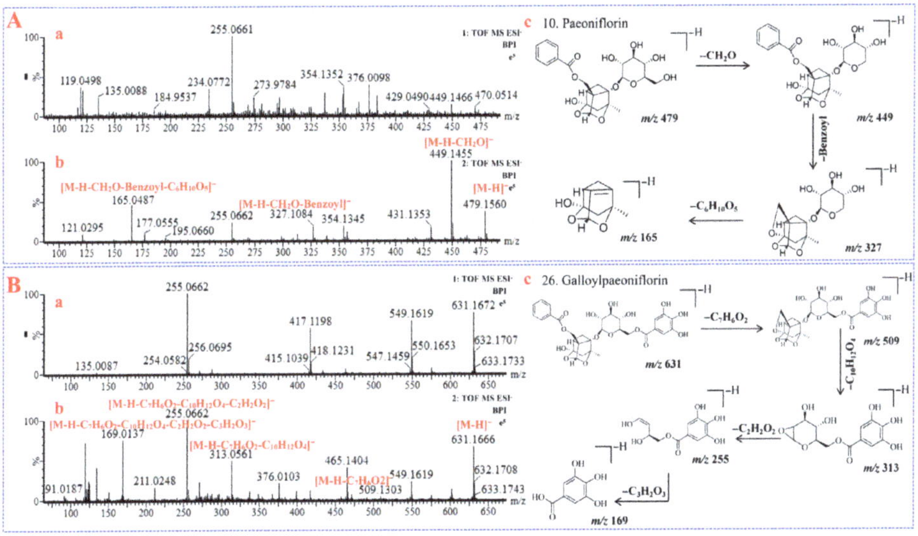

Figure 10. The MS1 (**a**), MS2 spectra (**b**) and characteristic fragmentation pathways (**c**) of paeoniflorin (**A**) and galloylpaeoniflorin (**B**).

2.2.4. Characterization of Organic Acids Constituents in HQD

Organic acids, as core components of natural products, not only participate in energy metabolism within biological systems but also exhibit a variety of bioactivities, including antioxidant, antibacterial, and anti-inflammatory functions [44]. The presence of characteristic functional groups such as carboxyl and hydroxyl groups within their molecular structures results in distinct fragmentation patterns during mass spectrometric analysis, including the elimination of carboxyl groups and the cleavage of aromatic ring moieties. These characteristic fragmentation pathways are not only valuable for compound identification but also provide crucial insights into the mechanisms underlying their bioactivities. The *p*-Hydroxybenzene propanoic acid was used to illustrate the identification process (Figure 11). The precursor ion $[M-H]^-$ observed at m/z 165.0556 at 2.40 min, yields a characteristic fragment ion at m/z 121.0293 through the loss of CO_2 (44 Da). This decarboxylation is a characteristic fragmentation pathway for aromatic carboxylic acids [45]. Based on this fragmentation pattern, together with retention time and literature comparison, compound **7** was putatively annotated as *p*-hydroxybenzene propanoic acid.

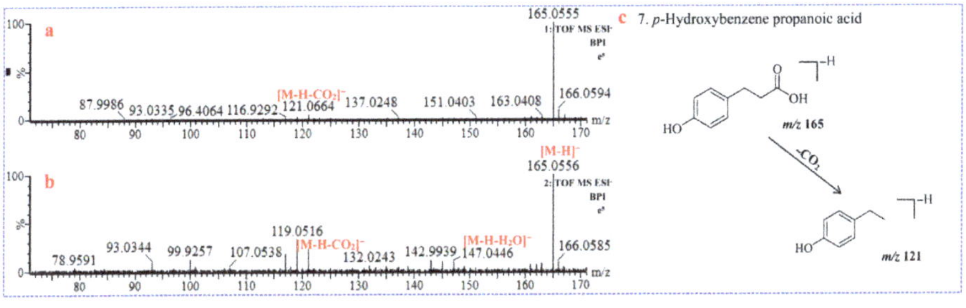

Figure 11. The MS1 (**a**), MS2 spectra (**b**) and characteristic fragmentation pathways (**c**) of *p*-Hydroxybenzene propanoic acid.

https://doi.org/10.3390/separations13040110

3. Discussion

The off-line 2D LC and UHPLC-QTOF/MS strategy established in this study effectively overcame the challenge in comprehensive component profiling of complex traditional Chinese prescription HQD—the masking of trace components by dominant constituents. It is worth noting that the off-line 2D LC and UHPLC-QTOF/MS strategy proposed in this study offers significant advantages for elucidating the comprehensive chemical profile of complex TCPs characterized by complex compositions and large disparities in component content. However, its application may be limited for samples that already exhibit sufficient detection sensitivity using conventional injection analysis. Compared with the traditional analytical strategy, this strategy significantly improved the detection sensitivity of trace components, successfully identifying 192 compounds, including 88 flavonoids, 80 triterpenoids, 7 monoterpene glycosides, 3 fatty acid amides, 3 phenylethanoid glycosides, 4 coumarins, 3 saccharides, 1 organic acid, and 3 others, a substantial supplement from previous studies. The identification of these compounds provides a foundation for understanding the chemical basis of HQD's pharmacological activities. However, it was important to acknowledge the limitations and confidence levels associated with these identifications. Following the MSI guidelines, of the 192 compounds characterized in this study, **11** compounds were identified at confidence Level 1 through comparison with reference standards using retention time and MS/MS fragmentation pattern matching. A total of 146 compounds were assigned Level 2.1 based on matching with publicly or commercially available MS/MS spectral libraries, while **15** compounds were assigned Level 2.2 based on characteristic product ions and neutral losses, supported by structure elucidation tools and network analysis. While Level 2 identifications provided valuable information for chemical profiling, they should be considered tentative pending confirmation with certified reference standards.

It was noteworthy that this study successfully differentiated some isomeric compounds in HQD based on characteristic mass spectrometric fragmentation ions and differences in reversed-phase chromatographic retention behavior. By elucidating the cleavage rules of positional isomers of functional groups and glycosyl linkage positions in flavonoid glycosides, some flavonoid glycoside isomers in HQD were distinguished. Specifically, 7-*O*-glycosides underwent direct heterolytic cleavage of the glycosidic bond due to the absence of a neighboring proton source, characterized by the neutral loss of 176 Da, whereas 8-*O*-glycosides, through intramolecular proton transfer driven by the adjacent phenolic hydroxyl group, triggered a sugar ring rearrangement leading to the preferential neutral loss of 148 Da. The flavanone glycosides are typically characterized by the RDA cleavage, resulting in the characteristic fragment ion m/z 135 $[C_8H_7O_2]^-$. The chalcone glycosides are typically characterized by the β-fragmentation and rearrangement of the enone framework, generating the characteristic ion m/z 147 $[C_9H_7O_2]^-$. The triterpenoid saponins in HQD were primarily represented by oleanane-type pentacyclic triterpenoid saponins derived from GRR. Different isomers could be inferred by observing characteristic fragmentation pathways of their aglycones, such as decarboxylation ($-CO_2$) and hydroxymethyl rearrangement loss ($-CH_2O$) [46]. These regularities provided support for clarifying the complex component groups in HQD. However, for those isomeric pairs that exhibited highly similar fragmentation pathways or lacked characteristic fragment ions, definitive characterization and identification could not be achieved with the current methodology. Future research could integrate additional orthogonal techniques to address this challenge.

The comprehensive chemical profiling conducted in this study provided a solid foundation for understanding the "holistic view" and synergistic effects underlying HQD. The results demonstrated that the chemical constituents of HQD were predominantly

https://doi.org/10.3390/separations13040110

flavonoids and triterpenoids. Among these, albiflorin, paeoniflorin, baicalin, wogonoside, wogonin, baicalein, and glycyrrhizic acid were identified as the seven principal components with the highest content, confirmed at MSI Level 1. Flavonoids exhibited significant anti-inflammatory, antibacterial, and antioxidant activities, contributing to heat-clearing and detoxification, while triterpenoids displayed anti-inflammatory and immunomodulatory functions. Beyond these major constituents, the pharmacological significance of many trace components identified in this study warrants consideration. For example, fatty acid amides are known to function as endogenous signaling molecules with anti-inflammatory and immunomodulatory properties [47–49]. Specifically, stearamide (compound **187**) modulates membrane stability and immune function through mechanisms distinct from flavonoid-mediated COX-2 inhibition. Erucamide (compound **191**) exhibits significant antisecretory effects in intestinal models, while also upregulating anti-inflammatory cytokines (IL-6, TNF-α) in myeloid cells, thereby complementing the anti-inflammatory actions of baicalin and glycyrrhizic acid. Docosanamide (compound **192**) has been characterized as an angiogenic lipid with potential roles in tissue repair, aligning with HQD's cooling-blood effects. Although present at low abundance, these fatty acid amides may contribute to the formula's heat-clearing and detoxifying effects through mechanisms distinct from those of the major flavonoids and triterpenoids. Similarly, trace phenolic acids, such as p-hydroxybenzene propanoic acid (compound **7**), have been reported to exhibit antioxidant and anti-inflammatory activities [50,51], potentially augmenting the overall efficacy of HQD. Collectively, the chemical components of HQD exhibit substantial diversity, ranging from abundant flavonoids and triterpenoids to trace fatty acid amides and phenolic acids. The complexity of their functional networks underpins the multi-target integrated regulation achieved through the emperor-minister-adjuvant-courier compatibility principle. This compositional and mechanistic complexity constitutes the basis for the core efficacy of the prescription in clearing heat, drying dampness, cooling blood, and relieving diarrhea.

4. Materials and Methods

4.1. Reagents, Chemicals, and Plant Materials

LC/MS-grade acetonitrile, formic acid and HPLC-grade acetonitrile were purchased from Thermo Fisher Scientific (Thermo Fisher, Waltham, MA, USA). Distilled water was obtained from Watson's Food & Beverage Co., Ltd. (Guangzhou, China). Other analytical-grade reagents were procured commercially from Sinopharm Chemical Reagent Co., Ltd. (Beijing, China).

SR, PL, GRR, and JF were sourced from Shanxi, Inner Mongolia, and Shandong respectively, with batch numbers 2309044, 2312075, 2304043, and 2401124 in sequence. A total of 11 compounds (purity > 98%), involving baicalin (1), baicalein (2), wogonin (3), wogonoside (4), oroxylin A (5), apigenin (6), liquiritin (7), glycyrrhizic acid (8), paeoniflorin (9), albiflorin (10), galloylpaeoniflorin (11) used as the reference compounds were purchased from Shanghai yuanye Bio-Technology Co., Ltd. (Shanghai, China) or Shanghai Acmec Biochemical Technology Co., Ltd. (Shanghai, China).

4.2. Preparation of Sample Solutions

The SR, PRR, GRR, and JF were ground to a fine powder (<40 mesh). Precisely weighed quantities of 12 g of SR powder, 8 g of PRR powder, 8 g of GRR powder, and 8 g of JF powder were placed in a 250 mL round-bottom flask. Eight times the weight of the powders of 50% ethanol (*v*/*v*) was added, and the mixture was subjected to heating reflux extraction once for 2 h. After filtration, the residue was then subjected to a second heating reflux extraction with eight times the weight of the powders of 50% ethanol

https://doi.org/10.3390/separations13040110

for 2 h, and the two filtrates were combined. The extracted solution was then concentrated under reduced pressure (maintaining a temperature below 45 °C) until a dry residue was obtained. Subsequently, 100 mg of the concentrated extract was re-dissolved in 5 mL of 50% ethanol, filtered through a 0.22 μm filter, and the resulting filtrate was collected for further analysis.

4.3. Sample Preparation

The ^{1}D LC system used in this work was the Thermo Scientific Vanquish Core Analytical Purification System (Thermo Fisher Scientific, USA), which integrates a valve-based VFC into the high-performance Vanquish Core HPLC platform. The integrated platform enables the execution of a complete workflow—from method development and sample analysis to target fraction collection—within a single system. First, the elution conditions were systematically optimized to ultimately establish the following preparative conditions: an Agilent ZORBAX SB-C18 column (4.6 mm × 250 mm, 5 μm) was employed for separation, with the column oven temperature maintained at 35 °C. The mobile phase consisted of eluent A (0.1% formic acid in water) and eluent B (0.1% formic acid in acetonitrile), utilizing a gradient program as follows: 0−10 min, 15−20% B; 10−15 min, 20−23% B; 15−45 min, 23−23% B; 45−50 min, 23−35% B; 50−70 min, 35−35% B; 70−74 min, 35−95% B; 74−76 min, 95−95% B; 76−77 min, 95−15% B; and 77−80 min, 15−15% B, with a flow rate of 1.0 mL/min. The autosampler was set at 18 °C for optimal performance. The injection volume was 100 μL, the injection was repeated 45 times, and the UV detection wavelengths were set at 210 nm and 365 nm. The CAD operated with a collection frequency of 10 Hz, an evaporation temperature of 35 °C, and a filter time of 1 s. The chromatographic peaks with RTs of 8.10−8.75 min, 9.35−9.95 min, 24.20−25.50 min, 31.75−32.75 min, 37.5−39.00 min, 44.50−46.50 min, and 65.00−66.00 min were removed. The sample eluent was collected and simultaneously concentrated to dryness under reduced pressure, with the distillation temperature controlled not to exceed 45 °C, re-dissolved in 50% ethanol, filtered through a 0.22 μm filter, and the subsequent filtrate was obtained.

4.4. UHPLC-Q-TOF/MS Analysis Conditions

The ^{2}D UHPLC-Q-TOF/MS analysis was performed using the Waters ACQUITY UPLC system equipped with Synapt G2-S mass spectrometry (Waters Corp., Milford, MA, USA). A Waters ACQUITY BEH C18 column (2.1 × 100 mm, 1.7 μm, USA) maintained at 35 °C was used for chromatographic separation. The mobile phase system comprised water with 0.1% formic acid (A) and acetonitrile (B), with a 0.4 mL/min flow rate. The gradient was set as follows: 0–2 min, 5–21% B; 2–11 min, 21–32% B; 11–20 min, 32–42% B; 20–23 min, 42–80% B; 23–25 min, 80–95% B; 25–27 min, 95–95% B; 27–28 min, 95–5% B; 28–30 min, 5–5% B. The autosampler temperature was set at 18 °C. The injection volume was 1 μL.

High-accuracy mass spectrometric data were recorded using the QTOF mass spectrometer in both positive and negative electrospray ionization (ESI) modes employing the MS^E. The LockSpray ion source was configured with optimized parameters: mass range of m/z 100–1200, scan time of 0.2 s, source temperature set to 120 °C, desolvation gas temperature at 450 °C, cone gas flow (N_2) of 50 L/h, desolvation gas flow (N_2) at 850 L/h, cone voltage at 30 V, and capillary voltage of 3.0 kV for positive ion mode and −2.5 kV for negative ion mode. In MS^E mode, the low collision energy was set to 6.0 eV, with the high collision energy ramp ranging from 35 to 55 eV for ESI^+ analysis and 30 to 50 eV for ESI^- analysis. Ultra-high purity argon (Ar, purity $\geq$ 99.997%) was used as the collision-induced dissociation (CID) gas for all MS/MS experiments. To ensure mass accuracy and reproducibility, leucine-enkephalin was utilized as the lock mass, with m/z

https://doi.org/10.3390/separations13040110

values of 556.2766 in positive ion mode and 554.2620 in negative ion mode, respectively. The mass spectrometer was operated at a working resolution of 50,000~60,000 FWHM (full width at half maximum) for both positive and negative ion modes.

4.5. Date Processing and Analysis

The acquired MS data were analyzed using Masslynx (Version 4.1, Waters Corp.), while UHPLC-CAD data were processed with Chromeleon 7 software (Version 4.1 Thermo Fisher Scientific, USA). An in-house database was established to compile the primary published phytochemical constituents (including compound name, molecular formula, exact mass, structure, and MS fragmentation) found in the four herbs of HQD. This was accomplished by exploring the TCMID platform (https://ngdc.cncb.ac.cn/databasecommons/database/id/437, accessed between 5 November 2025 and 30 December 2025), TCMSP databases (https://www.tcmsp-e.com/load_intro.php?id=43, accessed between 5 November 2025 and 30 December 2025), and supplemented by the relevant literature from PubMed, Web of Science, and CNKI (https://www.cnki.net/, accessed between 5 November 2025 and 30 December 2025). The fragment ion information obtained from reference standard analysis was summarized, and the structurally relevant fragments were designated as characteristic fragments and neutral loss fragments, respectively. The adduct ions were set as $[M-H]^-$, $[M-H_2O-H]^-$, and $[M+HCOO]^-$ in negative ion mode, while $[M+H]^+$ and $[M+Na]^+$ were used in positive ion mode.

The confidence levels for the characterization of these components are defined according to the MSI as follows: Level 1: structure confirmed by comparison with a standard; 2.1: structure annotated based on publicly or commercially available MS/MS spectral libraries; 2.2: structure putatively annotated using characteristic product ions and neutral losses, supported by structure elucidation tools and network analysis.

5. Conclusions

This study developed and applied an off-line 2D LC and UHPLC-QTOF/MS strategy for the in-depth chemical profiling of HQD, a complex TCP characterized by diverse constituents and significant variations in component content. This strategy enabled the characterization of 192 compounds in HQD, including 88 flavonoids, 80 triterpenoids, 7 monoterpene glycosides, 3 fatty acid amides, 3 phenylethanoid glycosides, 4 coumarins, 3 saccharides, 1 organic acid, and 3 others. Compounds were classified according to MSI confidence levels, with 11 compounds confirmed at Level 1 using reference standards, 146 annotated at Level 2.1 via spectral library matching, and 15 putatively annotated at Level 2.2 based on characteristic fragmentation patterns. While 99 compounds could be identified using a traditional analytical strategy, our approach enabled the detection of 192 peaks in the base peak chromatogram. Notably, this study successfully differentiated isomers of flavonoid glycosides and triterpenoid saponins based on their diagnostic fragmentation behaviors. Flavonoid glycoside isomers were distinguished by characteristic neutral losses, while flavanones and chalcones were identified through RDA cleavage and β-rearrangement, respectively. Furthermore, isomers of triterpenoid saponins were inferred via aglycone-specific fragmentation pathways.

Supplementary Materials: The following supporting information can be downloaded at: https://www.mdpi.com/article/10.3390/separations13040110/s1. Figure S1: The UHPLC-CAD chromatograms of HQD. (A): the injection volume is 1 μL, and the response signal has not undergone amplification processing, (B): the injection volume is 1 μL, and the response signal has undergone amplification processing, (C): the injection volume is 10 μL, and the response signal has undergone amplification processing; Figure S2: Optimization of chromatographic conditions for HQD. (A) Comparison of different chromatographic columns (a for YMC-Triart C18, b for NanoChrom

https://doi.org/10.3390/separations13040110

ChromCore Phenyl, c for COSMOSIL 5C18-AR-II, d for Agilent ZORBAX SB-C18). (B) Comparison of elution conditions on an Agilent ZORBAX SB-C18 column (c for the optimal elution condition). The main seven constituents are indicated by red dots in Figure S3. The number of annotated compounds of HQD in the UHPLC-Q-TOF/MS chromatogram before and after ^{1}D LC system processing; Figure S4. Comparison of Q-TOF peak areas for trace-level components **19**, **21**, **49**, **63**, **76**, **93**, **101**, **130**, and **175** of different structural types before and after ^{1}D LC system processing; Figure S5. The prepared chromatogram of the sample's repeated injection in the ^{1}D LC system. Table S1. Identification of 192 compounds in HQD by UHPLC-QTOF/MS in positive and negative ion modes; Table S2. Relative standard deviations (RSDs) of retention times (RTs) and peak areas for peaks 1~7 in replicate injections. Table S3. Stability recovery of representative compounds during repeated injections and concentration.

Author Contributions: Conceptualization, X.T.; methodology, Y.F., H.L. and B.M.; software, J.Z.; validation, X.T., X.C. and J.S.; formal analysis, H.L.; investigation, X.C. and J.S.; resources, Y.N.; data curation, B.M.; writing—original draft preparation, Y.F.; writing—review and editing, Y.F., H.L. and Y.N.; visualization, J.Z.; supervision, Y.N.; project administration, B.M.; funding acquisition, B.M. All authors have read and agreed to the published version of the manuscript.

Funding: This work was funded by the National Key Research and Development Program of China (2023YFC3502804).

Institutional Review Board Statement: Not applicable.

Informed Consent Statement: Not applicable.

Data Availability Statement: The original contributions presented in this study are included in the article. Further inquiries can be directed to the corresponding authors.

Acknowledgments: The authors have reviewed and edited the output and take full responsibility for the content of this publication.

Conflicts of Interest: The authors declare no conflicts of interest.

Abbreviations

The following abbreviations are used in this manuscript:

TCPs	traditional Chinese prescriptions
TCM	traditional Chinese medicine
HQD	Huangqin decoction
SR	Scutellariae Radix
PRR	Paeoniae Radix Rubra
GRR	Glycyrrhizae Radix et Rhizoma
JF	Jujubae Fructus
CAD	charged aerosol detector
VFC	vanquish fraction collector
UV	ultraviolet
UHPLC-Q-TOF/MS	ultra-high performance liquid chromatography-quadrupole time-of-flight mass spectrometry
ESI	electrospray ionization
BPI	base peak ion
RT	retention time
CID	collision-induced dissociation
FWHM	full width at half maximum

References

1. Wang, Z.B.; Tang, W.; Tong, K.; Huang, W.J.; Chen, B.B.; Qiao, S.J.; Han, H.R.; Li, X.D.; Wang, Y. Thinking on compatibility of prescription and drugs based on perspective of combination of disease, syndrome, and symptom. *J. Beijing Univ. Tradit. Chin. Med.* **2024**, *47*, 929–933.

2. Ding, Y.; Hong, Y.L.; Qi, F.J.; Gan, S.Y.; Wang, Z.J.; Zuo, X.H.; Zhao, Y. The innovative application of "Jun (emperor)—Chen (minister)—Zuo (adjuvant)—Shi (courier)" in clinical practice. *Lishizhen Med. Mater. Medica Res.* **2022**, *33*, 1178–1179. [CrossRef]
3. Liu, L.W.; Tang, M.; Zhang, Z.B.; Zhou, P.P.; Xue, L.P.; Jia, Q.Q.; Zhao, L.G.; Zuo, L.H.; Sun, Z. A stepwise integrated strategy to explore quality markers of Qishen Yiqi dripping pills against myocardial ischemia. *Phytomedicine* **2024**, *135*, 156182. [CrossRef]
4. Wang, X.; Zhang, X.; Li, J.; Fu, J.; Zhao, M.; Zhang, W.; Weng, W.; Li, Q. Network pharmacology and LC-MS approachs to explore the active compounds and mechanisms of Yuanjiang decoction for treating bradyarrhythmia. *Comput. Biol. Med.* **2023**, *152*, 106435. [CrossRef]
5. Guo, T.J.; Yuan, X.; Zou, Y.H.; Li, H.J.; Chen, Y.H.; Qing, Y.; Zheng, Y.F.; Liao, W. Modern research progress on classic formula Huangqin Decoction. *Chin. Tradit. Herb. Drugs* **2025**, *56*, 1414–1427.
6. Wang, L.F. Research progress on classical prescription of Huangqin Decoction. *Chin. Med. Mod. Distance Educ. China* **2020**, *18*, 191–194. [CrossRef]
7. Gu, S.H.; He, M.; Zhu, R.H.; Li, L.; Zhang, T. Historical evolution and key information research of classic Chinese formula Huangqin Decoction. *Shanghai J. Tradit. Chin. Med.* **2024**, *58*, 23–30. [CrossRef]
8. Wang, E.; Bussom, S.; Chen, J.G.; Quinn, C.; Bedognetti, D.; Lam, W.; Guan, F.; Jiang, Z.; Mark, Y.; Zhao, Y.D.; et al. Interaction of a traditional Chinese medicine (PHY906) and CPT-11 on the inflammatory process in the tumor microenvironment. *BMC Med. Genom.* **2011**, *4*, 38. [CrossRef]
9. Lam, W.; Jiang, Z.; Guan, F.; Hu, R.; Liu, S.H.; Chu, E.; Cheng, Y.C. The number of intestinal bacteria is not critical for the enhancement of antitumor activity and reduction of intestinal toxicity of irinotecan by the Chinese herbal medicine PHY906 (KD018). *BMC Complement. Altern. Med.* **2014**, *14*, 490. [CrossRef]
10. Lam, W.; Bussom, S.; Guan, F.; Jiang, Z.; Zhang, W.; Gullen, E.A.; Liu, S.H.; Cheng, Y.C. The four-herb Chinese medicine PHY906 reduces chemotherapy-induced gastrointestinal toxicity. *Sci. Transl. Med.* **2010**, *2*, 45–59. [CrossRef]
11. Lam, W.; Jiang, Z.; Guan, F.; Huang, X.; Hu, R.; Wang, J.; Bussom, S.; Liu, S.H.; Zhao, H.; Yen, Y.; et al. PHY906(KD018), an adjuvant based on a 1800-year-old Chinese medicine, enhanced the anti-tumor activity of Sorafenib by changing the tumor microenvironment. *Sci. Rep.* **2015**, *5*, 9384. [CrossRef]
12. Yang, X.; Lam, W.; Jiang, Z.; Guan, F.; Han, X.; Hu, R.; Cai, W.; Cheng, W.; Liu, S.H.; Cheng, P.; et al. YIV-906 potentiated anti-PD1 action against hepatocellular carcinoma by enhancing adaptive and innate immunity in the tumor microenvironment. *Sci. Rep.* **2021**, *11*, 13482. [CrossRef]
13. Lam, W.; Hu, R.; Liu, S.H.; Cheng, P.; Cheng, Y.C. YIV-906 enhances nuclear factor of activated T-cells (NFAT) activity of T cells and promotes immune checkpoint blockade antibody action and CAR T-cell activity. *Front. Pharmacol.* **2023**, *13*, 1095186. [CrossRef]
14. Li, M.X.; Li, M.Y.; Lei, J.X.; Wu, Y.Z.; Li, Z.H.; Chen, L.M.; Zhou, C.L.; Su, J.Y.; Huang, G.X.; Huang, X.Q.; et al. Huangqin decoction ameliorates DSS-induced ulcerative colitis: Role of gut microbiota and amino acid metabolism, mTOR pathway and intestinal epithelial barrier. *Phytomedicine* **2022**, *100*, 154052. [CrossRef]
15. Huang, S.; He, J.; Chen, Y.; Wang, X.; Li, Y.; Su, Y.; Wen, R.; Li, X.; Yang, G.; Luo, S.; et al. Effect of Huangqin decoction on regulating intestinal flora in colitis mice characterized as inhibition of the NOD2-dependent pathway. *Pharm. Biol.* **2022**, *60*, 108–118. [CrossRef]
16. Li, M.Y.; Li, M.X.; Xu, N.; Li, Z.H.; Zhang, Y.M.; Gan, Y.X.; Luo, H.J.; Zhou, C.L.; Liu, Y.H.; Su, Z.R.; et al. Effects of Huangqin Decoction on ulcerative colitis by targeting estrogen receptor alpha and ameliorating endothelial dysfunction based on system pharmacology. *J. Ethnopharmacol.* **2021**, *271*, 113886. [CrossRef] [PubMed]
17. Wang, D.; Ma, X.; Guo, S.; Wang, Y.; Li, T.; Zou, D.; Song, H.; Yang, W.; Ge, Y. Effect of Huangqin Tang on urine metabolic profile in rats with ulcerative colitis based on UPLC-Q-Exactive Orbitrap MS. *Evid.-Based Complement. Altern. Med.* **2020**, *2020*, 1874065. [CrossRef]
18. Yan, B.F.; Pan, L.F.; Quan, Y.F.; Sha, Q.; Zhang, J.Z.; Zhang, Y.F.; Zhou, L.B.; Qian, X.L.; Gu, X.M.; Li, F.T.; et al. Huangqin decoction alleviates lipid metabolism disorders and insulin resistance in nonalcoholic fatty liver disease by triggering Sirt1/NF-κB pathway. *World J. Gastroenterol.* **2023**, *29*, 4744–4762. [CrossRef]
19. Yan, B.F.; Wang, Y.; Wang, W.B.; Ding, X.J.; Wei, B.; Liu, S.J.; Fu, T.M.; Chen, L.; Zhang, J.Z.; Liu, J.; et al. Huangqin decoction mitigates hepatic inflammation in high-fat diet-challenged rats by inhibiting TLR4/NF-κB/NLRP3 pathway. *J. Ethnopharmacol.* **2023**, *303*, 115999. [CrossRef]
20. Wang, J.; Li, R.; Zhang, M.; Gu, C.; Wang, H.; Feng, J.; Bao, L.; Wu, Y.; Chen, S.; Zhang, X. Influence of Huangqin Decoction on the immune function and fecal microbiome of chicks after experimental infection with Escherichia coli O78. *Sci. Rep.* **2022**, *12*, 16632. [CrossRef]
21. Xu, X.; Fang, C.; Wang, Y.; Lu, F.; Liu, S. Integrating network pharmacology and metabolomics to elucidate the mechanism of action of Huang Qin Decoction for treament of diabetic liver injury. *Front. Pharmacol.* **2022**, *13*, 899043. [CrossRef]

https://doi.org/10.3390/separations13040110

22. Gao, L.; Zhang, A.P.; Fu, L.; Li, Q.W.; Qin, X.M.; Zhao, J. Huangqin decoction attenuates spared nerve injury (SNI)-induced neuropathic pain by modulating microglial M1/M2 polarization partially mediated by intestinal nicotinamide metabolism. *Phytomedicine* **2024**, *129*, 155594. [CrossRef]
23. Gao, L.; Cao, M.; Du, G.H.; Qin, X.M. Huangqin Decoction exerts beneficial effects on rotenone-induced rat model of Parkinson's disease by improving mitochondrial dysfunction and alleviating metabolic abnormality of mitochondria. *Front. Aging Neurosci.* **2022**, *14*, 911924. [CrossRef] [PubMed]
24. Schilling, K.; Holzgrabe, U. Recent applications of the charged aerosol detector for liquid chromatography in drug quality control. *J. Chromatogr. A* **2020**, *1619*, 460911. [CrossRef]
25. Wu, L.; Gong, X.; Qu, H. An optimization strategy for charged aerosol detection to linearize the detector response in the multicomponent quantitative analysis of Qishen Yiqi dripping pills. *J. Sep. Sci.* **2024**, *47*, e2300784. [CrossRef]
26. Rosentreter, U.; Huber, U. Optimal fraction collecting in preparative LC/MS. *J. Comb. Chem.* **2004**, *6*, 159–164. [CrossRef] [PubMed]
27. Song, J.; Liu, S.; Sun, M.; Tu, G.; Zhang, Z.; Liu, Y.; Nan, Y.; Shen, G.; Chen, X.; Liang, H.; et al. Isolation and structural elucidation of immunomodulatory oligosaccharides from Trillium tschonoskii. *Glycoconj. J.* **2026**, *43*, 8. [CrossRef] [PubMed]
28. Yao, L.; Wang, X.; Nan, Y.; Liang, H.Z.; Wang, M.Y.; Song, J.; Chen, X.J.; Ma, B.P. Exploring the chemical compositions of Fufang Yinhua Jiedu granules based on ultra-high-performance liquid chromatography coupled to quadrupole time-of-flight mass spectrometry combined with multistage intelligent data annotation strategy. *J. Chromatogr. A* **2024**, *1728*, 465010. [CrossRef]
29. Wang, X.; Wu, H.; Li, M.; Guo, X.; Cheng, X.; Jing, W.; Wei, F. A comprehensive analysis of Fel Ursi and its common adulterants based on UHPLC-QTOF-MSE and chemometrics. *Molecules* **2024**, *29*, 3144. [CrossRef]
30. Sun, Z.R.; Jia, D.; Chen, X.F.; Chai, Y.F. Identification on components of Huangqin Tang and rats blood components after administration based on UHPLC-QTOF MS analysis. *J. Instrum. Anal.* **2023**, *42*, 960–967. [CrossRef]
31. Tang, X.; Cui, Y.; Feng, B. The chemical constituents and metabolite profiles of Huangqin decoction in normal and ulcerative colitis rats by UHPLC-Q-TOF/MS analysis. *J. Pharm. Biomed. Anal.* **2024**, *237*, 115763. [CrossRef]
32. Jia, L.Y.; Zheng, F.Y.; Ma, Y.; Zhang, Y.F. Studies on chemical composition, pharmacological effect, clinical application of Huangqin Decoction and its quality marker prediction analysis. *J. Liaoning Univ. Tradit. Chin. Med.* **2025**, *27*, 149–156. [CrossRef]
33. Teng, X.; Wu, B.; Liang, Z.; Zhang, L.; Yang, M.; Liu, Z.; Liang, Q.; Wang, C. Three bioactive compounds from Huangqin decoction ameliorate Irinotecan-induced diarrhea via dual-targeting of Escherichia coli and bacterial β-glucuronidase. *Cell Biol. Toxicol.* **2024**, *40*, 88. [CrossRef]
34. Wu, Y.; Liu, X.; Li, G. Integrated bioinformatics and network pharmacology to identify the therapeutic target and molecular mechanisms of Huangqin decoction on ulcerative Colitis. *Sci. Rep.* **2022**, *12*, 159. [CrossRef] [PubMed]
35. Zhang, B.W.; Dong, Y.S.; Yu, N.; Sun, Y.M.; Xing, Y.; Yang, F.; Yua, X.X.; Sun, W.L.; Sun, J.; Lia, X.; et al. Intestinal metabolism of baicalein after oral administration in mice: Pharmacokinetics and mechanisms. *J. Funct. Foods* **2019**, *54*, 53–63. [CrossRef]
36. Łukasz, M.; Maciej, S.; Michał, J.; Wieslaw, O.; Piotr, K. Fragmentation pathways of acylated flavonoid diglucuronides from leaves of Medicago truncatula. *Phytochem. Anal.* **2010**, *21*, 224–233. [CrossRef]
37. Zhu, H.D.; Tian, P.; Guan, L.J.; Chen, L.M.; Liu, X.Q.; Gao, H.M.; Wang, Z.M. Analysis and structural identification of relevant substances in Breviscapine for Injection. *Zhongguo Zhong Yao Za Zhi* **2020**, *45*, 1350–1356. [CrossRef]
38. Kitagawa, I.; Hori, K.; Uchida, E.; Chen, W.Z.; Yoshikawa, M.; Ren, J. Saponin and sapogenol. L. On the constituents of the roots of *Glycyrrhiza uralensis* Fischer from Xinjiang, China. Chemical structures of licorice-saponin L3 and isoliquiritin apioside. *Chem. Pharm. Bull.* **1993**, *41*, 1567–1572. [CrossRef]
39. Batiha, G.E.-S.; Beshbishy, A.M.; El-Mleeh, A.; Abdel-Daim, M.M.; Devkota, H.P. Traditional Uses, Bioactive Chemical Constituents, and Pharmacological and Toxicological Activities of *Glycyrrhiza glabra* L. (Fabaceae). *Biomolecules* **2020**, *10*, 352. [CrossRef] [PubMed]
40. Yang, R.; Yuan, B.C.; Ma, Y.S.; Zhou, S.; Liu, Y. The anti-inflammatory activity of licorice, a widely used Chinese herb. *Pharm. Biol.* **2017**, *55*, 5–18. [CrossRef]
41. Yin, R.; Yu, X.Y.; Wei, X.X.; Shi, C.Z.; Zhao, S.; Wang, X.; Sun, Y.F.; Han, F. Analysis of chemical components and components migrating to blood of Guizhi Gancao Decoction by UHPLC-Q-Exactive Orbitrap MS method. *Chin. Arch. Tradit. Chin. Med.* **2025**, *43*, 198–207. [CrossRef]
42. Yang, S.; Zhang, X.; Dong, Y.; Sun, G.; Jiang, A.; Li, Y. Cleavage rules of mass spectrometry fragments and rapid identification of chemical components of Radix Paeoniae Alba using UHPLC-Q-TOF-MS. *Phytochem. Anal.* **2021**, *32*, 836–849. [CrossRef] [PubMed]
43. He, X.; Wu, H.M.; Xiong, L.L.; Jin, H.L.; Liu, Y.F.; Shen, A.J.; Liang, X.M. Monoterpenoids from Paeoniae Radix Rubra based on UPLC-Q-TOF-MS. *China J. Chin. Mater. Medica* **2023**, *48*, 1005–1013. [CrossRef]
44. Ehteshami, S.; Abdollahi, F.; Ramezanian, A.; Rahimzadeh, M.; Dastjerdi, A.M. Maintenance of quality and bioactive compounds of cold stored pomegranate (*Punica granatum* L.) fruit by organic acids treatment. *Food Sci. Technol. Int.* **2021**, *27*, 151–163. [CrossRef] [PubMed]

https://doi.org/10.3390/separations13040110

45. Hossain, M.B.; Rai, D.K.; Brunton, N.P.; Martin-Diana, A.B.; Barry-Ryan, C. Characterization of phenolic composition in Lamiaceae spices by LC-ESI-MS/MS. *J. Agric. Food Chem.* **2010**, *58*, 10576–10581. [CrossRef] [PubMed]
46. Zheng, Y.F.; Qi, L.W.; Zhou, J.L.; Li, P. Structural characterization and identification of oleanane-type triterpene saponins in *Glycyrrhiza uralensis* Fischer by rapid-resolution liquid chromatography coupled with time-of-flight mass spectrometry. *Rapid Commun. Mass Spectrom.* **2011**, *24*, 3261–3270. [CrossRef]
47. Dang, H.T.; Kang, G.J.; Yoo, E.S.; Hong, J.; Choi, J.S.; Kim, H.S.; Chung, H.Y.; Jun, J.H. Evaluation of endogenous fatty acid amides and their synthetic analogues as potential anti-inflammatory leads. *Bioorganic Med. Chem.* **2011**, *19*, 1520–1527. [CrossRef]
48. Raboune, S.; Stuart, J.M.; Leishman, E.; Takacs, S.M.; Rhodes, B.; Basnet, A.; Jameyfield, E.; McHugh, D.; Widlanski, T.; Bradshaw, H.B. Novel endogenous N-acyl amides activate TRPV1-4 receptors, BV-2 microglia, and are regulated in brain in an acute model of inflammation. *Front. Cell. Neurosci.* **2014**, *8*, 195. [CrossRef]
49. Tahara, K.; Kobayashi, M.; Yoshida, S.; Onodera, R.; Inoue, N.; Takeuchi, H. Effects of cationic liposomes with stearylamine against virus infection. *Int. J. Pharm.* **2018**, *543*, 311–317. [CrossRef]
50. Zhang, Y.Y.; Li, X.L.; Li, T.Y.; Li, M.Y.; Huang, R.M.; Li, W.; Yang, R.L. 3-(4-Hydroxyphenyl)propionic acid, a major microbial metabolite of procyanidin A2, shows similar suppression of macrophage foam cell formation as its parent molecule. *Rsc Adv.* **2018**, *8*, 6242–6250. [CrossRef]
51. Kiriyama, Y.; Tokumaru, H.; Sadamoto, H.; Kobayashi, S.; Nochi, H. Effects of phenolic acids produced from food-derived flavonoids and amino acids by the gut microbiota on health and disease. *Molecules* **2024**, *29*, 5102. [CrossRef] [PubMed]

https://doi.org/10.3390/separations13040110

foods

MDPI

Article

Sex- and Regio-Specific Lipid Profiling of Shishamo and Capelin Fish by Nontargeted Liquid Chromatography/Mass Spectrometry

Yusuke Minami [1], Siddabasave Gowda B. Gowda [2,3,*], Divyavani Gowda [2], Hitoshi Chiba [4] and Shu-Ping Hui [2,*]

1 Graduate School of Health Sciences, Hokkaido University, Kita-12, Nishi-5, Kita-ku, Sapporo 060-0812, Japan; minami.yusuke.t2@elms.hokudai.ac.jp
2 Faculty of Health Sciences, Hokkaido University, Kita-12, Nishi-5, Kita-ku, Sapporo 060-0812, Japan; divyavani@hs.hokudai.ac.jp
3 Graduate School of Global Food Resources, Hokkaido University, Kita-9, Nishi-9, Kita-Ku, Sapporo 060-0809, Japan
4 Department of Nutrition, Sapporo University of Health Sciences, Nakanuma, Nishi-4-3-1-15, Higashi-ku, Sapporo 007-0894, Japan; chiba-h@sapporo-hokeniryou-u.ac.jp
* Correspondence: gowda@gfr.hokudai.ac.jp (S.G.B.G.); keino@hs.hokudai.ac.jp (S.-P.H.)

Abstract

Shishamo smelt (*Spirinchus lanceolatus*), which is endemic to Hokkaido, Japan, is frequently substituted in markets with morphologically similar capelin (*Mallotus villosus*) imported from abroad. Lipids are essential nutrients that play important roles in fish authenticity, validation, and nutritional assessment. Although shishamo has long been consumed in Japan, its region-specific lipid distribution and composition are different from those of capelin and have not been well explored. To overcome these gaps, we used untargeted liquid chromatography–mass spectrometry to profile sex- and region-specific lipids in the Japanese *S. lanceolatus* and Nordic *M. villosus*. The results revealed that female *S. lanceolatus* heads exhibited high triacylglycerol (TAG) content, closely resembling roe composition. Docosahexaenoic acid and eicosapentaenoic acid were enriched in the female *S. lanceolatus* viscera. Multivariate analysis identified monounsaturated fatty acids, such as fatty acid (FA) 22:1 and FA 20:1, as robust discriminatory markers between *S. lanceolatus* and *M. villosus*. Overall, sex- and regiospecific differences in lipid composition between the two species were correlated and compared. These lipidomic signatures provide a basis for verifying species authenticity and geographic origin, while highlighting the nutritional lipid potential of *S. lanceolatus*, particularly from the female viscera.

Keywords: liquid chromatography; fish lipids; mass spectrometry; lipidomics; fatty acid; shishamo

1. Introduction

Shishamo smelt (*Spirinchus lanceolatus*) is an endemic species restricted to Hokkaido, the northernmost island of Japan [1]. In early winter, it has an anadromous life cycle, migrating from the coastal ocean to the lower reaches of rivers to spawn [2]. Owing to its restricted distribution and seasonally limited landings, shishamo is a culturally and economically important fishery resource in Hokkaido. However, shishamo populations have declined markedly, leading to its classification as an endangered species on Japan's Red List [1,3]. Capelin (*Mallotus villosus*), a closely related osmerid often referred to in Japan as "karafuto-shishamo," occupies sub-Arctic pelagic habitats throughout the Northern Hemisphere [4]. Although a large proportion of the global capelin catch is processed into fishmeal, prespawning individuals are also valued for direct human consumption because of their favorable nutritional profiles [5]. In Japan, the majority of "shishamo" marketed is imported frozen capelin, yet the biological and nutritional differences between shishamo and capelin are poorly understood [6].

Fish lipids are a well-recognized dietary source of bioactive compounds with significant health benefits [7]. These benefits are largely attributed to their high content of ω-3 polyunsaturated fatty acids (PUFAs), particularly eicosapentaenoic acid (EPA) and docosahexaenoic acid (DHA) [8]. ω-3 PUFAs lower circulating triacylglycerol levels and reduce the risk of neurodegenerative disorders, including Alzheimer's disease, dementia, and certain cancers [9]. By contrast, saturated fatty acids (SFAs), especially long-chain SFAs (LCSFAs), are associated with elevated serum cholesterol and increased cardiovascular risk, as suggested by previous studies [10]. To evaluate the nutritional quality of fish-derived lipids, free fatty acid (FFA)-based indices such as the PUFA/SFA ratio, index of atherogenicity (IA), and hypocholesterolemic/hypercholesterolemic (HH) ratio, are widely employed [11].

Advances in lipidomics now allow for the comprehensive characterization of food lipidomes and nutritional profiles beyond FFA-based assessments [12]. Lipid composition in fish is modulated by species, age, diet, and environment [13]. Lipidomics provides valuable insights into species-specific nutritional properties, feeding ecology, and even geographic origin [14]. For instance, in capelin (*Mallotus villosus*), ω-3 PUFAs account for ~71% of total lipids, with EPA and DHA constituting over 16% [5]. Long-chain PUFAs in food typically occur in triacylglycerols (TAGs) and phospholipids (PLs) [15], and evidence suggests that ω-3 PUFAs in TAG and PL forms enhance tissue incorporation more effectively than free forms [16]. Therefore, comprehensive lipidomic analyses, including TAGs, PLs, and other complex lipid classes, are crucial for evaluating the nutritional and functional potential.

Sex-and region-specific differences also influence fish lipid profiles. For example, ayu (*Plecoglossus altivelis*) roe contains nearly three times more phospholipids than muscle [17], and female capelin accumulates greater lipid reserves in the gonads than males [18]. Recently, organ-specific lipidomic maps of capelin revealed that ceramides (Cer) are enriched in the viscera and phosphatidylcholine (PC) is enriched in the roe [5]. Our previous work analyzing Japanese marine products demonstrated that female shishamo smelt (*Spirinchus lanceolatus*) meat is particularly rich in ω-3 PUFAs [19]. However, sex- and region-specific differences in lipid composition between shishamo and capelin have not yet been fully elucidated. In the current study, we performed a comprehensive lipidomic comparison of shishamo from Hokkaido, Japan, and capelin from Nordic countries, two species widely consumed in Japan but often confused in markets. Using liquid chromatography/mass spectrometry (LC/MS)-based untargeted lipidomics, we assessed the lipid composition across multiple body parts (head, meat, viscera, and roe) and sexes. This approach enabled us to identify species-, sex-, and region-specific lipid signatures, discuss their nutritional implications, and propose candidate molecular markers to distinguish shishamo from capelin.

2. Materials and Methods

2.1. Materials

LC-MS-grade solvents, including methanol, isopropanol, and chloroform, were purchased from Wako Pure Chemical Industries (Osaka, Japan). Ammonium acetate (1 M), used as the mobile phase additive, was obtained from Sigma-Aldrich (St. Louis, MO, USA). For sample homogenization, 1.4 mm ceramic beads (item no. 15-340-159) were purchased from Thermo Fisher Scientific (Tokyo, Japan). Pierce™ LTQ ESI Positive Ion Calibration Solution and Pierce™ Negative Ion Calibration Solution were obtained from Thermo Fisher Scientific (Tokyo, Japan). To prepare the internal standards, EquiSPLASH LIPIDOMIX and

https://doi.org/10.3390/foods15020298

oleic acid (d9) were obtained from Avanti Polar Lipids, Inc. (Alabaster, AL, USA). A mixed internal standard solution was prepared in methanol (1 μg/mL) containing: phosphatidylcholine (PC 15:0/18:1(d7)), phosphatidylethanolamine (PE 15:0/18:1(d7)), phosphatidylglycerol (PG 15:0/18:1(d7)), phosphatidylserine (PS 15:0/18:1(d7)), phosphatidylinositol (PI 15:0/18:1(d7)), lysophosphatidylethanolamine (LPE 18:1(d7)), lysophosphatidylcholine (LPC 18:1(d7)), sphingomyelin (SM d18:1/18:0(d9)), ceramide (Cer d18:1/15:0(d7)), monoacylglycerol (MAG) (18:1(d7)), diacylglycerol (DAG), triacylglycerol (TAG) (15:0–18:1(d7)-15:0), and cholesterol ester (18:1(d7)) along with 10 μg/mL oleic acid (d9).

2.2. Sample Preparation

Male and female shishamo smelt (*Spirinchus lanceolatus*) from Hokkaido, Japan, were obtained from Nomoto Suisan Co., Ltd. (Hokkaido, Japan). Male capelin (*Mallotus villosus*) from Iceland was purchased from Taikai Co., Ltd. (Tottori, Japan), and female capelin from Norway was purchased from Kanenaka Food Industry Co., Ltd. (Ishikawa, Japan). Sex and geographical origin were completely confounded in the current study. This mixed design makes it difficult to disentangle the independent contributions of sex-specific effects (e.g., hormonal differences) and region-specific effects (e.g., feeding grounds or environmental factors), which may complicate the interpretation of their interactions. Head, meat, viscera, and roe (female only) were dissected within 10 min of arrival and stored at −80 °C until extraction. Four individuals of each fish species were prepared, dissected to obtain specific tissues, and subjected to four replicate analyses.

2.3. Lipid Extraction

Approximately 100 mg of tissue was homogenized with 1.4 mm ceramic beads (Beadmill 4, Fisher Scientific, Tokyo, Japan). One milliliter of methanol containing 0.01% butylated hydroxytoluene was added, followed by the extraction of total lipids using a modified Folch method [19,20]. Each homogenate was spiked with an internal standard, mixed with chloroform and Milli-Q water, and then centrifuged. The chloroform phase was collected, re-extracted, dried under vacuum, and reconstituted in 100 μL methanol. After centrifugation at 15,000 rpm (10 min, 4 °C), the supernatant was transferred to autosampler vials for LC/MS analysis.

2.4. LC/MS Analysis

For LC-MS analysis, a protocol similar to that reported in our previous study was followed [21]. Lipid profiling was performed using a Prominence UFLC system (Shimadzu, Kyoto, Japan) coupled to an LTQ Orbitrap mass spectrometer (Thermo Fisher Scientific, San Jose, CA, USA) with an Atlantis T3 C18 column (2.1 mm × 150 mm, 3 μm, Waters). The column temperature was maintained at 40 °C. Lipids were separated using (A) Milli-Q (10 mM ammonium acetate), (B) isopropanol, and (C) methanol with gradient elution in both positive and negative mode at 200 μL/min over 30 min. Eluent distribution in positive mode was 30% B and 35% C (0–1 min), 82.5% B and 15% C (1–9 min), 95% B and 5% C (9–15 min), 95% B and 5% C (15–25 min), 30% B and 35% C (25–26 min) and equilibrated for 4 min. Eluent distribution in negative mode was 30% B and 35% C (0–1 min), 75% B and 15% C (1–9 min), 82.5% B and 15% C (9–21 min), 95% B and 5% C (21–25 min), 30% B and 35% C (25–26 min), and equilibrated for 4 min. Data were acquired in electrospray ionization (ESI) positive and negative modes, with the following conditions: capillary temperature 330 °C; nitrogen sheath gas flow rate 50 units; auxiliary gas flow rate 20 units (positive) and 30 units (negative); source voltage 4 kV (positive) and 3 kV (negative); and capillary voltage 25 V (positive) and 10 V (negative). The MS^1 scan ranges were m/z 100–1750 (positive), and m/z 160–1900 (negative) in Fourier transform mode with a resolution of 60,000. MS^2 and MS^3 spectra were acquired using collision-induced dissociation (CID) at

https://doi.org/10.3390/foods15020298

collision energies of 40 and 45 eV, respectively. Calibration was performed using Pierce™ LTQ ESI Positive Ion Calibration Solution and Pierce™ Negative Ion Calibration Solution (both from Thermo Fisher Scientific, Tokyo, Japan). Before sample analysis, blanks were used to confirm the absence of contamination and quality control (QC) samples were used to monitor the performance of the analytical instrument.

2.5. Data Processing for Identification of Lipids

Alignment, lipid annotation, and peak area integration of the raw data acquired by LC-MS analysis were performed using MS-DIAL ver. 5.1. The parameters set in MS-DIAL were the same as previously reported [22], specifically employing a minimum peak height of 1000 amplitude units, 0.1 Da mass slice width, smoothing across 3–5 scans with a 0.5 sigma window, signals at least fivefold above blank levels, and tolerances of 0.5 min for retention time alongside 0.015 Da for MS^1. To confirm the identification of lipid molecular species, MS^1, MS^2, and MS^3 spectra were confirmed using Xcalibur 2.2 (Thermo Fisher Scientific, Waltham, MA, USA). Relative quantification was performed using internal standards of the same or analogous lipid subclasses. Analyte levels were determined by multiplying the peak area ratio (analyte/internal standard) by the internal standard concentration and normalized to sample weight, following the Level 2 or 3 guidelines of the Lipidomics Standard Initiative.

2.6. Determination of Lipid Nutritional Indices

For nutritional assessment based on FFA, the following formulas were used to calculate the nutritional indices: fish lipid quality (FLQ), hypo/hypercholesterolemic ratio (HH), and index of atherogenicity (IA) [11].

$$\text{FLQ} = 100 \times (\text{C20:5} + \text{C22:6})/\Sigma\text{FA}$$

$$\text{HH} = (\text{C18:1} + \Sigma\text{PUFA})/(\text{C12:0} + \text{C14:0} + \text{C16:0})$$

$$\text{IA} = [\text{C12:0} + (4 \times \text{C14:0}) + \text{C16:0}]/\Sigma\text{UFA}$$

2.7. Statistical Analysis

Multivariate analyses (principal component analysis, heatmaps, and orthogonal partial least squares discriminant analysis) were performed using MetaboAnalyst 6.0, and additional graphs/statistics were generated using GraphPad Prism 8. Data are expressed as mean $\pm$ standard error (SE). Group comparisons for head, meat, and viscera were assessed using one-way analysis of variance (ANOVA) with Tukey's post hoc test; roe lipids were compared using Student's *t*-test. Distinct letters in the figures denote significant group differences ($p < 0.05$).

3. Result & Discussion

3.1. Lipid Annotation and Distribution of Major Lipid Classes

Untargeted LC-MS analysis annotated 329 lipid molecular species in *Spirinchus lanceolatus* (shishamo) and *Mallotus villosus* (capelin) in both positive and negative ionization modes (Supplementary Table S1). Scatter plots (m/z vs. retention time) representing the elution of different lipid subclasses are shown in Figure 1A. This illustrates the distribution of the detected lipid species, which were classified into five major categories: fatty acyls (FA), sphingolipids (SP), glycerophospholipids (GP), glycerolipids (GL), and sterols (ST). The relative abundances of these categories differed considerably according to the species, sex, and body region (Figure 1B). In capelin, GP predominated in male head (43.9%) and meat (44.2%), whereas FA dominated viscera (63.1%). Female capelins also exhibited high GP

https://doi.org/10.3390/foods15020298

in the head (53.7%), high FA in viscera (43.8%), and high GL in meat (39.0%) and roe (62.5%). By contrast, male shishamo heads were GP-rich (60.1%), while meat contained elevated GL (39.6%), and viscera showed balanced GP and FA (~33% each), respectively. Elevated levels of GPs were detected in the lipid profile of the golden threadfin bream head [23]. Interestingly, female shishamo heads were GL-dominated (71.8%), a unique lipid feature absent in the other head samples. Previous studies have reported that TAG, a type of GL, is the predominant lipid class in salmon head paste [24]. Female shishamo meat (49.9%) and viscera (53.8%) were FA-rich, while roe was dominated by GL (56.2%). The relative abundance of lysophospholipids (LPL) along with other major lipid classes by body region (Figure 1C) further highlighted differences: female shishamo heads contained significantly higher GL and PL levels ranked as capelin males < capelin females < shishamo males < shishamo females, and male shishamo meat exhibited the highest levels across nearly all classes except ST. In the viscera, females of both species had significantly elevated FA, LPL, SP, and ST compared with males, and the storage and distribution of lipids were influenced by sex-related differences in energy allocation to reproduction and timing of migration [25]. The roe was uniformly dominated by GL, with no major interspecific differences.

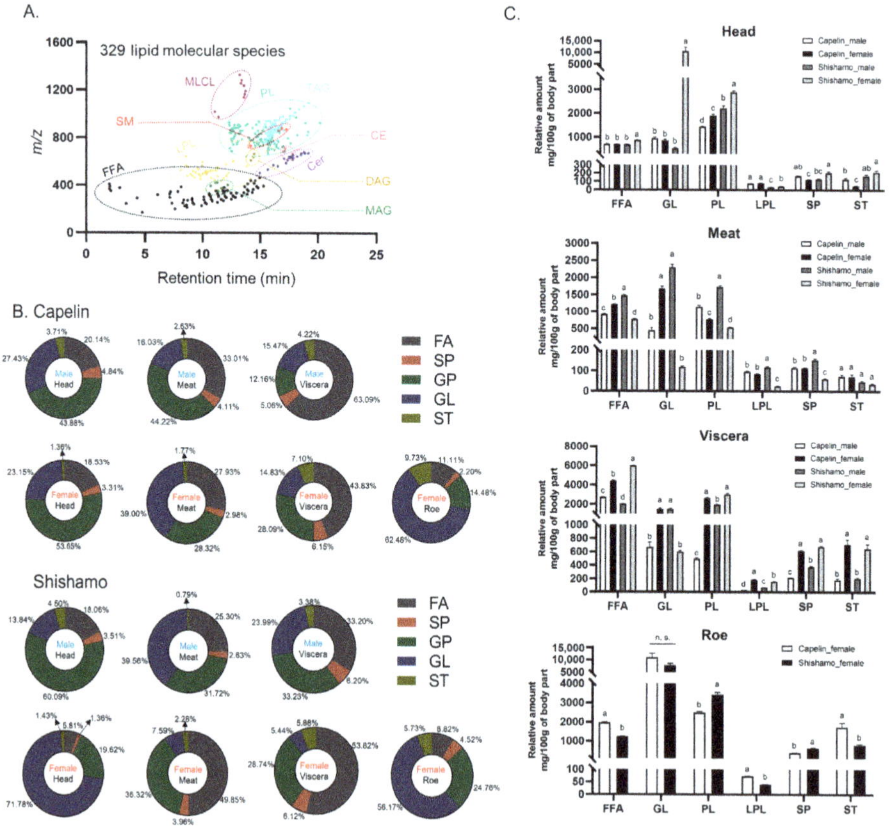

Figure 1. (**A**) m/z vs. retention time (RT) plot of detected lipids in shishamo and capelin. (**B**) Species, sex, body parts-specific percentage distribution of five lipid classes (FA: fatty acids, SP: sphingolipids, GP: glycerophospholipids, GL: glycerolipids, ST: sterol lipids). (**C**) Relative amount of six lipid classes based on body part (FFA: free fatty acids, PL: phospholipids, LPL: lyso-phospholipids). Different letters (a–d) indicate significant differences ($p < 0.05$); n.s. indicates no significant difference.

https://doi.org/10.3390/foods15020298

3.2. Region-Specific Free Fatty Acid Profiles

Free fatty acids (FFAs) are key regulators of energy metabolism, gene expression, and signaling [26]. Tissue FFA composition was classified into saturated fatty acids (SFA), monounsaturated fatty acids (MUFA), and polyunsaturated fatty acids (PUFA) (Figure 2A). SFAs predominated in head tissues (57.6–66.5%), particularly in female shishamo meat (59.9%). By contrast, PUFAs were enriched in viscera across all groups (40.2–50.0%), and this trend largely reflects lipid storage in the liver and intestine, the major visceral organs. Semi-quantitative analysis confirmed that visceral FFA levels were substantially higher in females than in male-most notably, visceral PUFA reached 2873 ± 30 mg/100 g in female shishamo and 1981 ± 29 mg/100 g in female capelin (Figure 2B).

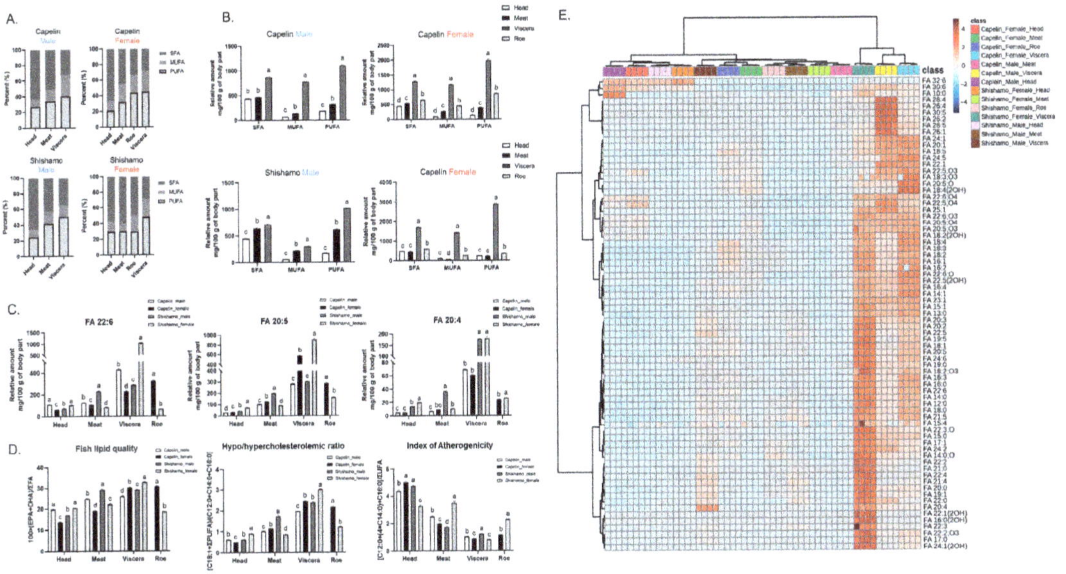

Figure 2. Free fatty acids (FFA) profiling in shishamo and capelin. (**A**) Percentage distribution of FFA based on unsaturation. (**B**) Relative amount of SFA, MUFA, PUFA. (**C**) Relative amount of FA 22:6, FA 20:5, and FA 20:4. (**D**) Lipid nutritional assessment by FFA (Fish lipid quality, Hypo/hypercholesterolemic ratio, index of atherogenicity). (**E**) Hierarchical Clustering Heat Map of free fatty acid, Data were quantile normalized followed by autoscaling for features (i.e., z-scores for each feature) using MetaboAnalyst 6.0. Different letters (a–d) indicate significant differences ($p < 0.05$).

Among nutritionally relevant long-chain PUFAs (FA 22:6, FA 20:5, FA 20:4; Figure 2C), DHA (FA 22:6) was highest in female shishamo viscera (1048 ± 7 mg/100 g), followed by male capelin viscera (426 ± 7 mg/100 g). EPA (FA 20:5) was likewise elevated in shishamo female viscera (904 ± 15 mg/100 g) compared with capelin females (568 ± 8 mg/100 g). Arachidonic acid (FA 20:4; ω-6) showed no significant sex differences in shishamo, with similar levels in males and females (~179 mg/100 g). DHA and EPA are ω-3 PUFAs with well-documented cardioprotective and anti-inflammatory effects, whereas arachidonic acid (ω-6 PUFA) contributes to pro-inflammatory signaling, particularly when the ω-6/ω-3 ratio is elevated [27,28]. The high ω-3 PUFA content of both shishamo and capelin viscera suggests that these tissues may represent valuable dietary sources of health-promoting lipids. Chinese capelin roe contains higher levels of FA 22:6 and FA 20:5 than the viscera [5], and this difference may be related to the maturity of roe [29].

To assess the potential health effects of shishamo and capelin, lipid-based nutritional indices were calculated (Figure 2D). The fish lipid quality index (FLQ), which reflects the proportion of EPA and DHA relative to total FFA, is a widely used indicator of the nutritional value of fish lipids. FLQ values ranged from 13.8 to 32.7, consistent with previous reports for fish (13.0–36.4) [11]. The highest FLQ was observed in shishamo female viscera (32.7 $\pm$ 0.2), followed by capelin female roe (30.6 $\pm$ 0.2), while capelin female head yielded the lowest value (13.8 $\pm$ 0.2). The hypocholesterolemic/hypercholesterolemic ratio (HH), which predicts the effect of FA composition on serum cholesterol, was greatest in shishamo female viscera (3.01 $\pm$ 0.03), with lower values in capelin female viscera (2.43 $\pm$ 0.04) and shishamo male viscera (2.37 $\pm$ 0.04). Among roe, capelin exhibited a significantly higher HH ratio (2.15 $\pm$ 0.02) than shishamo (1.21 $\pm$ 0.01). These values align with literature ranges, where fish generally display ratios of 1.5–4.8, higher than meat (1.3–2.8) and dairy products (0.3–1.3) [11], supporting the relative nutritional advantage of fish lipids.

The index of atherogenicity (IA), used to evaluate the potential risk of lipid intake promoting atherosclerosis [30], was highest in capelin female head (5.0 $\pm$ 0.1) and shishamo male head (4.7 $\pm$ 0.1). Female shishamo meat also exhibited elevated IA (3.5 $\pm$ 0.1). By contrast, viscera across all groups showed comparatively low IA values (0.8–1.2), suggesting a reduced atherogenic risk from these tissues. Reported IA ranges were 0.2–1.4 for fish, 1.4–5.1 for dairy, and 0.2–1.3 for meat [11], indicating variability across food groups but underscoring the favorable profile of visceral lipids. All indices were calculated based on the amount of free fatty acids using formulas reported in our earlier studies [19,21]. Hierarchical clustering heatmaps (Figure 2E) highlighted the distinct differences in the FFA profiles between shishamo and capelin. Viscera consistently exhibited a high relative FFA abundance, particularly in females, except for shishamo males. In the heatmap, although samples from the same body part cluster together and the differences among body parts are much greater than those among species, these trends are primarily driven by fatty acid composition rather than the overall lipid composition. Moreover, our objective is to identify region- and sex-specific lipid profiles, which do not necessarily reflect drastic compositional differences. Lipid reserves serve as critical energy resources during gonadal development and sexual maturation and are known to vary with sex and reproductive stage [31]. Notably, very long-chain fatty acids (VLC-FAs; >C28), such as FA 32:6 and FA 30:6, were detected at relatively high levels in head tissues. The high VLC-FA content in head tissues may reflect contributions from specialized tissues, such as the retina [32,33], although specific functional attribution is difficult due to the use of composite head tissues. The identification of region-specific and sex-dependent FFA distribution in shishamo and capelin provides a valuable reference for future studies on fish nutrition, reproductive physiology, and biochemical adaptations.

3.3. Principal Component Analysis of Shishamo and Capelin by Sex

Principal component analysis (PCA) was used to explore sex- and region-specific lipidomic variations in shishamo and capelin (Figure 3). The score plots demonstrated clear clustering by body part across all samples, reflecting a distinct lipid composition. The corresponding loading plots identified the lipid molecular species that contributed the most strongly to these separations. In male capelin, phosphatidylcholine (PC) species containing very long-chain FAs (e.g., PC(18:1/24:1), PC(30:6/22:6)) differentiated the head tissues, whereas omega-3-rich PCs (PC(16:0/22:6), PC(20:5/20:5), PC(20:5/22:6)) differentiated the meat. FFAs (FA 22:6, FA 20:5, FA 18:1) were the primary contributors to the viscera. Notably, VLC-FA-containing PCs are enriched in the head. This lipid is retinal-specific and supports normal visual function in mammals [34]. In female capelins, the head and meat profiles were more similar, with PCs (PC(18:1/24:0), PC(18:1/24:1)) as key discriminators.

FFAs (FA 22:6, FA 20:5, FA 18:1) distinguish the viscera, whereas triacylglycerols (TAGs) and cholesterol esters (CE 22:6, CE 20:5, CE 18:1) are the defining features of roe. The enrichment of PC species in the head and meat is consistent with previous observations in other marine fish, such as saury [35]. Lipid profiling of Chinese female capelins revealed PC as a characteristic of roes [5], suggesting that this difference may be due to their geographic origin [36]. In male shishamo, heads were separated by PC species containing 24-carbon FAs (PC(18:1/24:0), PC(18:1/24:1)), whereas meat was differentiated by TAGs (TAG(16:0/18:1/18:1), TAG(16:0/18:0/18:1)), and PC(20:5/20:5). Viscera were characterized by FFAs (FA 22:6, FA 20:5, FA 18:1), with FA 20:4 showing strong negative loading on PC1. In female shishamo, PCA revealed a close association between the head and roe profiles, with the TAG species driving separation. This observation suggests the metabolic mobilization of TAGs from somatic tissues (e.g., muscle and viscera) into developing oocytes, as reported for other marine fishes [37,38]. The accumulation in the cranial tissues may represent an energy reserve that indirectly contributes to oogenesis. Furthermore, TAG distribution patterns in fish vary by order, species, and swimming mode [39]. Collectively, these PCA findings highlight species-, sex-, and region-specific lipidomic partitioning, reinforcing the distinctive head–roe link in female shishamo and the enrichment of VLC-FA-containing PCs in the capelin head tissue. Although lipidomic maps of capelin have recently been developed, this study provides, for the first time, a comprehensive region-resolved lipidomic profile of shishamo smelt.

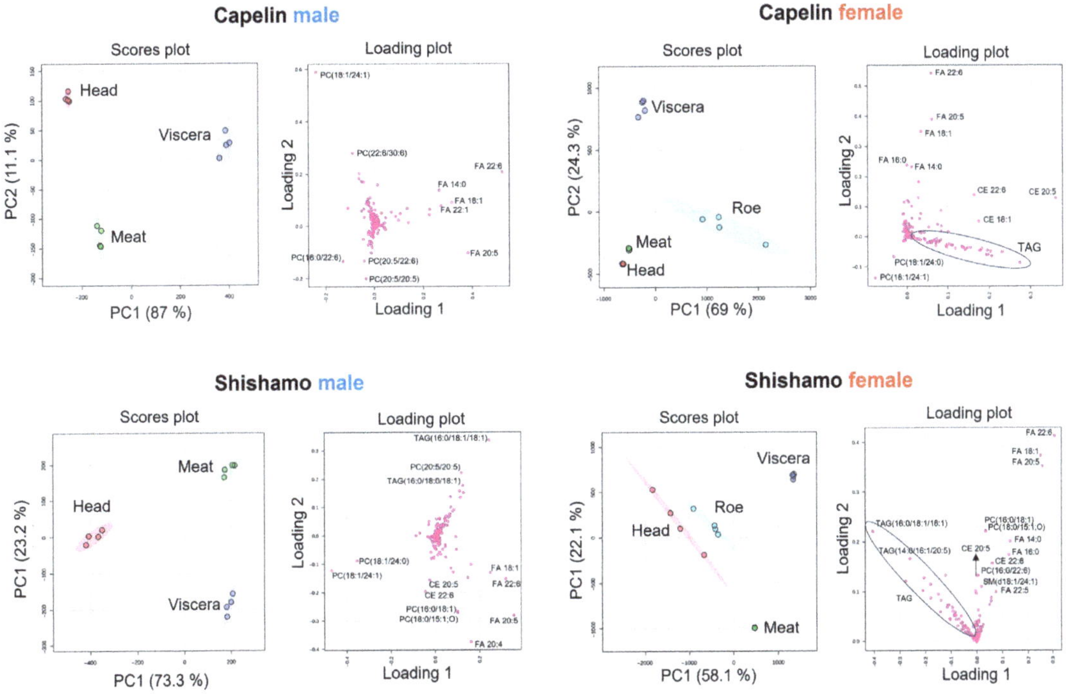

Figure 3. Principal component analysis (PCA) by species and sex (Score plot and Loading plot). Colored circles display 95% confidence regions.

3.4. Lipid Markers Distinguishing Shishamo from Capelin

To identify the anatomical regions most suitable for the exploration of lipid-based species markers, Figure 4A,B depict PCA score plots for the head, meat, viscera, and roe (females only), comparing male shishamo vs. capelin and female shishamo vs. capelin. The

https://doi.org/10.3390/foods15020298

PCA score plots for males revealed a distinct separation of muscle and viscera between shishamo and capelin. Loading plot analysis revealed that FFAs were the primary contributors to species separation in the viscera, whereas PC and TAG played key roles in the meat. In the female PCA score plots, the head and viscera were separated between species. Loading plot analysis revealed that in females, TAG contributed to separation in shishamo head tissue, whereas PC was the primary contributor to the capelin head. By contrast, FFA drives visceral separation in both species. The viscera facilitates marker discovery owing to the sole contribution of FFAs, while also enabling simplified analysis for practical applications. By contrast, in male meat, multiple factors, including PC and TAG, make marker identification challenging and complicate practical analysis. Orthogonal partial least squares discriminant analysis (OPLS-DA) of male meat tissues (Figure S1) further confirmed the multifactor complexity of the FA-TAG reversal patterns. Although the head tissue initially appeared promising, OPLS-DA (Figure S1) showed TAG reversal, which was abundant in shishamo female heads yet equally enriched in capelin female meat. This may reflect reproductive status [40], but the uncontrolled reproductive status in this study limits interpretation and deems the head unsuitable for species identification marker exploration. Furthermore, comparison of the roes (Figure S1), multiple lipid molecular species were intermixed, rendering interpretation challenging and imposing a high analytical burden for practical applications. These findings suggest that the viscera are an optimal tissue for lipid-based species identification.

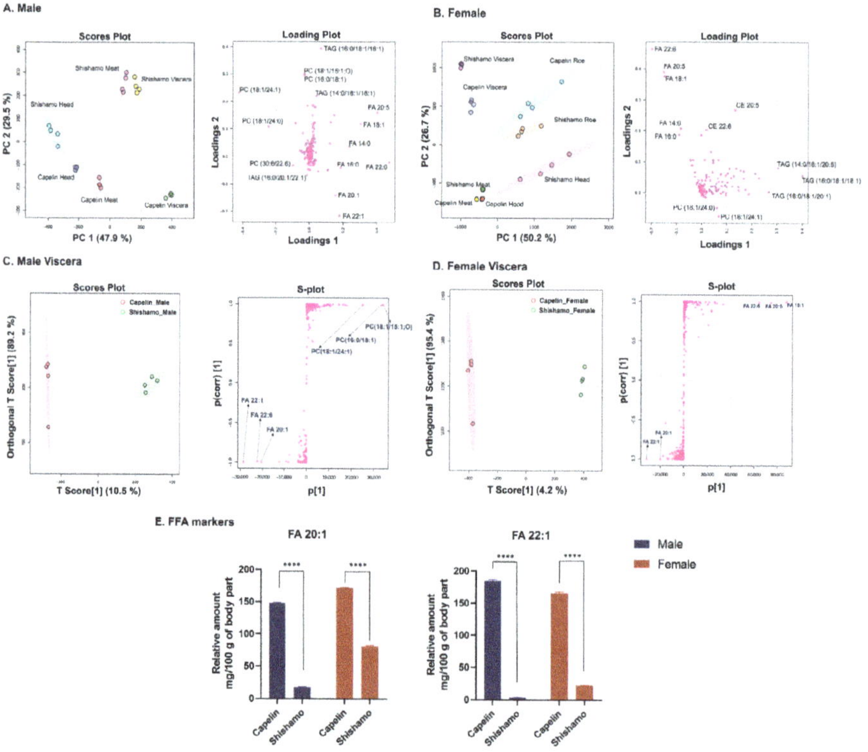

Figure 4. Multi-dimensional assessment of tissue-specific lipid markers for shishamo–capelin discrimination. (**A**) PCA for male, (**B**) PCA for female, (**C**) Orthogonal partial least squares discriminant analysis (OPLS-DA) for male, (**D**) OPLS-DA for female, (**E**) Semi-quantitative levels (means ± SE, n = 4) of FA 20:1 and FA 22:1. **** $p < 0.0001$, Student's t-test.

https://doi.org/10.3390/foods15020298

For species discrimination, OPLS-DA analyses targeting visceral tissues were conducted separately for males and females (Figure 4C,D). Score plots showed a visual separation between shishamo and capelin for both sexes, with model robustness validated by 100 permutation tests confirming lipid composition differences (males: $R^2Y = 0.992$, $Q^2 = 0.985$; females: $R^2Y = 0.999$, $Q^2 = 0.994$). In female viscera, FA 18:1, FA 20:5, and FA 22:6 showed strong positive loading scores, whereas FA 20:1 and FA 22:1 showed strong negative loading scores. In male visceral tissues, several PC molecular species exhibited positive loadings, whereas FA 20:1 and FA 22:1 displayed negative loadings. Semi-quantitative levels of FA 20:1 and FA 22:1 were compared between the species for each sex in Figure 4E. FA 20:1 and FA 22:1, which are consistently elevated in capelin in both sexes, are promising species-specific lipid identification markers. Capelin primarily feeds on zooplankton [41], which inhabit the water off Iceland and Norway and are rich in C20:1 and C22:1 fatty acids [42], suggesting that dietary differences are the cause of lipidome variation. The lipid markers FA 20:1 and FA 22:1 identified in the viscera offer potential practical advantages for species discrimination. From the perspective of analysis time reduction, quantification of two FFA species promotes interspecies separation, simplifying the process compared to multivariate analysis required for other tissues, and contributing to determinations within tens of minutes. Furthermore, for food safety and traceability applications, this method supports the detection of imported capelin masquerading as shishamo, with the diet-derived FFA composition serving as a promising candidate marker suitable for simplified authentication during commercial distribution.

Overall, TAG enrichment in female shishamo heads and their compositional similarity with the roe suggest that the head serves as a TAG storage site and nutrient source for oogenesis. Meanwhile, high ω-3 PUFA concentrations in viscera reflect TAG hydrolysis-mediated release and reallocation, providing a biological basis for systemic energy supply networks. Additionally, FFAs generated during this visceral process may function as interspecific discriminant markers.

This study had several limitations. First, the reported lipid concentrations were semi-quantitative, as absolute quantification was not performed, and the values were derived relative to internal standards. Second, all specimens were sourced from commercial providers, introducing potential variability in handling, storage, and transport before analysis. Important biological and environmental factors that influence fish lipid composition, including diet, spawning stage, age, seasonality, and water temperature, have not been controlled for or systematically assessed. Additionally, post-harvest factors such as storage conditions and duration, particularly for imported capelin, may affect lipid stability, including the susceptibility of PUFAs to oxidation. Future studies should integrate controlled sampling, longitudinal monitoring of environmental and dietary influences, and absolute quantification approaches to validate and expand these findings.

4. Conclusions

This study used untargeted LC-MS to characterize the sex- and region-specific lipidomes of shishamo (*Spirinchus lanceolatus*) and capelin (*Mallotus villosus*). Female shishamo heads were enriched in glycerolipids, with a composition resembling roe, and uniquely contained very long-chain fatty acids (>C30) and ester-linked PCs. Both species exhibited elevated DHA and EPA levels in their viscera. Multivariate analysis highlighted clear region-specific lipid signatures; PCs predominated in head and meat, TAGs and CEs in the roe, and FFAs in the viscera. The lipid markers used for species discrimination were FA 20:1 and FA 22:1, which are abundant in capelin. Overall, these findings provide new insights into the nutritional lipid composition of shishamo and capelin, reveal region- and

https://doi.org/10.3390/foods15020298

sex-specific lipid distributions, and identify lipidomic markers with potential applications in the authentication, nutrition, and tracing of geographic origin.

Supplementary Materials: The following supporting information can be downloaded at: https://www.mdpi.com/article/10.3390/foods15020298/s1, Table S1: List of lipid metabolites and their relative amount (mg/100 g of body parts). Figure S1: OPLS-DA by sex and body part between shishamo and capelin, excluding viscera.

Author Contributions: Conceptualization, Y.M. and H.C.; methodology, Y.M., S.G.B.G. and D.G.; validation, Y.M., S.G.B.G. and D.G.; formal analysis, Y.M.; investigation, Y.M. and S.G.B.G.; resources, S.G.B.G. and S.-P.H.; data curation, Y.M., S.G.B.G. and D.G.; writing—original draft preparation, Y.M.; writing—review and editing, Y.M., S.G.B.G., D.G., H.C. and S.-P.H.; visualization, Y.M.; supervision, S.G.B.G.; funding acquisition, Y.M. and S.G.B.G. All authors have read and agreed to the published version of the manuscript.

Funding: This work was supported by JSPS KAKENHI Grant Number JP24KJ0261 (to Y.M.), 25K00258 (to S.G.B.G), and next-generation researcher support from Hokkaido University (to S.G.B.G).

Institutional Review Board Statement: Not applicable.

Informed Consent Statement: Not applicable.

Data Availability Statement: The original contributions presented in this study are included in the article/Supplementary Materials. Further inquiries can be directed to the corresponding authors.

Acknowledgments: The authors would like to thank Hokkaido University for supporting the research expense and the staff of the Laboratory for Advanced Lipid Analysis (LALA) at Hokkaido University for their cooperation in this study. English language editing was provided by a professional service.

Conflicts of Interest: The authors declare no conflicts of interest.

References

1. Yatsuyanagi, T.; Ishida, R.; Sakata, M.K.; Kanbe, T.; Mizumoto, H.; Kobayashi, Y.; Kamada, S.; Namba, S.; Nii, H.; Minamoto, T.; et al. Environmental DNA Monitoring for Short-Term Reproductive Migration of Endemic Anadromous Species, Shishamo Smelt (*Spirinchus lanceolatus*). *Environ. DNA* **2020**, *2*, 130–139. [CrossRef]
2. Yatsuyanagi, T.; Araki, H. Understanding Seasonal Migration of Shishamo Smelt in Coastal Regions Using Environmental DNA. *PLoS ONE* **2020**, *15*, e0239912. [CrossRef]
3. Mizuno, S.; Sasaki, Y.; Omoto, N.; Imada, K. Elimination of Adhesiveness in the Eggs of Shishamo Smelt Spirinchus Lanceolatus Using Kaolin Treatment to Achieve High Hatching Rate in an Environment with a High Iron Concentration. *Aquaculture* **2004**, *242*, 713–726. [CrossRef]
4. Bragadóttir, M.; Pálmadóttir, H.; Kristbergsson, K. Composition and Chemical Changes during Storage of Fish Meal from Capelin (*Mallotus villosus*). *J. Agric. Food Chem.* **2004**, *52*, 1572–1580. [CrossRef]
5. Yin, M.; Chen, M.; Matsuoka, R.; Song, X.; Xi, Y.; Zhang, L.; Wang, X. UHPLC-Q-Exactive Orbitrap MS/MS Based Untargeted Lipidomics Reveals Fatty Acids and Lipids Profiles in Different Parts of Capelin (*Mallotus villosus*). *J. Food Compos. Anal.* **2023**, *116*, 105096. [CrossRef]
6. Hayashi, K. Comparative Study on the Feature and Evaluation of the Production and Trade Activity on Capelin for Human Consumption Exported to Japan. *J. Reg. Fish.* **2007**, *47*, 99–124. [CrossRef]
7. Jamshidi, A.; Cao, H.; Xiao, J.; Simal-Gandara, J. Advantages of Techniques to Fortify Food Products with the Benefits of Fish Oil. *Food Res. Int.* **2020**, *137*, 109353. [CrossRef]
8. Coniglio, S.; Shumskaya, M.; Vassiliou, E. Unsaturated Fatty Acids and Their Immunomodulatory Properties. *Biology* **2023**, *12*, 279. [CrossRef] [PubMed]
9. Dicks, L.M.T. How Important Are Fatty Acids in Human Health and Can They Be Used in Treating Diseases? *Gut Microbes* **2024**, *16*, 2420765. [CrossRef]
10. Perna, M.; Hewlings, S. Saturated Fatty Acid Chain Length and Risk of Cardiovascular Disease: A Systematic Review. *Nutrients* **2023**, *15*, 30. [CrossRef] [PubMed]
11. Chen, J.; Liu, H. Nutritional Indices for Assessing Fatty Acids: A Mini-Review. *Int. J. Mol. Sci.* **2020**, *21*, 5695. [CrossRef] [PubMed]

12. Wu, B.; Wei, F.; Xu, S.; Xie, Y.; Lv, X.; Chen, H.; Huang, F. Mass Spectrometry-Based Lipidomics as a Powerful Platform in Foodomics Research. *Trends Food Sci. Technol.* **2021**, *107*, 358–376. [CrossRef]
13. Všetičková, L.; Suchý, P.; Straková, E. Factors Influencing the Lipid Content and Fatty Acids Composition of Freshwater Fish: A Review. *Asian J. Fish. Aquat. Res.* **2019**, *5*, 1–10. [CrossRef]
14. Harlina, P.W.; Maritha, V.; Geng, F.; Subroto, E.; Yuliana, T.; Shahzad, R.; Sun, J. Lipidomics: A Comprehensive Review in Navigating the Functional Quality of Animal and Fish Products. *Int. J. Food Prop.* **2023**, *26*, 3115–3136. [CrossRef]
15. Liu, L.; Bartke, N.; Daele, H.V.; Lawrence, P.; Qin, X.; Park, H.G.; Kothapalli, K.; Windust, A.; Bindels, J.; Wang, Z.; et al. Higher Efficacy of Dietary DHA Provided as a Phospholipid than as a Triglyceride for Brain DHA Accretion in Neonatal Piglets. *J. Lipid Res.* **2014**, *55*, 531–539. [CrossRef]
16. Ding, N.; Xue, Y.; Tang, X.; Sun, Z.-M.; Yanagita, T.; Xue, C.-H.; Wang, Y.-M. Short-Term Effects of Different Fish Oil Formulations on Tissue Absorption of Docosahexaenoic Acid in Mice Fed High- and Low-Fat Diets. *J. Oleo Sci.* **2013**, *62*, 883–891. [CrossRef] [PubMed]
17. Jeong, B.-Y.; Moon, S.-K.; Jeong, W.-G.; Ohshima, T. Lipid Classes and Fatty Acid Compositions of Wild and Cultured Sweet Smelt Plecoglossus Altivelis Muscles and Eggs in Korea. *Fish. Sci.* **2000**, *66*, 716–724. [CrossRef]
18. Henderson, R.J.; Sargent, J.R.; Hopkins, C.C.E. Changes in the Content and Fatty Acid Composition of Lipid in an Isolated Population of the Capelin Mallotus Villosus during Sexual Maturation and Spawning. *Mar. Biol.* **1984**, *78*, 255–263. [CrossRef]
19. B. Gowda, S.G.; Minami, Y.; Gowda, D.; Chiba, H.; Hui, S.-P. Detection and Characterization of Lipids in Eleven Species of Fish by Non-Targeted Liquid Chromatography/Mass Spectrometry. *Food Chem.* **2022**, *393*, 133402. [CrossRef]
20. Folch, J.; Lees, M.; Stanley, G.H.S. A simple method for the isolation and purification of total lipides from animal tissues. *J. Biol. Chem.* **1957**, *226*, 497–509. [CrossRef]
21. Minami, Y.; B. Gowda, S.G.; Gowda, D.; Chiba, H.; Hui, S.-P. Regio-Specific Lipid Fingerprinting of Edible Sea Cucumbers Using LC/MS. *Food Res. Int.* **2024**, *184*, 114253. [CrossRef]
22. B. Gowda, S.G.; Minami, Y.; Gowda, D.; Furuko, D.; Chiba, H.; Hui, S.-P. Lipidomic Analysis of Non-Esterified Furan Fatty Acids and Fatty Acid Compositions in Dietary Shellfish and Salmon by UHPLC/LTQ-Orbitrap-MS. *Food Res. Int.* **2021**, *144*, 110325. [CrossRef]
23. Cui, J.; Cao, J.; Zeng, S.; Ge, J.; Li, P.; Li, C. Comprehensive Evaluation of Lipidomics Profiles in Golden Threadfin Bream (*Nemipterus virgatus*) and Its by-Products Using UHPLC-Q-Exactive Orbitrap-MS. *LWT* **2022**, *165*, 113690. [CrossRef]
24. Monteiro, J.P.; Sousa, T.; Melo, T.; Pires, C.; Marques, A.; Nunes, M.L.; Calado, R.; Domingues, M.R. Unveiling the Lipid Features and Valorization Potential of Atlantic Salmon (*Salmo salar*) Heads. *Mar. Drugs* **2024**, *22*, 518. [CrossRef] [PubMed]
25. House, A.H.; Debes, P.V.; Kurko, J.; Erkinaro, J.; Käkelä, R.; Primmer, C.R. Sex-Specific Lipid Profiles in the Muscle of Atlantic Salmon Juveniles. *Comp. Biochem. Physiol. Part D Genomics Proteomics* **2021**, *38*, 100810. [CrossRef] [PubMed]
26. Kimura, I.; Ichimura, A.; Ohue-Kitano, R.; Igarashi, M. Free Fatty Acid Receptors in Health and Disease. *Physiol. Rev.* **2020**, *100*, 171–210. [CrossRef]
27. Djuricic, I.; Calder, P.C. Beneficial Outcomes of Omega-6 and Omega-3 Polyunsaturated Fatty Acids on Human Health: An Update for 2021. *Nutrients* **2021**, *13*, 2421. [CrossRef] [PubMed]
28. Jeong, H.Y.; Moon, Y.S.; Cho, K.K. ω-6 and ω-3 Polyunsaturated Fatty Acids: Inflammation, Obesity and Foods of Animal Resources. *Food Sci. Anim. Resour.* **2024**, *44*, 988–1010. [CrossRef]
29. Quirós-Pozo, R.; Robaina, L.; Calderón, J.A.; Filgueira, J.R. Reproductive Management of the Mugilid *Liza aurata* and Characterization of Proximate and Fatty Acid Composition of Broodstock Tissues and Spawnings. *Aquaculture* **2023**, *564*, 739055. [CrossRef]
30. Ulbricht, T.L.V.; Southgate, D.A.T. Coronary Heart Disease: Seven Dietary Factors. *Lancet* **1991**, *338*, 985–992. [CrossRef]
31. Nogueira, N.; Fernandes, I.; Fernandes, T.; Cordeiro, N. A Comparative Analysis of Lipid Content and Fatty Acid Composition in Muscle, Liver and Gonads of *Seriola fasciata* Bloch 1793 Based on Gender and Maturation Stage. *J. Food Compos. Anal.* **2017**, *59*, 68–73. [CrossRef]
32. He, Y.; Phan, K.; Bhatia, S.; Pickford, R.; Fu, Y.; Yang, Y.; Hodges, J.R.; Piguet, O.; Halliday, G.M.; Kim, W.S. Increased VLCFA-Lipids and ELOVL4 Underlie Neurodegeneration in Frontotemporal Dementia. *Sci. Rep.* **2021**, *11*, 21348. [CrossRef]
33. Nwagbo, U.; Bernstein, P.S. Understanding the Roles of Very-Long-Chain Polyunsaturated Fatty Acids (VLC-PUFAs) in Eye Health. *Nutrients* **2023**, *15*, 3096. [CrossRef] [PubMed]
34. Kuroha, S.; Katada, Y.; Isobe, Y.; Uchino, H.; Shishikura, K.; Nirasawa, T.; Tsubota, K.; Negishi, K.; Kurihara, T.; Arita, M. Long Chain Acyl-CoA Synthetase 6 Facilitates the Local Distribution of Di-Docosahexaenoic Acid- and Ultra-Long-Chain-PUFA-Containing Phospholipids in the Retina to Support Normal Visual Function in Mice. *FASEB J.* **2023**, *37*, e23151. [CrossRef] [PubMed]
35. Tao, X.; Yin, M.; Lin, L.; Song, R.; Wang, X.; Tao, N.; Wang, X. UPLC-ESI-MS/MS Strategy to Analyze Fatty Acids Composition and Lipid Profiles of Pacific Saury (*Cololabis saira*). *Food Chem. X* **2024**, *23*, 101682. [CrossRef]

https://doi.org/10.3390/foods15020298

36. Han, H.; Lin, X.; Wu, Y.; Li, H.; Jia, Q.; Hu, C.; Lin, Y.; Liao, G.; Qian, Y.; Qiu, J. Lipid Fingerprinting and Geographical Origin Identification of Large Yellow Croaker (*Larimichthys crocea*) Using UHPLC-QTOF-MS. *Food Res. Int.* **2024**, *196*, 115048. [CrossRef]
37. Cejas, J.R.; Almansa, E.; Villamandos, J.E.; Badía, P.; Bolaños, A.; Lorenzo, A. Lipid and Fatty Acid Composition of Ovaries from Wild Fish and Ovaries and Eggs from Captive Fish of White Sea Bream (*Diplodus sargus*). *Aquaculture* **2003**, *216*, 299–313. [CrossRef]
38. Hou, Z.; Fuiman, L.A. Incorporation of Dietary Lipids and Fatty Acids into Red Drum *Sciaenops ocellatus* Eggs. *Comp. Biochem. Physiol. B Biochem. Mol. Biol.* **2022**, *258*, 110694. [CrossRef] [PubMed]
39. Kaneko, G.; Shirakami, H.; Hirano, Y.; Oba, M.; Yoshinaga, H.; Khieokhajonkhet, A.; Nagasaka, R.; Kondo, H.; Hirono, I.; Ushio, H. Diversity of Lipid Distribution in Fish Skeletal Muscle. *Zoolog. Sci.* **2016**, *33*, 170–178. [CrossRef]
40. Torsabo, D.; Ishak, S.D.; Noordin, N.M.; Koh, I.C.C.; Abduh, M.Y.; Iber, B.T.; Kuah, M.-K.; Abol-Munafi, A.B. Enhancing Reproductive Performance of Freshwater Finfish Species through Dietary Lipids. *Aquac. Nutr.* **2022**, *2022*, 7138012. [CrossRef]
41. Dalpadado, P.; Mowbray, F. Comparative Analysis of Feeding Ecology of Capelin from Two Shelf Ecosystems, off Newfoundland and in the Barents Sea. *Prog. Oceanogr.* **2013**, *114*, 97–105. [CrossRef]
42. Huenerlage, K.; Graeve, M.; Buchholz, F. Lipid Composition and Trophic Relationships of Krill Species in a High Arctic Fjord. *Polar Biol.* **2016**, *39*, 1803–1817. [CrossRef]

https://doi.org/10.3390/foods15020298

foods

MDPI

Article

Comparative Study of Key Aroma Components in Rice of Different Aroma Types Using Flavor Metabolomics

Shengmin Qi [1,2,3,4], Haibin Ren [4], Haiqing Yang [4], Lianhui Zhang [4] and Min Zhang [1,2,3,*]

1 School of Food and Health, Beijing Technology and Business University, Beijing 100048, China; 13720035676@163.com
2 Beijing Engineering and Technology Research Center of Food Additives, Beijing 100048, China
3 Beijing Advanced Innovation Center for Food Nutrition and Human Health, Beijing 100048, China
4 Grain Research and Development Center, Nutrition & Health Research Institute, COFCO Corporation, Beijing 102209, China; renhaibin@cofco.com (H.R.); yanghaiqing1@cofco.com (H.Y.); zhanglianhui@cofco.com (L.Z.)
* Correspondence: zmin@th.btbu.edu.cn

Abstract

This study aimed to analyze the volatile organic compounds (VOCs) for different rice aroma types using sensory evaluation, headspace solid-phase microextraction gas chromatography mass spectrometry (HS-SPME-GC-MS), and gas chromatography-ion mobility spectrometry (GC-IMS) techniques, and to explore the material basis for the flavor differences. Based on the sensory evaluation results, rice aroma was categorized into three types, distinguished by their unique aroma compounds. Type A was characterized by a prominent sweet, popcorn aroma, type B by a more prominent cereal and starchy flavor, and type C by a more complex aroma. Untargeted metabolomics analysis using HS-SPME-GC-MS identified and characterized 74 volatile compounds. A comparison of A versus B versus C revealed 8 key aroma compounds, primarily alkanes, aldehydes, ketones, alcohols, and heterocyclic compounds. (E)-2-Octenal, 6-Undecanone, 2-Acetyl-1h-pyrrole, and P-menthan-1-ol in type A gave it a better sweet aroma, Dodecane, 2,6,10-trimethyl-, 1-Octen-3-one, and 2-Methyldecane in type B gave it a better starchy and cereal flavor. 2-Acetyl-1h-pyrrole, Heptacosane, and 1-Propanol in type C contributed to a complex aroma. GC-IMS analysis showed that the fingerprints of rice with different aroma types were significantly different. The VOCs of aroma type A contained (+)-limonene, 2-methylpyrazine, 2-pentanone, ethyl butanoate, n-pentanal, styrene, 1-butanol, 3-methyl-, acetate, 1-hexanal, 1-pentanol, and 2-heptanone, which gave it a better sweet aroma. The VOCs of aroma type C contained 1-octen-3-ol, 2,6-dimethyl pyrazine, 2-acetylpyridine, and ethyl hexanoate, which gave it a better complex aroma.

Keywords: rice aroma types; GC-IMS; GC-MS; aroma compounds

1. Introduction

Rice, one of the major food crops in the world today, serves as the primary food source for more than half the world's population, particularly in Asia and Africa [1,2]. Rice, domesticated by the Chinese between 13,500 and 8200 years ago [3,4], is now a staple crop worldwide [5]. The United Nations Food and Agriculture Organization reports that rice production is expected to reach a record high of 555.5 million tons (converted into refined rice) in the 2025/26 fiscal year [6]. Changes in the consumer market and advancements in rice breeding have made the flavor of rice a critical indicator of its market price and consumer acceptance. Therefore, the study of rice flavor is particularly important.

Current rice flavor research mainly focuses on two key areas: analyzing volatile compounds in rice and characterizing the aroma quality [7]. The quantitative descriptive analysis (QDA) method is a subjective evaluation technique mainly used to establish descriptive words, which are then used to analyze the aroma attributes of the fragrant rice [8–10]. Gas chromatography, supplemented by related pretreatment and detection techniques, is currently the main objective flavor evaluation method [11]. Numerous studies have confirmed that the flavor of rice is jointly determined by the concentrations, thresholds, and interactions of different types of volatile compounds. Currently, more than 500 volatile components have been detected and identified in rice [12,13], mainly including hydrocarbons, alcohols, aldehydes, esters, ketones, phenols, heterocyclic compounds, and so forth [14,15]. Multiple studies using sensory evaluation and flavor metabolomics have analyzed the volatile organic compounds (VOCs) in rice/millet [16,17], confirming that only a select few are responsible for the distinct and significant flavor profile of rice [18]. Studies have also reported the effects of different processing conditions on the flavor of rice [14]. These key aroma components undergo concentration changes during breeding, planting, drying, storage, processing, preservation, cooking, and oral chewing processes, and the compounds exhibit different flavors due to varying concentrations, thereby influencing the overall flavor of rice [19].

Currently, rice is classified into two main categories based on aroma: fragrant rice and nonfragrant rice. The key differences in volatile components between fragrant and nonfragrant rice include 2-acetyl-1-pyrrolidine (2-AP), 2-octene, isopentanol, 2-ethyl-1-hexanol, 1-octen-3-ol, and so forth [20]. 2-AP, first identified in 1982 [21], has a strong popcorn aroma [22–24]. It is responsible for the characteristic aroma compound of fragrant rice [25]. Several studies have also reported on the aroma types of rice. Lu et al. [12] used gas chromatography–mass spectrometry (GC-MS) to analyze the volatile components in fragrant rice from China and Thailand. They divided fragrant rice into three types (popcorn flavor, corn flavor, and lotus seed flavor) and determined the main aroma-active substances of each type. Dong et al. [26] used GC-MS, Gas chromatography–olfactometry (GC-O), and multivariate analysis to analyze the aroma chemical components of six different types of rice from India, Thailand, and South Korea. They identified the main compounds causing aroma differences. Zhou et al. [19] collected nine samples of rice with different aromas, established an aroma description word bank based on QDA, identified four aroma types, and analyzed the flavor substances in each aroma type. At present, the rice aroma research is mainly focused on fragrant rice. However, fragrant and nonfragrant rice have their own specific aromas for consumers, and the differences in their flavor substances result in certain differences in their aroma types. Thus, the classification of rice with different aroma types due to different flavor substances needs further systematic research.

Taking into account the significant influence of the aroma of raw rice on consumers' purchasing decisions, this study focuses on raw rice as the research object, aiming to verify the feasibility of rice aroma differentiation based on the main varieties available in China. The material basis of rice aroma differences was systematically analyzed using sensory evaluation, GC-MS, and GC-IMS analytical methods. The aromas of different Chinese rice varieties were differentiated and the key flavor substances were explored, providing a reference for the quantitative evaluation and standard upgrade of rice aroma and offering high-quality products to consumers.

https://doi.org/10.3390/foods15020200

2. Materials and Methods

2.1. Chemicals and Reagents

2-Methyl-3-heptanone, 4-Hydroxy-2,5-dimethylfuran-3-one, Butyraldehyde, 1-Octanol, 2-Acetyl-1-pyrrolidine were procured from Tokyo Chemical Industry Co., Ltd. (Shanghai, China). A mixed standard of n-alkanes (C9–C23) (chromatographic grade, purity $\geq$ 99.5%) was bought from Shanghai Anpu Laboratory Technology Co., Ltd. (Shanghai, China). n-Ketones [2-butanone, 2-pentanone, 2-hexanone, 2-heptanone, 2-octanone, and 2-nonanone (all analytical grade)], phenethyl alcohol were procured from Shanghai Aladdin Biochemical Technology Co., Ltd. (Shanghai, China). Starch, Toothpick, pasteurized milk, quaker oatmeal, cream style sweet corn were procured from Wal-Mart Stores (Beijing, China). Ultrapure water was prepared by a Hokee Company (Hefei, China, HK-DI-10/20/30 model).

2.2. Sample Collection

11 different types of commercial raw rice, all with distinct aromas, were purchased between July and October 2024. Prior to the experiments, each sample was divided into three portions for subsequent research. These samples were vacuum-packed and stored at 4 °C. All experiments were completed within 2 months of rice purchase. All samples were prepared before each test, kept fresh in the refrigerator, and then brought to room temperature for evaluation. The samples were assigned three random numbers. Table 1 presents the numbers and variety information of the samples.

Table 1. Information on commercially available rice samples.

No.	652	953	574	429	691	724	860	836	398	560	951
varieties	Nanjing46	Daohuaxiang	Kongyu131	Suijing18	jasmine rice	Longyang16	Suxiu	Taixiangjing1402	Jihong6	Ningxiangjing9	Zhongkefa5

2.3. Sensory Evaluation of Rice Flavor

According to the requirements of GB/T 16291.1-2012 (National Standard, 2012) [27], a rice sensory evaluation panel was formed based on previous findings [28–30]. 10 participants regularly engaged in sensory analysis of cereal products, primarily adults over the 20 years of age with at least two years of experience in the sensory evaluation of rice products, were enrolled in the study. Three sensory training sessions were conducted. In the first one, panelists with rich experience in rice descriptive sensory analysis generated aroma descriptors for raw rice samples studied. For the second session, the panelists performed a sensory evaluation of the samples based on their own subjective perceptions and recorded some different aroma descriptors. In the last session, all sensory panelists were asked to participate in a discussion of sensory descriptors screening and deletion. Ultimately, ten aroma attributes (starchy, grainy, floral, Popcorn, sweet, grassy, woody, oatmeal, dairy, corn) were determined to be most effective in characterizing the rice and highlighting key differences among them. After the training period, the quantitative description analysis was performed using the screened sensory descriptors and the provided standard references (Table 2) [19]. The strength of each descriptor ranged from 0 to 9 [31], with 0 representing no sensation and 9 representing a strong sensation (0.5 is the minimum increment).

The sensory experiment was conducted in an air-conditioned laboratory at 25 °C with good ventilation. The rice samples (20 g) were placed in a plastic cup and marked with a three-digit code according to Table 1. The participants were asked to quickly select the descriptors they thought were suitable to describe the flavor attributes of the rice. Samples were presented at intervals of approximately 5 min. The results of each aroma descriptor of samples were averaged for statistical analysis.

https://doi.org/10.3390/foods15020200

Table 2. The rice aroma descriptors and definitions.

Descriptor	Definition	Reference Substance
Starchy	Aroma of bread and flour	Boiled starch:water (1:5) and cool
Grainy	a flavor reminiscent of raw grain	28 mg/kg Butyraldehyde
Floral	The smell produced by a non-specific type of flower	1600 mg/kg phenethyl alcohol
Popcorn	aroma of popcorn	1300 mg/kg 2-Acetyl-1-pyrrolidine
Sweet	aroma of Demerara sugar	255 mg/kg 4-Hydroxy-2,5-dimethylfuran-3-one
Grassy	A green, slightly earthy, and slightly sweet flavor	26.3 mg/kg 1-Octanol
Woody	aroma of the dry, freshly cut wood smell	Toothpick
Oatmeal	aroma of cooked oatmeal	Quaker Oatmeal (Quaker, London, UK)
Dairy	aroma of uncooked milk	High-quality 2% Brand pasteurized milk
Corn	aroma of canned creamed corn	Cream Style Sweet Corn

2.4. *Flavor Metabolomic Analysis by HS-SPME-GC-MS*

Flavor metabolomic analysis was performed on 11 different types of commercial raw rice. The method used headspace solid-phase microextraction gas chromatography mass spectrometry (HS-SPME-GC-MS), with an Agilent system consisting of a 7890B chromatograph and a 5977B mass spectrometer (Agilent Technologies, Santa Clara, CA, USA). The aroma compounds were extracted using a Supelco solid-phase microextraction fiber [50/30 μm (Divinylbenzene (DVB)/Carboxen (CAR)/Polydimethylsiloxane (PDMS), StableFlex (2 cm)], and the volatile compounds were separated using a DB-Wax capillary column (30 m × 250 μm × 0.25 μm, Agilent Technologies, Santa Clara, CA, USA).

2.4.1. Metabolite Extraction

First, 0.816 μg/μL of 2-methyl-3-heptanone standard solution was prepared using deionized water as the solvent. The rice samples (2000 ± 40 mg) were placed in 20 mL headspace bottles, and 1 μL of 2-methyl-3-heptanone was added as an internal standard. Headspace injection conditions included incubation at 60 °C, preheating for 15 min, adsorbing for 30 min, and desorption for 4 min.

2.4.2. GC-MS Analysis

The volatile substances were analyzed using a DB-Wax chromatographic column and an Agilent 7890B gas chromatography–5977B mass spectrometry system (Agilent Technologies, Santa Clara, CA, USA). The system used a nonsplit injection mode, with high-purity helium as the carrier gas. The pre-injection port purge flow rate was 3 mL/min, and the column flow rate was 1 mL/min. The initial temperature was controlled at 40 °C for 4 min, and then increased at a rate of 5 °C/min to 245 °C and maintained for 5 min. The injection port temperature was 250 °C, the transfer line temperature was 250 °C, the ion source temperature was 230 °C, and the quadrupole temperature was 150 °C. The electron impact ionization energy was 70 eV. The mass spectrometry scan range was set at m/z 20–400, and the solvent delay time was 2.37 min. After the measurement, the mass spectrum of the volatile compounds were then compared with data from the NIST 2017 database. Then, the actual retention index (RI) value is calculated according to the relevant research methods of the same laboratory [32,33]. The RI were compared with the values reported in the literature. When the difference is <50, it was inferred that the two match. The volatile compounds were quantified (semi-quantitative analysis) by dividing the peak areas of the compounds of interest by the peak area of 2-methyl-3-heptanone as internal standard. Each experiment with the samples was conducted triplicately.

https://doi.org/10.3390/foods15020200

2.4.3. Calculation of Relative Odor Activity Value

The relative odor activity values (ROAV) were performed to evaluate the specific contribution of each compound to the overall aroma, which was well reported in previous studies [34]. ROAV was mainly used to quantitatively evaluate the contribution of each volatile substance to the overall flavor of the test sample and thus identify key aroma-active compounds. A high ROAV is also indicative of the great contribution of a component to the overall flavor of the sample. The compounds with ROAV ≥ 1 were the principal aroma compounds of the sample, whereas those with $0.1 \leq$ ROAV < 1 had a important modifying effects on its overall flavor. ROAV was calculated as described previously [35,36].

2.5. GC-IMS Analysis

GC-IMS analysis of volatiles was performed using an Agilent 490 GC (Agilent Technologies, CA, USA) and a FavourSpec IMS instrument (Gesellschaft für analytische Sensorsysteme (G.A.S.), Dortmund, Germany) as described previously [37,38]. First, 5000 ± 100 mg of rice sample was ground at low temperature and placed in a 20 mL headspace vial. The headspace injection conditions included incubation at 60 °C for 20 min and splitless injection with an injection volume of 500 μL, an injection speed of 500 rpm, and an injector temperature maintained at 85 °C. A capillary column (15 m × 0.53 mm × 1.0 μm, MXT-5; Restek, Bellefonte, PA, USA) was used, with a column temperature of 60 °C and 99.99% nitrogen as the carrier gas. A programmed pressure was employed with an initial flow rate of 2.0 mL/min, which was maintained for 2 min and then linearly increased to 10.0 mL/min for 8 min, 100.0 mL/min for 10 min, and 150.0 mL/min for 10 min. The chromatographic run time was set at 30 min, and the inlet temperature was maintained at 80 °C. The volatile compounds were identified based on the drift time and RI according to the National Institute of Standards and Technology (NIST) and G.A.S. databases. These measurements were performed in triplicate. The IMS conditions were as follows: ionization source, tritium (3H); drift tube length, 53 mm; electric field strength, 500 V/cm; drift tube temperature, 45 °C; drift gas, high-purity nitrogen (≥99.999%); flow rate, 75 mL/min; and positive ion mode. Each experiment with the samples was conducted in triplicate.

The retention index (RI) was calculated using n-ketones C4–C9 (Sinopharm Chemical Reagent Beijing Co., Ltd., Beijing, China) as external references by the Laboratory Analysis View (LAV) software(LAV version 2.0.0; G.A.S.) in the GC-IMS device. The volatile compounds were tentative identified based on comparison of RI and the drift time with the NIST library and IMS database retrieval software obtained from G.A.S (Dortmund, Germany, version 2.0.0). Finally, the intensities of the volatile compounds were analyzed according to the peak volumes of the selected signal peaks by Gallery Plot analysis (v.1.0.7, G.A.S.) [37]. Visual and quantitative comparisons of volatility differences between samples were performed using the Gallery plot plug-in. Two-dimensional top views and three-dimensional fingerprints were constructed using the Reporter plug-in.

2.6. Statistical Analysis

Chroma Time-of-Flight (TOF) 4.3X software of LECO Corporation (St. Joseph, MI, USA) and NIST database were used for raw peak extraction, data baseline filtering and calibration of the baseline, peak alignment, deconvolution analysis, peak identification, integration, and spectrum matching of the peak area [39].

The Reporter, Gallery Plot, and Dynamic Principal component analysis (PCA) plug-ins in LAV software were used to generate two- and three-dimensional spectra, difference spectra, fingerprints, and PCA plots of volatile components for comparing VOCs between samples. In order to avoid deviations between measurements, the drift time of sample spectra was normalized relative to RIP drift time, which proceed automatically in the LAV software.

Excel and Origin 2021 software were used for statistical analysis, including principal component analysis (PCA), cluster analysis, spider plots, and histograms.

3. Results

3.1. Sensory Evaluation and Aroma Classification of Rice

The flavor sensory properties of 11 different rice types were evaluated, and their aroma profiles were clustered using the Ward hierarchical clustering method. The number of clusters was determined by the elbow method using the Within-Cluster Sum of Squares (WCSS). The results are presented in Figure 1a. As the number of clusters increases, the WCSS value decreases. The decrease is rapid from class 1 to class 3, and after the number of clusters reaches 3, the curve of WSS changes gradually and becomes flat, which is identified as the "elbow" point. To maintain the tightness within the clusters while ensuring the degree of segmentation, cluster 3 is selected as the best clustering number. The 11 rice samples were divided into 3 major groups. The first group comprised three samples, and the second and third groups comprised four samples each. The results are shown in Figure 1b.

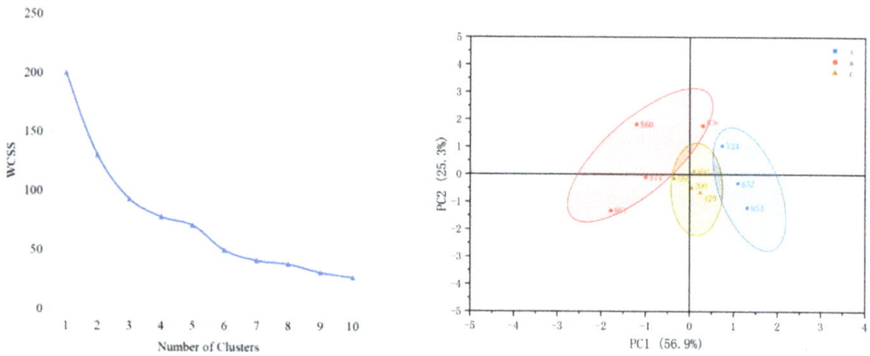

(a) (b)

Figure 1. Ward hierarchical cluster analysis of sensory evaluation: (**a**) relationship between the number of clusters and WCSS, (**b**) cluster diagram of sensory evaluation of 11 rice samples, A: aroma type A rice; B: aroma type B rice; C: aroma type C rice.

The aroma characteristics of each group of rice were analyzed, and a flavor profile spider plots was drawn (Figure 2), illustrating different aromas. (1) The first group included samples 652, 953, and 724. Their sweet, popcorn, cereal, and grassy aromas were more prominent, and no obvious floral aroma was observed. This group of samples was defined as the sweet aroma type (aroma type A). (2) The second group included samples 574, 951, 860, and 836. The most prominent aromas of this group were cereal and starchy, whereas their sweet and popcorn aromas were significantly reduced compared with other samples. Other sensory attributes were not obvious, so this group of samples was defined as the cereal aroma type (aroma type B). (3) The third group included samples 429, 691, 398, and 560. Their cereal, sweet, and starchy aromas were higher, and the aroma of this group was complex. Therefore, this group of samples was defined as a complex aroma type (aroma type C).

https://doi.org/10.3390/foods15020200

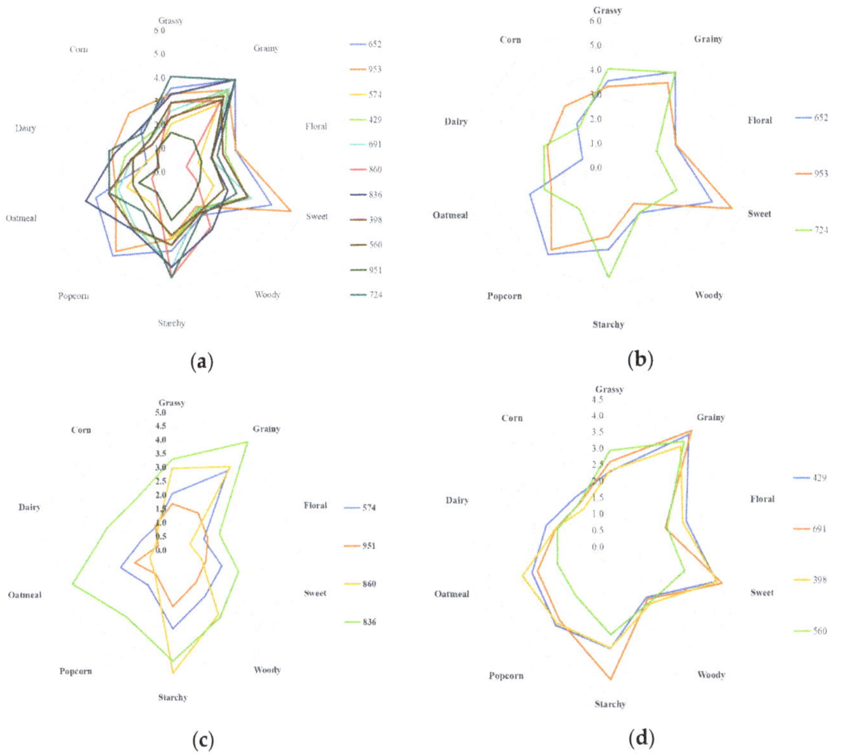

Figure 2. Aroma profile spider plots of rice of different aroma types: (**a**) all rice samples; (**b**) aroma type A rice; (**c**) aroma type B rice; and (**d**) aroma type C rice.

3.2. Analysis of Volatile Aroma Compounds in Different Rice Aromas Based on HS-SPME-GC-MS

3.2.1. VOCs in Rice with Different Aroma Types

HS-SPME-GC-MS was used to analyze the volatile components of 11 aromatic rice samples. Table 3 presents the volatile compounds of the three aromatic rice types. HS-SPME-GC-MS analysis of 11 samples identified 74 volatile compounds. The number of volatile compounds varied among rice with different aromas. In addition, 52 volatile compounds with flavor characteristics were identified, which constituted 70.27% of the identified aroma compounds.

HS-SPME-GC-MS analysis of 11 samples characterized 74 volatile compounds into 8 categories, as shown in Figure 3a. Of these, alcohols accounted for 27%, followed by ketones molecules for 20.2%, aldehydes for 17.5%, esters, alkanes and heterocyclic for 8.1%, and aromatics compounds for 6.7%, and alkenes for 4%.

https://doi.org/10.3390/foods15020200

Table 3. Comparison of the odor profiles and odor threshold values of the volatile compounds found in the three rice cultivars.

No.	Volatile Compounds	RI	Formula	CAS	Category	Odor Description [1]	Odor Threshold [2] (μg/L)	Mean Content(μg/L) [3] ± Standard Deviation A	B	C
1	Butanal	884	C_4H_8O	123-72-8	Aldehydes	Bready, Chocolate, Creamy, Pear, Vegetable, Fatty, Earthy, Woody, Pungent, Nutty, Wine-Like, Meaty, Spicy, Citrus, Herbaceous, Green, Apple, Floral, Musty, Cocoa, Fruity, Malty, Characteristic, Aldehyde Odor, Banana, Cocoa, Musty, Bready	9.0	1.87 ± 0.16	1.17 ± 0.48	2.20 ± 2.10
2	2-Butanone	898	C_4H_8O	78-93-3	Ketones	Ethereal, Ether, Fruity, Acetone, Camphor, Acetone-Like Odor, Sweet, Pleasant, Pungent, Moderately Sharp, Mint- Or Acetone-Like Odor, Pleasant, Camphoraceous	50,000.0	23.01 ± 2.21	17.87 ± 6.34	23.02 ± 12.50
3	2-Methylbutanal	907	$C_5H_{10}O$	96-17-3	Aldehydes	Musty, Coffee, Cocoa, Nutty, Almond, Fermented, Hazelnut, Malt, Malt, Burnt, Choking, Estery Apple, Fruity, Green Grass, Sour	1.0	0.45 ± 0.03	0.42 ± 0.12	0.31 ± 0.09
4	3-Methylbutanal	910	$C_5H_{10}O$	590-86-3	Aldehydes	Peach, Sour, Ethereal, Fatty, Aldehydic, Apple-Like Odor, Powerful Penetrating, Acrid Odor, Almond, Cocoa, Cheese, Chocolate, Malty	n.f.	7.57 ± 0.70	7.05 ± 3.64	6.08 ± 1.75
5	Pentanal	968	$C_5H_{10}O$	110-62-3	Aldehydes	Bready, Fermented, Berry, Malt, Fruity, Nutty, Acrid, Almond, Bitter, Oil, Pungent, Aldehyde, Banana, Grass, Lower Aldehyde,	12.0	3.14 ± 0.87	1.66 ± 1.06	4.39 ± 4.48
6	Decane	1000	$C_{10}H_{22}$	124-18-5	Alkanes	Oily, Green, Beany, Berry, Coffee, Green, Fatty Alkane	n.f.	0.35 ± 0.10	0.67 ± 0.09	0.74 ± 0.39
7	2-Butanol	1023	$C_4H_{10}O$	78-92-2	Alcohols	Oily, Sweet, Wine, Apricot, Has A Strong Vinous Odor, Sweet, Wine Odor Liquid, Strong, Pleasant Odor, Fruit, Medicine, Alcoholic	43,000.0	3.39 ± 0.91	2.84 ± 0.25	2.32 ± 1.87
8	Toluene	1029	C_7H_8	108-88-3	Aromatics	Paint, Sweet, Pungent, Benzene-Like Odor	2140.0	14.78 ± 2.55	14.12 ± 6.70	12.95 ± 4.60
9	1-Propanol	1035	C_3H_8O	71-23-8	Alcohols	Alcoholic, Fermented, Alcohol, Musty, Fusel, Pungent, Peanut, Similar To Ethanol, Alcohol, Candy	30,000.0	3.45 ± 2.14	3.97 ± 2.23	2.63 ± 0.63
10	2-Methyl-decane	1053	$C_{11}H_{24}$	6975-98-0	Alkanes	—	n.f.	1.13 ± 0.18	ND	1.23 ± 0.44
11	Undecane, 5-methyl-	1064	$C_{12}H_{26}$	1632-70-8	Alkanes	—	n.f.	1.04 ± 0.60	ND	1.87 ± 0.40
12	Hexanal	1072	$C_6H_{12}O$	66-25-1	Aldehydes	Leafy, Sweaty, Tallow, Fresh, Fatty, Fruity, Aldehydic, Green, Characteristic Fruity Odor (On Dilution), Strong, Green Grass Odor, Sharp, Aldehyde Odor, Apple, Fresh, Oil, Leaves, Vinous, Grassy	4.0	307.27 ± 33.46	ND	356.20 ± 210.81
13	Ethylbenzene	1111	C_8H_{10}	100-41-4	Aromatics	Aromatic, Pungent, Sweet, Gasoline-Like	140,000.0	1.95 ± 1.72	ND	3.28 ± 0.75
14	o-Xylene	1119	C_8H_{10}	95-47-6	Aromatics	Geranium, Sweet, Aromatic Odor	50.0	1.20 ± 0.19	1.66 ± 0.57	1.13 ± 0.94

https://doi.org/10.3390/foods1

Table 3. *Cont.*

No.	Volatile Compounds	RI	Formula	CAS	Category	Odor Description [1]	Odor Threshold [2] (μg/L)	Mean Content(μg/L) [3] ± Standard Deviation A	B	C
15	2-n-Butyl furan	1122	$C_8H_{12}O$	4466-24-4	Heterocyclic	Mild, Wine, Sweet, Fruity, Spicy, Wet Hay, Noncharacteristic, Weak	n.f.	3.27 ± 0.45	1.66 ± 1.19	3.77 ± 2.58
16	1-Methoxy-2-propanol	1125	$C_4H_{10}O_2$	107-98-2	Alcohols	Weak Pleasant Odor, Mild, Ethereal Odor	10,000.0	25.29 ± 6.71	29.70 ± 5.43	19.75 ± 16.89
17	Dodecane, 2,6,10-trimethyl-	1140	$C_{15}H_{32}$	3891-98-3	Alkanes	—	n.f.	0.78 ± 0.68	3.68 ± 1.73	1.82 ± 1.05
18	1-Butanol	1144	$C_4H_{10}O$	71-36-3	Alcohols	Oil, Vanilla, Fusel, Sweet, Balsam, Medicinal, Harsh Fusel Odor With Banana, Odor Similar To Amyl Alcohol, Rancid, Strong Characteristic, Mildly Alcoholic Odor, Wine, Floral, Fragrant, Fruity	500.0	8.84 ± 1.40	16.62 ± 17.70	12.43 ± 7.10
19	Undecane, 4-methyl-	1147	$C_{12}H_{26}$	2980-69-0	Alkanes	—	n.f.	0.89 ± 0.31	2.40 ± 1.12	1.07 ± 0.59
20	3-Hexanol	1157	$C_6H_{14}O$	623-37-0	Alcohols	Alcoholic, Ether, Medicinal, Solvent, Spice, Wine	820.0	2.27 ± 0.22	1.82 ± 1.10	1.52 ± 1.16
21	1-Penten-3-ol	1158	$C_5H_{10}O$	616-25-1	Alcohols	Butter, Pungent, Tropical, Horseradish, Green, Vegetable, Bitter, Fruity, Fish, Oxidized, Wet Earth, Burnt, Grass, Slightly Meaty, Mushroom	400.0	13.77 ± 3.29	6.53 ± 6.55	8.30 ± 5.88
22	2-Propenoic acid, butyl ester	1168	$C_7H_{12}O_2$	141-32-2	Esters	Sharp, Fragrant, Biting Characteristic Odor, Fruity	n.f.	0.99 ± 0.28	0.44 ± 0.51	0.78 ± 0.54
23	2-Heptanone	1173	$C_7H_{14}O$	110-43-0	Ketones	Coconut, Soap, Herbal, Sweet, Woody, Fruity, Spicy, Cinnamon, Banana, Blue Cheese, Fruit, Green, Nut, Musty, Peardrops, Soapy, Butter	20.0	11.46 ± 2.37	6.79 ± 1.93	16.27 ± 14.37
24	Heptanal	1175	$C_7H_{14}O$	111-71-7	Aldehydes	Citrus, Ozone, Herbal, Fresh, Wine-Lee, Rancid, Fatty, Aldehydic, Green, Pungent Odor, Penetrating Fruity Odor, Citrus, Nut, Heavy, Oily, Planty Green, Putty, Milky	3.0	10.31 ± 1.32	6.63 ± 2.71	10.44 ± 5.17
25	1-Butanol, 3-methyl-	1205	$C_5H_{12}O$	123-51-3	Alcohols	Oil, Alcoholic, Burnt, Whiskey, Malt, Banana, Fusel, Fruity, Characteristic, Disagreeable, Burnt, Cocoa, Floral	250.0	7.14 ± 3.52	7.45 ± 1.99	6.52 ± 0.87
26	2-Pentylfuran	1222	$C_9H_{14}O$	3777-69-3	Heterocyclic	Butter, Green Bean, Vegetable, Earthy, Beany, Fruity, Metallic, Green, Greenbean, Butter, Floral, Pungent, Sweet, Buttery, Caramel, Fishy, Roasted, Popcorn	6.0	12.80 ± 1.44	8.50 ± 1.80	15.35 ± 9.76
27	1-Decanol	1237	$C_{10}H_{22}O$	112-30-1	Alcohols	Orange, Rose, Fat, Waxy, Fatty, Floral, Sweet, Clean, Watery, Fruity, Orange, Fat, Oil, Citrus, Soapy, Unripe Fruit	6.0	0.58 ± 0.19	1.16 ± 0.56	0.65 ± 0.33
28	Styrene	1243	C_8H_8	100-42-5	Alkenes	Balsamic, Gasoline, Floral, Sweet, Balsam, Plastic, Aromatic, Sweet, Floral	3.6	0.40 ± 0.39	1.19 ± 1.10	0.35 ± 0.45

https://doi.org/10.3390/foods1

Table 3. *Cont.*

No.	Volatile Compounds	RI	Formula	CAS	Category	Odor Description [1]	Odor Threshold [2] (μg/L)	Mean Content(μg/L) [3] ± Standard Deviation A	B	C
29	1-Pentanol	1248	$C_5H_{12}O$	71-41-0	Alcohols	Oil, Balsamic, Vanilla, Fusel, Sweet, Fusel-Like, Fruit, Green, Pungent, Yeast, Fruit, Wax, Bread-Like	120.0	34.45 ± 4.85	28.07 ± 10.62	36.50 ± 17.34
30	1,2,3-Trimethylbenzene,	1266	C_9H_{12}	526-73-8	Aromatics	Distinctive, Aromatic Odor	n.f.	0.08 ± 0.13	0.17 ± 0.20	0.22 ± 0.16
31	2-Hexanol, 2,3-dimethyl-	1269	$C_8H_{18}O$	19550-03-9	Alcohols	——	n.f.	0.61 ± 0.33	0.21 ± 0.19	0.36 ± 0.45
32	3-Hydroxy-2-butanone	1271	$C_4H_8O_2$	513-86-0	Ketones	Butter, Cream, Milky, Fatty, Creamy, Sweet, Dairy, Buttery, Bland, Woody, Yogurt Odor, Green Pepper, Fermented, Milky, Fatty	n.f.	15.09 ± 9.24	4.89 ± 1.75	9.38 ± 6.60
33	1-Decanol, 2-hexyl-	1272	$C_{16}H_{34}O$	2425-77-6	Alcohols	——	n.f.	0.50 ± 0.15	0.96 ± 0.37	0.54 ± 0.34
34	2-Octanone	1276	$C_8H_{16}O$	111-13-7	Ketones	Natural, Earthy, Gasoline, Weedy, Herbal, Woody, Bitter, Soap, Fatty, Green Cheese Aroma, Floral, Bitter, Fruity (Unripe Apple), Pleasant, Fat, Fragrant, Mold, Butter, Herb, Resin, Floral, Ketone, Musty, Soapy	41.0	17.01 ± 3.06	20.55 ± 3.75	22.08 ± 5.74
35	Octanal	1279	$C_8H_{16}O$	124-13-0	Aldehydes	Lemon, Citrus, Soap, Orange Peel, Fat, Waxy, Fatty, Aldehydic, Green, Strong, Fruity Odor, Fatty, Honey Odor On Dilution, Pungent Odor, Citrus-Like On Dilution, Aldehyde, Fruity, Pungent, Slightly Fragment, Soapy, Fresh, Citrus-Like	1.4	2.27 ± 0.43	1.33 ± 0.42	2.28 ± 1.72
36	1-Nonadecene	1306	$C_{19}H_{38}$	18435-45-5	Alkenes	——	n.f.	0.08 ± 0.04	0.18 ± 0.16	0.08 ± 0.16
37	(E)-2-Heptenal	1311	$C_7H_{12}O$	18829-55-5	Aldehydes	Soap, Vegetable, Fresh, Fatty, Pungent, Almond, Green, Fruity, Melting Plastic, Soapy, Tallow	13.0	2.77 ± 0.43	2.07 ± 0.81	2.97 ± 2.04
38	2,3-Octanedione	1316	$C_8H_{14}O_2$	585-25-1	Ketones	Dill, Buttery, Broccoli, Cooked, Green, Roasted Nuts	n.f.	3.66 ± 0.78	2.92 ± 1.50	2.58 ± 0.72
39	5-Hepten-2-one, 6-methyl-	1328	$C_8H_{14}O$	110-93-0	Ketones	Apple, Mushroom, Citrus, Musty, Rubber, Nutty, Hazelnut, Bitter, Lemongrass, Powerful, Fatty, Green, Mushroom, Pepper, Strawberry, Sweet, Fruity, Lemongrass	50.0	1.54 ± 0.27	1.34 ± 0.37	1.16 ± 0.24
40	1-Hexanol	1350	$C_6H_{14}O$	111-27-3	Alcohols	Oil, Alcoholic, Ethereal, Resin, Fusel, Sweet, Fruity, Green, Characteristic, Sweet Alcohol, Pleasant, Banana, Herb, Fatty, Floral, Grassy	10.0	90.76 ± 14.00	120.41 ± 72.43	96.17 ± 19.66
41	2-Cyclopenten-1-one, 3-ethyl-2-hydroxy-	1377	$C_7H_{10}O_2$	21835-01-8	Ketones	Butterscotch, Chocolate, Caramel, Smoky, Sweet, Maple, Fenugreek, Savory, Strong Caramel	n.f.	0.37 ± 0.25	0.36 ± 0.42	0.51 ± 0.38
42	2-Decanone	1381	$C_{10}H_{20}O$	693-54-9	Ketones	Floral, Orange, Peach, Fatty, Fruity, Musty	n.f.	1.08 ± 0.12	0.65 ± 0.25	0.99 ± 0.83

https://doi.org/10.3390/foods1

Table 3. *Cont.*

No.	Volatile Compounds	RI	Formula	CAS	Category	Odor Description [1]	Odor Threshold [2] (μg/L)	Mean Content(μg/L) [3] ± Standard Deviation A	B	C
43	Nonanal	1385	$C_9H_{18}O$	124-19-6	Aldehydes	Citrus, Lime, Rose, Green, Fishy, Waxy, Fresh, Fatty, Peely, Aldehydic, Orris, Grapefruit, Orange-Rose Odor, Floral, Fat, Green, Lemon, Aldehyde, Floral, Grass, Slightly Pungent, Soapy, Tallow, Wax, Citrus-Like, Tallow, Fruity, Orange Peel	1.0	17.78 ± 4.03	13.83 ± 1.28	12.87 ± 2.67
44	3-Octanol	1392	$C_8H_{18}O$	589-98-0	Alcohols	Citrus, Nut, Moss, Herbal, Earthy, Woody, Melon, Minty, Mushroom, Spicy, Oil, Burnt, Chemical, Metal	18.0	0.68 ± 0.12	0.20 ± 0.40	0.33 ± 0.39
45	Formic acid, heptyl ester	1404	$C_8H_{16}O_2$	112-23-2	Esters	Apple, Herbal, Waxy, Floral, Cucumber, Plum, Green	n.f.	0.09 ± 0.16	0.27 ± 0.20	0.10 ± 0.20
46	P-menthan-1-ol	1413	$C_{10}H_{20}O$	3901-95-9	Alcohols	——	n.f.	2.25 ± 0.75	0.80 ± 0.59	1.77 ± 1.85
47	Furfural	1416	$C_5H_4O_2$	98-01-1	Aldehydes	Fragrant, Bread, Woody, Sweet, Baked, Almond, Peculiar Odor, Almond-Like Odor, Almond, Baked Potatoes, Burnt, Spice	280.0	0.42 ± 0.36	0.46 ± 0.32	0.20 ± 0.40
48	(E)-2-Octenal	1418	$C_8H_{14}O$	2548-87-0	Aldehydes	Nut, Fat, Herbal, Fresh, Green, Fatty, Banana, Waxy, Leaf, Cucumber, Green Leaf, Walnut, Almond, Fruity, Nutty	3.0	0.56 ± 0.07	0.40 ± 0.09	ND
49	1-Octen-3-one	1433	$C_8H_{14}O$	4312-99-6	Ketones	Herbal, Earthy, Metal, Musty, Mushroom, Dirty, Metallic	0.05	2.73 ± 0.45	3.88 ± 1.04	1.51 ± 1.48
50	Cis-linalool oxide	1436	$C_{10}H_{18}O_2$	5989-33-3	Alcohols	Musty, Alcohol, Fenchyl, Camphor, Earthy, Floral, Green, Herbal, Wood	320.0	2.02 ± 0.23	1.42 ± 0.63	1.61 ± 0.36
51	1-Tetradecene	1442	$C_{14}H_{28}$	1120-36-1	Alkenes	Pleasant	n.f.	0.60 ± 0.06	0.74 ± 0.33	0.53 ± 0.05
52	1-Octen-3-ol	1447	$C_8H_{16}O$	3391-86-4	Alcohols	Raw, Fishy, Oily, Earthy, Fungal, Chicken, Mushroom, Green, Cucumber, Floral, Plastic, Soap, Fatty, Fruity, Grass, Perfumy, Sweet, Balsamic	1.0	43.75 ± 4.49	38.82 ± 6.43	47.65 ± 19.96
53	1-Heptanol	1452	$C_7H_{16}O$	111-70-6	Alcohols	Leafy, Coconut, Herbal, Peony, Strawberry, Chemical, Musty, Sweet, Woody, Violet, Green, Fragrant, Faint, Aromatic, Fatty, Mushroom	3.0	5.71 ± 0.85	4.12 ± 1.23	5.76 ± 2.93
54	2-Propylfuran	1481	$C_7H_{10}O$	4229-91-8	Heterocyclic	——	n.f.	0.57 ± 0.49	0.57 ± 0.45	0.46 ± 0.55
55	2-Ethyl-1-hexanol	1487	$C_8H_{18}O$	104-76-7	Alcohols	Citrus, Rose, Fresh, Floral, Oily, Sweet, Mild, Green, Rose, Citrus Fresh Floral Oily Sweet, Rosy	n.f.	11.54 ± 2.17	14.15 ± 5.82	13.33 ± 2.83
56	Decanal	1491	$C_{10}H_{20}O$	112-31-2	Aldehydes	Citrus, Soap, Orange Peel, Tallow, Waxy, Floral, Sweet, Aldehydic, Floral-Fatty Odor, Penetrating, Floral, Fried, Fatty, Green	9.0	0.74 ± 0.14	0.58 ± 0.21	0.47 ± 0.38
57	6-Undecanone	1521	$C_{11}H_{22}O$	927-49-1	Ketones	——	85.0	0.51 ± 0.10	0.09 ± 0.18	0.29 ± 0.37
58	4-Ethylcyclohexanol	1535	$C_8H_{16}O$	4534-74-1	Alcohols	——	n.f.	0.86 ± 0.16	0.80 ± 0.07	0.67 ± 0.16
59	2,2-Dimethyl-3-heptanone	1539	$C_9H_{18}O$	19078-97-8	Ketones	——	n.f.	1.90 ± 0.34	1.05 ± 0.73	1.85 ± 1.37

https://doi.org/10.3390/foods1

Table 3. *Cont.*

No.	Volatile Compounds	RI	Formula	CAS	Category	Odor Description [1]	Odor Threshold [2] (μg/L)	Mean Content(μg/L) [3] ± Standard Deviation A	B	C
60	1-Octanol	1555	$C_8H_{18}O$	111-87-5	Alcohols	Burnt, Orange, Rose, Waxy, Chemical, Metal, Aldehydic, Mushroom, Green, Fresh, Aromatic, Bitter Almond, Burnt Matches, Fat, Floral, Moss, Nut, Green Herbaceous, Citrus-Like	42.0	3.68 ± 0.17	2.70 ± 0.76	3.47 ± 1.74
61	Isophorone	1580	$C_9H_{14}O$	78-59-1	Ketones	Cooling, Tobacco, Cedarwood, Leather, Musty, Woody, Sweet, Fruity, Camphoraceous, Camphor, Green, Peppermint-Like Odor, Spice	2000.0	1.26 ± 1.12	1.74 ± 1.29	0.96 ± 1.12
62	Myristic acid vinyl ester	1594	$C_{16}H_{30}O_2$	5809-91-6	Esters	——	n.f.	0.58 ± 0.06	0.52 ± 0.38	0.23 ± 0.29
63	Heptacosane	1622	$C_{27}H_{56}$	593-49-7	Alkanes	——	n.f.	ND	ND	0.22 ± 0.39
64	(E)-2-Decenal	1635	$C_{10}H_{18}O$	3913-81-3	Aldehydes	Orange, Coriander, Rose, Tallow, Waxy, Oily, Fatty, Earthy, Floral, Aldehydic, Mushroom, Green	n.f.	0.81 ± 0.09	0.43 ± 0.51	0.64 ± 0.45
65	Furan, 2-hexyl-	1755	$C_{10}H_{16}O$	3777-70-6	Heterocyclic	——	n.f.	0.24 ± 0.01	0.16 ± 0.04	0.11 ± 0.13
66	3-Tridecanone	1770	$C_{13}H_{26}O$	1534-26-5	Ketones	——	n.f.	0.27 ± 0.24	0.26 ± 0.17	0.31 ± 0.02
67	Phenol, 4-(1-methylpropyl)-	1832	$C_{10}H_{14}O$	99-71-8	Aromatics	——	n.f.	0.13 ± 0.06	0.03 ± 0.06	0.09 ± 0.05
68	Propanoic acid, 2-methyl-, 3-hydroxy-2,2,4-trimethylpentyl ester	1862	$C_{12}H_{24}O_3$	77-68-9	Esters	——	n.f.	0.29 ± 0.25	0.85 ± 0.32	0.44 ± 0.06
69	2,2,4-Trimethyl-1,3-pentanediol diisobutyrate	1873	$C_{16}H_{30}O_4$	6846-50-0	Esters	——	n.f.	3.29 ± 0.99	3.34 ± 1.52	1.97 ± 0.54
70	2-Acetyl-1h-pyrrole	1956	C_6H_7NO	1072-83-9	Heterocyclic	Licorice, Coumarin, Bread, Musty, Nutty, Walnut, Bread, Cocoa, Hazelnut, Caramel, Fatty, Mothball, Oily, Putty	170,000.0	0.06 ± 0.06	ND	0.02 ± 0.00
71	2(3H)-Furanone, dihydro-5-pentyl-	2013	$C_9H_{16}O_2$	104-61-0	Esters	Peach, Coconut, Waxy, Oily, Creamy, Sweet, Buttery	7.0	0.61 ± 0.16	0.65 ± 0.15	0.71 ± 0.24
72	2-Pyrrolidinone	2020	C_4H_7NO	616-45-5	Ketones	——	n.f.	0.07 ± 0.13	0.06 ± 0.12	0.29 ± 0.11
73	2-Tetradecanone	2122	$C_{14}H_{28}O$	2345-27-9	Ketones	——	n.f.	0.02 ± 0.03	0.07 ± 0.03	0.06 ± 0.02
74	Benzofuran, 2,3-dihydro-	2361	C_8H_8O	496-16-2	Heterocyclic	——	n.f.	0.18 ± 0.02	0.17 ± 0.12	0.12 ± 0.08

[1] Odor description found in the literature with database (Flavornet; FlavorDB2; The LRI and Odor Database). [2] All odor thresholds were obtained from: Odor & Flavor Detection Thresholds in Water (In Parts per Billion, lg/L) [11,15,26,35,40–50]. [3] The values are based on a semi-quantitative analysis using the area of 2-methyl-3-heptanone (IS). RI, retention index, which was calculated referred to the retention time of c5-c28 n-alkanes under the same conditions. "——", no odor description information was found in the literature. "n.f.", data was not found in the literature. "ND", not detected.

https://doi.org/10.3390/foods1

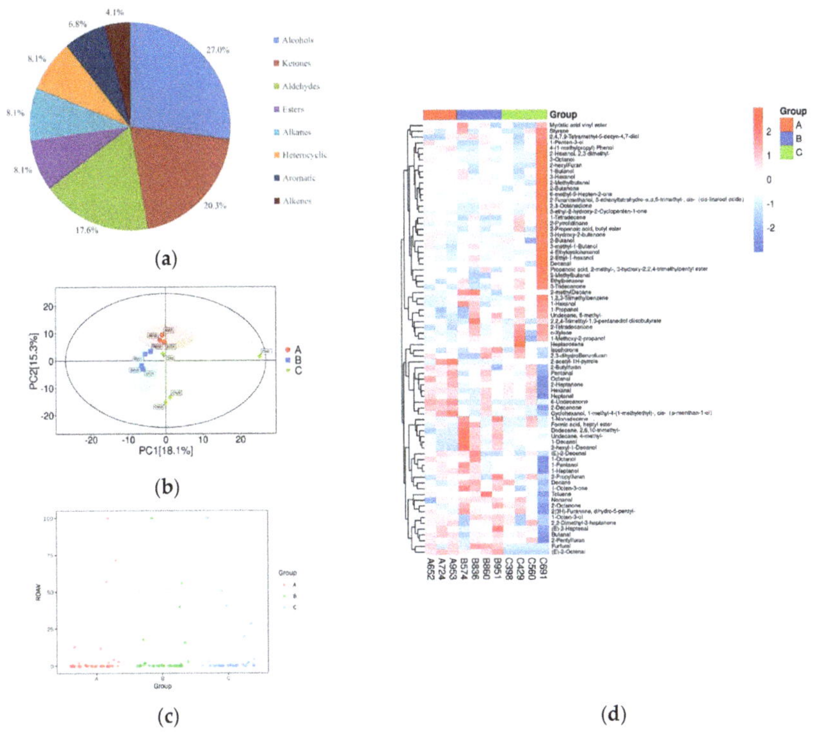

Figure 3. Multivariate statistical analysis of volatile compounds: (**a**) pie chart of metabolite classification proportions; (**b**) cluster analysis: A: aroma type A rice; B: aroma type B rice; C: aroma type C rice; (**c**) relative odor activity values; (**d**) heat map of hierarchical clustering analysis for all groups.

A PCA of 11 rice varieties was performed based on volatile aroma compounds, as shown in Figure 3b. All samples were within the 95% confidence interval (Hotelling's T2 ellipse), confirming clear differentiation between the three aroma types. Based on 74 volatile compounds, the samples were sorted into three distinct clusters: the first cluster comprised samples 652, 953, and 724; the second primarily comprised samples 574, 951, 860, and 836; and the third primarily comprised samples 429, 691, 398, and 560. This clustering was consistent with the cluster analysis results from the sensory evaluation. The ROAV of the volatile aroma compounds was calculated to determine which volatile compounds had the greatest impact on the overall flavor of the tested samples. This allowed the identification of key aroma compounds and more effective analysis of their contribution to the flavor of rice varieties, as shown in Figure 3c. An ROAV scatter plot was drawn based on the ROAV to visualize the distribution of the ROAV of flavor substances in different groups. Each point in the figure represents a substance. The horizontal axis of the ROAV scatter plot represents different sample groups, and the vertical axis represents the ROAV of flavor substances in these groups. A cluster analysis was conducted on all the volatile flavor substances of the samples, and a cluster heat map was drawn. Figure 3d shows the cluster heat map of the total content of volatile compounds in rice samples of different aroma types, indicating that volatile aroma compounds differed significantly among these samples.

https://doi.org/10.3390/foods15020200

3.2.2. Analysis of the Differential Volatile Profiling of Three Rice Cultivars

The 8 volatile compound values with the greatest differences were screened out by performing analysis of variance (ANOVA) on the volatile flavor compound data of the three rice types and based on the principle of ANOVA p-value < 0.05 (Table 4).

Table 4. Analysis of variance of differential volatile metabolites.

Volatile Compounds	RI	ANOVA p-Value	Q-Value	Mean Content(μg/kg) * ± Standard Deviation		
				A	B	C
2-Methyldecane	1053	0.012	0.363	1.13 ± 0.18	2.92 ± 1.26	1.23 ± 0.44
Dodecane, 2,6,10-trimethyl-	1140	0.035	0.623	0.78 ± 0.68	3.68 ± 1.73	1.82 ± 1.05
P-menthan-1-ol	1413	0.018	0.498	2.25 ± 0.75	0.80 ± 0.59	1.77 ± 1.85
(E)-2-Octenal	1418	0.001	0.333	0.56 ± 0.07	0.40 ± 0.09	ND
1-Octen-3-one	1433	0.044	0.623	2.73 ± 0.45	3.88 ± 1.04	1.51 ± 1.48
6-Undecanone	1521	0.022	0.498	0.51 ± 0.10	0.09 ± 0.18	0.29 ± 0.37
Heptacosane	1622	0.025	0.517	ND	ND	0.22 ± 0.39
2-Acetyl-1h-pyrrole	1956	0.007	0.498	0.06 ± 0.06	ND	0.02 ± 0.00

* The values are based on a semi-quantitative analysis using the area of 2-methyl-3-heptanone (IS). "ND", not detected.

Table 4 shows that the main differential volatile compounds were alkanes, aldehydes, ketones, alcohols, and heterocyclic compounds, and so forth. (E)-2-Octenal and 2-Acetyl-1h-pyrrole exhibited significant aroma expression in rice. Among these, (E)-2-Octenal mainly contributed to the fatty, herbal, fresh, green, fruity, and nutty flavors [19]. 2-Acetyl-1h-pyrrole contributed to a nutty flavor, 1-Octen-3-one contributed to a herbal flavor [19,42,50].

The volatile flavor data for the three aroma types were analyzed pairwise using Student t tests and Orthogonal Partial Least Squares-Discriminant Analysis (OPLS-DA) model analysis of variance. Figure 4 shows the differential volatile compounds across different rice aroma types based on the Student t test ($p < 0.05$) and the variable importance in the projection (VIP) of the OPLS-DA model greater than 1. In Figure 4a, different circles represent different comparison groups. The numbers in the overlapping area indicate the number of common differentially expressed metabolites between the comparison groups, and the numbers in the nonoverlapping area indicate the number of unique differentially expressed metabolites. A total of 7 common differentially expressed metabolites between groups A and C, 5 between groups A and B, and 3 between groups B and C. Across the comparison groups, 3 common differentially expressed metabolites between A versus C and A versus B groups, and 1 between A versus C and B versus C groups.

Figure 4b–d shows the results of the differential metabolite screening as volcano plots, with each dot representing a metabolite. The abscissa represents the fold change (logarithm to base 2) of each substance in the comparison group, and the ordinate represents the p value (negative logarithm to base 10) of the Student t test. The size of the scatter plot represents the VIP value of the OPLS-DA model, with larger scatter plots indicating higher VIP values. The significantly upregulated metabolites in Figure 4b–d are represented in red, the significantly downregulated metabolites are shown in blue, and the metabolites with no significant difference are indicated in gray. Figure 4b–d ashows that the A versus B comparison group had a higher number of differentially expressed metabolites, indicating a greater difference between aroma types A and B. In the three comparison groups, the flavor of the differential volatile substances in aroma type B was relatively weaker and not distinctive, indicating it had a less distinctive flavor than aroma types A and C, and A and C had better sweet, fruity, and other aromas compared with aroma type B.

https://doi.org/10.3390/foods15020200

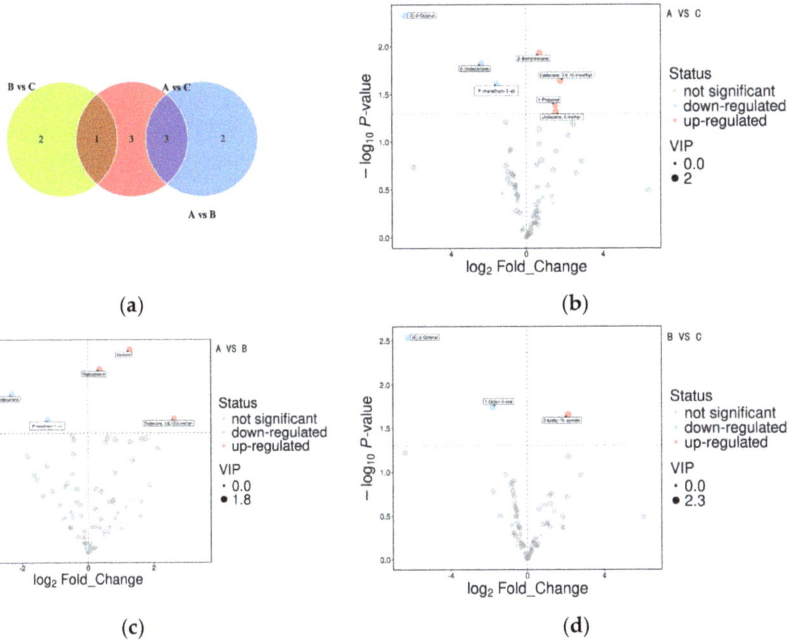

Figure 4. Differential flavor substances among aroma type A, B, C. (**a**) Venn analysis of differential metabolites among different comparison groups, (**b**) Volcano plots of differential metabolites of A versus C, (**c**) Volcano plots of differential metabolites of A versus B, (**d**) Volcano plots of differential metabolites of B versus C.

The Euclidean distance was used to quantify the differences in metabolites among different comparison groups. The data were clustered using the complete linkage method, and the results were presented as a heat map, shown in Figure 5.

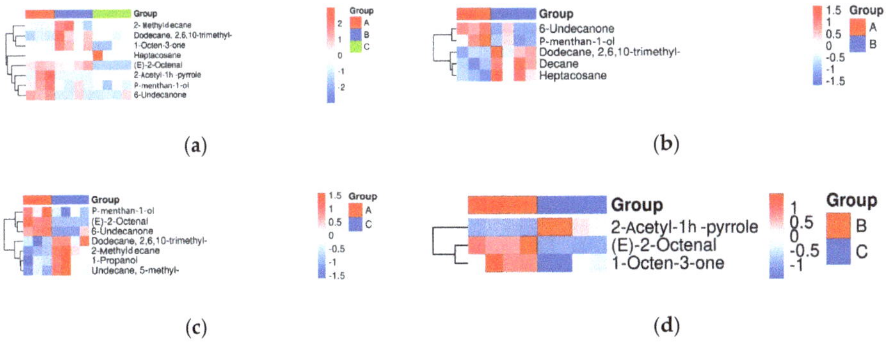

Figure 5. Hierarchical cluster analysis heat map among aroma type A, B, C. (**a**) A versus B versus C comparison group, (**b**) A versus B comparison group, (**c**) A versus C comparison group, and (**d**) B versus C comparison group.

The different colors in the graph on the horizontal axis represent different rice aroma types, the vertical axis represents the differential metabolites, and the different color blocks represent the relative expression levels of the corresponding metabolites. Red indicates a higher expression level of the substance, and blue indicates a lower expression level. Figure 5a presents the data of the A versus B versus C comparison group. The expression

https://doi.org/10.3390/foods15020200

levels of (E)-2-Octenal, 6-Undecanone, 2-Acetyl-1h-pyrrole, and P-menthan-1-ol were high in the aroma type A rice. Among these, (E)-2-Octenal had a nutty, fatty, herbal, fresh, and green aroma [19]. This also better explained the sensory results. Compared with aroma types B and C, aroma type A had a better sweet and grassy aroma. The expression levels of Dodecane, 2,6,10-trimethyl-, 1-Octen-3-one, and 2-Methyldecane were high in aroma type B, among which 1-Octen-3-one had a herbal, mushroom aroma [51]. This also better explained the sensory results that aroma type B had a more prominent starchy and grainy taste compared with aroma types A and C. In aroma type C, the expression levels of Heptacosane were high, and this group had richer differentiated volatile characteristics. This also confirmed that the aroma of this group was complex and was defined as a complex aroma type.

Figure 5b–d represents the data of the A versus B, A versus C, and B versus C comparison groups, respectively. In addition, the expression levels of 6-Undecanone, P-menthan-1-ol in type A were higher than in B in the A versus B comparison group. The expression levels of Dodecane, 2,6,10-trimethyl-, Decane and Heptacosane were higher in aroma type B than in aroma type A. However, the flavor characteristics of these volatile substances were not obvious, indicating that aroma type B had no prominent flavor characteristic. In the A versus C comparison group, the expression levels of 2-Methyldecane, 1-Propanol, and 5-Methylundecane were higher in aroma type C than in aroma type A. Among these, 1-Propanol had a better peanut aroma [51]. Therefore, type C had a better nutty and fruity aroma than types A. In aroma type B versus aroma type C comparison group, the expression levels of (E)-2-Octenal and 1-Octen-3-one were higher in aroma type B. (E)-2-Octenal contributed nutty, fatty, herbal, fresh, and green aroma characteristics [19]. The expression levels of 2-Acetyl-1h-pyrrole were higher in aroma type C. 2-Acetyl-1h-pyrrole had a bready, nutty, and caramel aroma [51]. Therefore, type C had a better nutty flavor than types B.

3.3. *Analysis of Volatile Odor Components of Rice by GC-IMS Spectroscopy*

3.3.1. Volatile Compounds Identified by GC-IMS

The volatile components of the three rice cultivars were determined using GC-IMS, and fingerprint patterns and cluster heat maps were obtained to more intuitively distinguish the differences in volatile components among different types of fragrant rice. As shown in Figure 6, the ion migration time and the position of the reactive ion peak (RIP) were normalized; the ordinate represents the retention time of the gas chromatography; the abscissa represents the ion migration time, and the vertical line at the abscissa 1.0 is the RIP. Figure 6a shows that most of the signals appeared in the retention time of 100–800 s and the drift time of 1.0–1.75. As shown in Figure 6b, the X, Y, and Z axes represent the ion drift time, retention time, and peak intensity of the volatile substances, respectively, which were used for identifying volatile substances [52]. In addition, some volatile compounds such as 1-pentanol, 1-propanol, and acetic acid ethyl ester produced multiple spots or signals, which were interpreted as protonated monomers, trimers, polymers, and proton-bound dimers in the fingerprint area, which could not be observed using GC-MS [53].

Figure 6a,b shows that the fingerprints of different aroma types of rice samples differed significantly after normalization of ion drift time and reactive ion peak position, with the signal intensities for A, B, and C showing considerable differences.

https://doi.org/10.3390/foods15020200

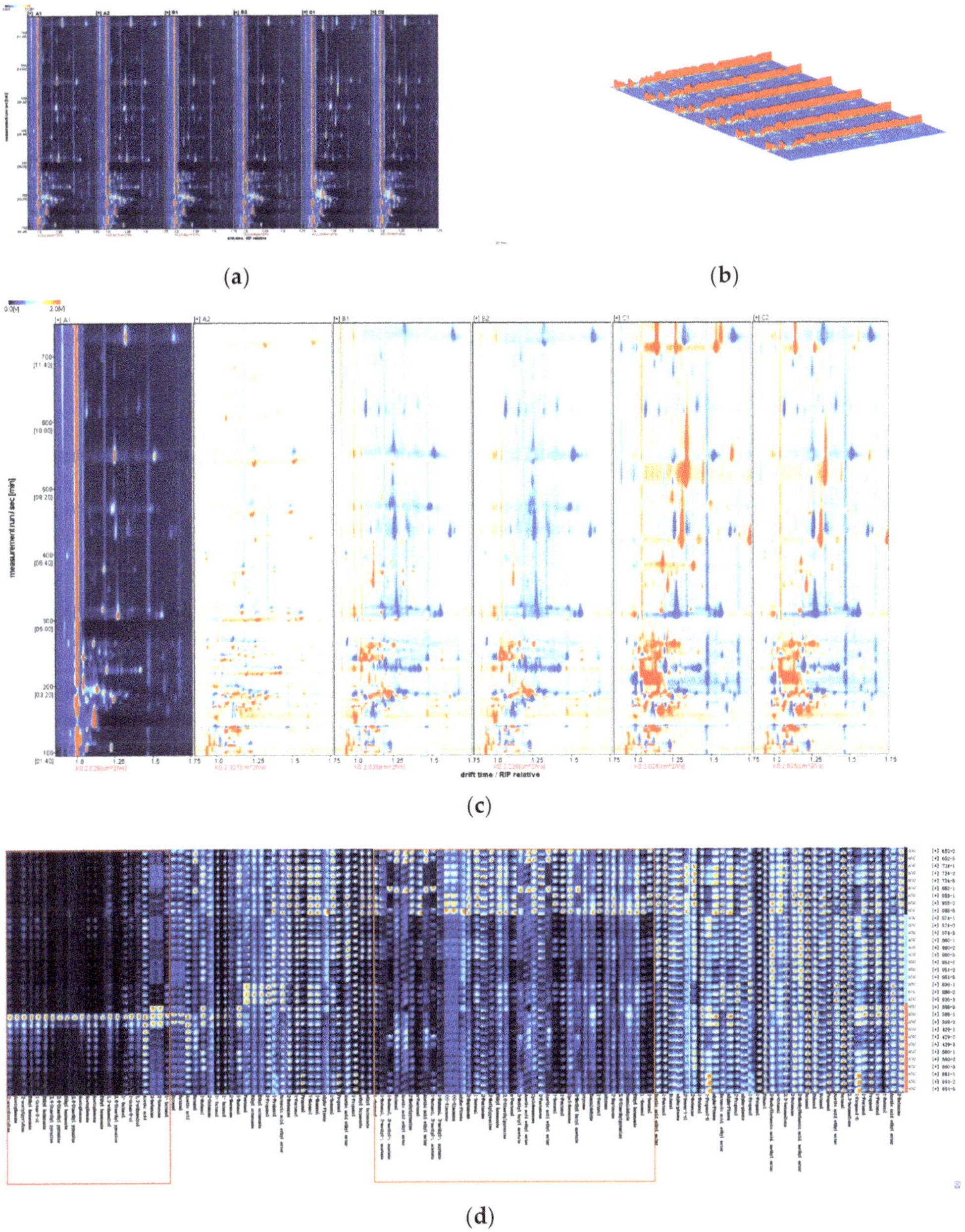

Figure 6. GC-IMS analysis of volatile compounds in 11 samples of the 3 aroma types A, B, and C. Volatile fingerprints (**a**), gallery plots (**b**), and comparison of differences using A1 as a reference (**c**). Fingerprint of volatile components of rice with different flavors (**d**).

The fingerprints of two samples were randomly selected for each aroma type, using A1 as a reference. This reference sample was compared with the fingerprints of the other aroma types to generate a differential spectrum, as shown in Figure 6c. Higher concentrations are displayed in red, concentrations equal to the reference value are displayed in white, and concentrations below the reference value are displayed in blue. Although VOCs signal intensities within the same aroma type (A1/A2, B1/B2, and C1/C2) were not significantly different, they were significantly different among different aroma types, with A and C showing a significant difference. Samples C1 and C2 displayed more red dots compared with samples A1 and A2, indicating a more intense aroma and flavor profile for aroma type C. Samples B1 and B2 displayed more blue dots, indicating a milder, more balanced flavor profile, consistent with the sensory evaluation and flavor metabolomics results.

The fingerprint identification method was adopted to display the volatile chemical information in the samples for a deeper understanding of the flavor characteristics of different aroma types of rice, as shown in Figure 6d. In this figure, each row represents the signal peak of the volatile components of each sample, and each column represents the signal peak of a specific volatile chemical substance in different samples. The color of the signal peak represents the substance. These graphs present the complete volatile compound profile of each sample in an easily understandable manner, highlighting the differences between samples [54,55]. The volatile chemical fingerprints of aroma types A, B, and C were significantly different. The two distinct regions in the figure represent the differences in the volatile chemical fingerprints between different flavor types. The red region represents primarily alcohols, heterocyclic compounds, and ketones, whereas the orange region represents primarily esters and aromatic hydrocarbons. In the red region, the volatile chemical fingerprint of flavor type C was significantly higher than that of A and B, indicating that C had a higher content of alcohols, heterocyclic compounds, and ketones, giving it a richer flavor profile. In the orange region, the volatile chemical fingerprint of flavor type A was significantly higher than that of B and C, indicating that A had a higher content of esters and aromatic hydrocarbons, giving it a better sweet and fragrant flavor [56]. Flavor type B had fewer volatile aroma compounds than flavor types A and C, which is why the flavor of type B was softer than the other two aroma types.

3.3.2. Differential Analysis of Volatile Compounds Based on GC-IMS

Table 5 lists all the volatile substances in the three fragrant rice types and their peak intensities. A total of 55 volatiles were detected and identified, including 14 ketones, 13 alcohols, 12 esters, 8 aldehydes, 4 heterocyclic compounds, 3 aromatic hydrocarbon compounds, and 1 other compound. The number of these categories differed from the results of GC-MS. A total of 14 VOCs were identified as monomers and dimers. These chemicals had high proton affinity or concentration, generating several signals in the analysis [57]. The aforementioned data showed that the volatile aroma compounds of the three different aromatic rice types had significant characteristics [51]. The samples of aroma type A contained higher contents of (+)-limonene (lemon- and orange-like flavors), 1-hexanol (grassy flavor), 2-methylpyrazine (nutty, coffee, and cocoa-like flavors), 2-pentanone (fruity, floral, and grassy flavors), 3-methyl butyl acetate, cyclohexanone, ethyl butanoate (a unique fresh fruity flavor), n-pentanal (fruity and floral flavors), styrene (floral and sweet flavor), 1-butanol (fruity and caramel flavors), 3-methyl-acetate 1-hexanal (grassy and fruity flavors), 1-pentanol (vanilla and sweet flavors), and 2-heptanon (coconut, sweet, and fruity flavors), indicating a better sweet aroma. In contrast, the samples of aroma type C contained higher contents of 1,2-ethanediol and 1-octen-3-ol (mushroom, lavender, and green-like flavors), 2,6-dimethyl pyrazine (peanut butter and peanut flavors), 2-acetylpyridine (popcorn flavor), acetophenone, and ethyl hexanoate (fruity and green flavors), indicating a better overall aroma. The various volatile flavor substances in aroma type B, were between those in aroma types A and C, without any significant outstanding features, which was consistent with the results of sensory evaluation and flavor metabolomics.

https://doi.org/10.3390/foods15020200

Table 5. GC-IMS analysis of volatile components of the three aromatic rice samples.

No.	Volatile Compounds	RI [1]	RT/s [2]	Drift Time/ms [3]	Peak Volume A	Peak Volume B	Peak Volume C
1	(+)-Limonene	1189.3	433.309	1.65766	47.32	21.00	18.45
2	1,2-Ethanediol	1611.1	1224.033	1.15889	193.75	156.70	307.64
3	1-Butanol-D	1150.3	376.987	1.3813	1348.39	1055.02	1285.03
4	1-Butanol-M	1150.8	377.739	1.18107	120.26	102.56	132.41
5	1-Butanol, 3-methyl-, acetate	1098.9	313.929	1.29261	94.11	55.98	48.65
6	1-Hexanal-D	1092.5	307.402	1.5605	2004.85	1105.25	812.02
7	1-Hexanal-M	1090.7	305.662	1.2717	410.76	168.15	199.22
8	1-Hexanol-D	1370.6	730.557	1.32995	2577.51	1732.20	1479.16
9	1-Hexanol-M	1368.9	727.847	1.6473	858.88	403.07	317.09
10	1-Octen-3-ol	1476.4	916.663	1.1603	140.36	152.57	455.21
11	1-Pentanol-D	1257.5	546.758	1.51302	3641.90	2330.38	2190.92
12	1-Pentanol-M	1258.7	548.929	1.25232	566.51	535.20	532.15
13	1-Propanol-D	1043.8	263.684	1.2563	1635.42	2108.75	2257.31
14	1-Propanol-M	1044.8	264.48	1.10846	77.39	82.29	62.42
15	2,3-Bbutanedione	927.4	190.575	1.17797	2020.23	2355.91	2300.38
16	2,6-Dimethyl pyrazine-D	1359.2	712.777	1.53432	96.20	87.06	325.36
17	2,6-Dimethyl pyrazine-M	1359.4	713.102	1.14292	42.39	39.75	82.61
18	2-Acetylpyridine	1569.9	1120.373	1.11215	138.31	152.23	372.34
19	2-Butanone	912	183.29	1.24516	1833.67	1415.23	1477.38
20	2-Heptanone-D	1192	437.464	1.62967	979.16	439.76	427.46
21	2-Heptanone-M	1188.7	432.325	1.26116	853.13	439.76	521.98
22	2-Hexanone-D	1097	311.827	1.50233	537.91	431.80	459.12
23	2-Hexanone-M	1095	309.868	1.19273	449.86	393.95	430.29
24	(E)-2-Hhexen-1-ol	1348.6	696.854	1.17499	186.52	159.96	159.59
25	2-Methylbutanoic acid, methyl ester	1028.1	250.954	1.18974	241.93	282.58	220.97
26	2-Methylpyrazine	1292.7	616.299	1.06994	894.63	288.30	321.60
27	2-Nonanone-D	1398	774.751	1.88186	49.58	47.39	49.36
28	2-Nonanone-M	1401.7	780.967	1.87851	43.77	42.12	39.74
29	2-Octanone-D	1295	621.087	1.75701	368.72	282.99	311.93
30	2-Octanone-M	1292.4	615.705	1.33009	88.85	88.59	90.09
31	2-Pentanone	996	226.78	1.3725	109.65	56.16	64.99
32	2-Propanol	920.1	187.091	1.22384	1142.23	1759.81	1455.53
33	3-Methyl butyl acetate	1149.9	376.536	1.31347	53.16	29.85	34.89
34	Acetic acid	1477.7	919.242	1.0518	2331.29	2504.44	4153.42
35	Acetic acid ethyl ester-D	914.3	184.354	1.32401	348.72	316.74	274.78
36	Acetic acid ethyl ester-M	918.1	186.15	1.09431	56.30	23.46	35.04
37	Acetophenone-D	1688.5	1445.218	1.17229	988.12	1180.18	4038.73
38	Acetophenone-M	1688	1443.765	1.38686	44.25	54.32	115.64
39	α-Pinene	992.3	224.692	1.29579	221.34	157.79	169.56
40	Benzaldehyde	1565.4	1109.612	1.14581	200.18	124.63	161.89
41	Butanal	894.9	175.517	1.28753	44.73	35.52	39.40
42	Butanoic acid, ethyl ester	1018.6	243.498	1.56902	171.15	179.31	126.95
43	Cyclohexanone	1292.1	614.912	1.15711	184.07	77.66	83.86
44	Ethanol	939.2	196.365	1.13225	3471.54	3298.02	3895.87
45	Ethyl acrylate	1025.7	249.08	1.41748	36.99	37.05	39.45
46	Ethyl butanoate	991	223.924	1.19944	1327.66	714.71	558.32
47	Ethyl hexanoate-D	1241.7	517.989	1.33922	361.29	366.11	920.86
48	Ethyl hexanoate-M	1242.7	519.833	1.78996	187.21	163.97	313.61
49	Ethyl octanoate	1401.9	781.337	1.47904	452.88	429.68	319.57
50	Ethyl pentanoate	1172.2	407.663	1.26934	121.34	88.84	99.17
51	Heptanal	1131.5	352.605	1.31628	30.45	24.15	26.67
52	n-Pentanal	990.2	223.486	1.41859	370.03	94.34	55.40

https://doi.org/10.3390/foods15020200

Table 5. *Cont.*

No.	Volatile Compounds	RI [1]	RT/s [2]	Drift Time/ms [3]	Peak Volume		
					A	B	C
53	Propanal-D	791.5	135.056	1.06433	422.97	489.08	569.58
54	Propanal-M	790.4	134.684	1.14053	601.24	489.08	435.88
55	Styrene	1296.4	623.002	1.41216	326.33	208.17	181.32

M, monomers; D, dimers. [1] The retention index (RI) calculated using n-ketones C4–C9 as external standard on FS-SE-54-CB column. [2] The retention time (RT) of identified volatiles in the capillary GC column. [3] The drift time in the drift tube.

3.4. Comparison of Different Volatile Compounds in Rice with Different Aroma Type Based on HS-SPME-GC-MS and GC-IMS

The differential flavor and compound analysis showed that the three rice aromas could be differentiated in sensory evaluation, HS-SPME-GC-MS, and GC-IMS.

From a sensory perspective, the main differential flavors of the three aromas included sweetness, woodiness, popcorn, cereal, and corn flavors. The main feature of aroma type A was the prominent sweetness, and the main feature of aroma type B was the prominent woodiness, cereal, and starch flavors. Aroma type C was more comprehensive than aroma types A and B, with a balanced focus on cereal, sweetness, starch, and popcorn flavors.

From the perspective of differential volatile substances, the main differential compounds included alkanes, aldehydes, ketones, alcohols, and heterocyclic compounds.

Aldehydes are an important type of volatile compounds in rice. These compounds have a low odor threshold and give rice pleasant grassy, citrus, and fresh grassy flavors at low levels. They are an important component of rice flavor. Aldehydes are decomposition products of lipid oxidation [58]. Among these aldehydes, the key volatile aroma compounds of rice are octanal [59], pentanal (valeraldehyde), and heptanal [48], which were also found in the differential compounds of the three aroma types. Studies have shown that heptanal, octanal, and so forth are formed from linoleic acid precursors [60]. In this study, the presence of aldehydes such as octanal and n-pentanal resulted in the prominent sweet flavor of aroma type A compared with aroma type B.

Alcohols are the second major VOCs identified in rice. Methanol, 1-propanol, 1-butanol, 1-pentanol [60], and 1-octen-3-ol are the key volatile aroma compounds of rice. They were all detected in the control sample and were the most abundant besides aldehydes [15]. Alcohols (1-octen-3-ol, hexanol, 1-octanol, etc.) have higher olfactory thresholds, contributing to rice's mild aromatic profile [61,62]. The floral and fruity aroma of rice is attributed to alcohols [1]. Among all alcohols, 1-octen-3-ol is produced during lipid oxidation. It has a high content, a low threshold, and the smell of wild mushrooms, and contributes the most to the aroma of rice [63]. Previous reports have shown that 1-octen-3-ol can be derived from the peroxide of linoleic acid [64]. In this study, the contents of alcohols such as 1-octen-3-ol and 1-propanol were higher in aroma type C than in aroma type A, which is why aroma type C exhibited a richer flavor than aroma type A.

Ketones are the third largest class of volatile components found in japonica rice. These compounds contribute a pleasant fruity flavor to the rice [61,65]. The main ketones identified in this study were 2-butanone and 2-heptanone, which gave cooked rice a unique fruity and pleasant aroma [66]. 2-Heptanone is produced from oleic acid [60]. In this study, the presence of ketones such as 2-heptanone resulted in a prominent sweet aroma in aroma type A compared with aroma types B and C.

Esters can give a fruity and floral aroma to rice [67]. However, they generally have a relatively high taste threshold, thus contributing little to the flavor of rice [68]. Ethyl acetate, octanoate acetate, and other esters affect the flavor of rice, making it more intense [15]. Therefore, the esters in type A contributed more to its sweet aroma.

https://doi.org/10.3390/foods15020200

Other volatile compounds contributing to odor include heterocyclic compounds such as pyridines and pyrazines. Heterocyclic compounds are mainly formed through nonenzymatic browning reactions or Maillard reactions [69]. Although these compounds have low concentrations, their thresholds are also low, so they contribute significantly to rice flavor, often imparting sweetness, nutty flavors, and beany notes to rice [1,15]. Aroma type C contains more heterocyclic compounds such as pyridines and pyrazines than aroma type A, which is one of the reasons why aroma type C has a more complex aroma.

Because hydrocarbon compounds have a higher threshold, their contribution to the overall aromatic properties of rice is limited [70]. Therefore, although some aroma type contain higher levels of hydrocarbons, their impact on their distinctive flavors is minimal.

4. Discussion

This study used sensory evaluation, HS-SPME-GC-MS, GC-IMS, and data analysis to systematically investigate the differences in flavor and volatile aroma compounds among different rice samples. The study also explored the key aroma components differentiating rice aroma types, dissecting the molecular basis for these flavor differences. The main findings were as follows:

1. Feasibility of distinguishing rice aromas based on sensory evaluation: Cluster analysis based on sensory evaluation data categorized rice aromas into three types: sweet aroma type (A), cereal aroma type (B), and complex aroma type (C). Type A exhibited a prominent sweet and popcorn aroma, type B exhibited a more prominent grainy and starchy aroma, and type C exhibited a more complex aroma.
2. Nontargeted metabolomics analysis using HS-SPME-GC-MS: The findings revealed the diversity and differences in volatile components of rice with different aroma types. A total of 74 volatile compounds were identified and classified into three aroma types through PCA, consistent with the sensory results. In the A versus B versus C comparison group, 8 major differential aroma compounds were identified, primarily including alkanes, aldehydes, ketones, alcohols, and heterocyclic compounds. (E)-2-Octenal, 6-Undecanone, 2-Acetyl-1h-pyrrole, and P-menthan-1-ol in type A gave it a better sweet aroma, Dodecane, 2,6,10-trimethyl-, 1-Octen-3-one, and 2-Methyldecane in type B gave it a better starchy and cereal flavor. 2-Acetyl-1h-pyrrole, Heptacosane, and 1-Propanol in type C contributed to a complex aroma.
3. GC-IMS analysis: The results revealed the diversity and differences in volatile components of rice with different aroma types. The fingerprints of rice samples with different aroma types showed significant differences. Compared with aroma type A, aroma type C was stronger and aroma type B was milder and more balanced, consistent with the results of sensory evaluation and flavor metabolomics. Samples of aroma type A contained (+)-limonene, 2-methylpyrazine, 2-pentanone, ethyl butanoate, n-pentanal, styrene, 1-butanol, 3-methyl-, acetate, 1-hexanal, 1-pentanol, and 2-heptanone, contributing to a more sweet aroma. Samples of aroma type C contained 1-octen-3-ol, 2,6-dimethyl pyrazine, 2-acetylpyridine, and ethyl hexanoate, contributing to a more complex aroma. The volatile aroma compounds in aroma type B fell between those in aroma types A and C, with no significant standout characteristics, consistent with the results of sensory evaluation and flavor metabolomics.

This study attempted to validate the feasibility of rice aroma differentiation. A comprehensive analysis was conducted on mainstream commercially available rice varieties to determine the molecular factors behind their distinct aromas, using sensory evaluation, flavor metabolomics. Combining complementary methods such as HS-SPME-GC-MS and GC-IMS, a more complete profile of volatile compounds was obtained, which helped distinguish different rice aromas and revealed their molecular basis. Future investigations

https://doi.org/10.3390/foods15020200

should build on this study by increasing sample diversity, integrating metabolomics and machine learning technologies to further analyze rice flavor differences and their formation pathways, thereby guiding processing innovation and quality improvement.

Author Contributions: Conceptualization, M.Z. and S.Q.; methodology, H.R.; software, H.Y.; validation, S.Q., H.R. and H.Y.; formal analysis, H.Y.; investigation, S.Q. and H.Y.; resources, M.Z. and S.Q.; data curation, H.Y.; writing—original draft preparation, S.Q.; writing—review and editing, S.Q.; visualization, H.R.; supervision, L.Z. and M.Z.; project administration, S.Q.; funding acquisition, L.Z., M.Z. and S.Q. All authors have read and agreed to the published version of the manuscript.

Funding: This research received no external funding.

Institutional Review Board Statement: 1. Nature of the Research Qualifies for Exemption: This study constitutes a scientific evaluation of raw rice based on sensory analysis. It does not fall under the category of research involving invasive human procedures, or exposing participants to significant physiological or psychological risk. 2. Adherence to National Standards and Safety Protocols: The design and execution of this sensory evaluation experiment strictly followed the relevant regulations of the Chinese National Standards, specifically GB/T 15682-2008 [71] "Inspection of grain and oils-Method for sensory evaluation of paddy or rice cooking and eating quality" and GB/T 16291.1-2012 "Sensory analysis-General guidance for the selection, training and monitoring of assessors-Part 1:Selected assessors". All testing procedures were standardized, low-risk sensory perception activities. 3. Participant Informed Consent and Voluntariness: All participants were fully informed prior to the test regarding the nature, procedures, duration, and information about the samples they would encounter. Participation was based entirely on voluntariness. Participants retained the right to withdraw from the test unconditionally at any stage without any adverse consequences. In light of the fact that this study presented minimal risk, followed testing procedures per authoritative national standards, and fully implemented informed consent and privacy protection measures, we hereby declare that this research qualified for an ethics exemption.

Informed Consent Statement: Informed consent was obtained from all subjects involved in the study.

Data Availability Statement: The original contributions presented in this study are included in the article. Further inquiries can be directed to the corresponding author.

Acknowledgments: We are grateful to all the participants of this tasting experiment who gave a fair and objective assessment of the different aroma types of rice.

Conflicts of Interest: Authors Shengmin Qi, Haibin Ren, Haiqing Yang and Lianhui Zhang were employed by the COFCO Corporation. They participated in study design, data collection, analysis, or decision to publish in the study. This relationship is disclosed here and has not influenced the content or conclusions of this work. The remaining authors declare that the research was conducted in the absence of any commercial or financial relationships that could be construed as a potential conflict of interest.

Abbreviations

The following abbreviations are used in this manuscript:

VOCs	Volatile organic compounds
HS-SPME-GC-MS	Headspace solid-phase microextraction gas chromatography mass spectrometry
HS-GC-IMS	Headspace gas chromatography-ion mobility spectrometry
GC-O	Gas chromatography-olfactometry
RI	Retention index
ROAV	Relative odor activity value
PCA	Principal component analysis
ANOVA	Analysis of variance
OPLS-DA	Orthogonal Partial Least Squares-Discriminant Analysis
VIP	Variable Importance in the Projection

https://doi.org/10.3390/foods15020200

)

RT	Retention time
QDA	Quantitative descriptive analysis
LAV	Laboratory Analysis View software

References

1. Verma, D.K.; Srivastav, P.P. A paradigm of volatile aroma compounds in rice and their product with extraction and identification methods: A comprehensive review. *Food Res. Int.* **2020**, *130*, 108924. [CrossRef] [PubMed]
2. Murtaza, G.; Huma, N.; Sharif, M.K.; Zia, M.A. Probing a best suited brown rice cultivar for the development of extrudates with special reference to Physico-Chemical, microstructure and sensory evaluation. *Food Sci. Technol.* **2022**, *42*, e103521. [CrossRef]
3. Molina, J.; Sikora, M.; Garud, N.; Flowers, J.M.; Rubinstein, S.; Reynolds, A.; Huang, P.; Jackson, S.; Schaal, B.A. Molecular evidence for a single evolutionary origin of domesticated rice. *Proc. Natl. Acad. Sci. USA* **2011**, *108*, 8351–8356. [CrossRef] [PubMed]
4. Huang, X.; Kurata, N.; Wang, Z.X.; Wang, A.; Zhao, Q.; Zhao, Y.; Lu, Y. A map of rice genome variation reveals the origin of cultivated rice. *Nature* **2012**, *490*, 497–501. [CrossRef]
5. Alvaro, D.M.; Willy, M. International Rice Outlook:International Rice Baseline Projections 2023–2033. *AAES Res. Rep.* **2024**, *5*, 1015. Available online: https://aaes.uada.edu/communications/publications/ (accessed on 12 October 2025).
6. FAO. Food and Agriculture Organization. 2025. Available online: https://www.fao.org/worldfoodsituation/csdb/en?amp= (accessed on 12 October 2025).
7. Zhang, X.; Dai, Z.; Fan, X.; Liu, M.; Ma, J.; Shang, W.; Zhou, Z. A study on volatile metabolites screening by HS-SPME-GC-MS and HS-GC-IMS for discrimination and characterization of white and yellowed rice. *Cereal Chem.* **2020**, *97*, 496–504. [CrossRef]
8. Goodwin, H.L.; Koop, L.A.; Riste, M.E.; Miller, R.K.; Maca, J.V.; Chambers, E.; Hollingsworth, M.; Bett, K.; Webb, B.D.; McClung, A.M. Developing a common language for the U.S. rice industry: Linkages among breeders, producers, processors and consumers. In *TAMRC Consumer Product Market Research*; Texas A&M: College Station, TX, USA, 1996.
9. Meullenet, J.F.; Marks, B.P.; Griftin, K.; Daniels, M.J. Effects of rough rice drying and storage conditions on sensory profiles of cooked rice. *Cereal Chem.* **1999**, *76*, 438–486. [CrossRef]
10. Park, J.K.; Kim, S.S.; Kim, K.O. Effect of milling ratio on sensory properties of cooked rice and on physicochemical properties of milled and cooked rice. *Cereal Chem.* **2001**, *78*, 151–156. [CrossRef]
11. Setyaningsih, W.; Majchrzak, T.; Dymerski, T.; Namiesnik, J.; Palma, M. Key-Marker volatile compounds in aromatic rice (*Oryza sativa*) grains: An HS-SPME extraction method combined with GCxGC-TOFMS. *Molecules* **2019**, *24*, 4180. [CrossRef]
12. Lu, L.; Hu, Z.; Fang, C.; Hu, X. Characteristic Flavor Compounds and Functional Components of Fragrant Rice with Different Flavor Types. *Foods* **2023**, *12*, 2185. [CrossRef]
13. Champagne, E.T. Rice aroma and flavor: A literature review. *Cereal Chem.* **2008**, *85*, 445–454. [CrossRef]
14. Hu, X.; Lu, L.; Guo, Z.; Zhu, Z. Volatile compounds, affecting factors and evaluation methods for rice aroma: A review. *Trends Food Sci. Technol.* **2020**, *97*, 136–146. [CrossRef]
15. Yang, D.S.; Lee, K.S.; Jeong, O.Y.; Kim, K.J.; Kays, S.J. Characterization of volatile aroma compounds in cooked black rice. *J. Agric. Food Chem.* **2008**, *56*, 235–240. [CrossRef] [PubMed]
16. Hoffmann, J.F.; Bassinello, P.Z.; Colombari Filho, J.M.; Lindemann, I.S.; Elias, M.C.; Takeoka, G.R.; Vanier, N.L. Volatile compounds profile of Brazilianaromatic brown rice genotypes and its cooking quality characteristics. *Cereal Chem.* **2018**, *96*, 292–301. [CrossRef]
17. Zheng, X.Z.; Lan, Y.B.; Zhu, J.M.; Westbrook, J.; Hoffmann, W.C.; Lacey, R.E. Rapid identification of rice samples using an electronic nose. *J. Bionic Eng.* **2009**, *6*, 290–297. [CrossRef]
18. Wakte, K.; Zanan, R.; Hinge, V.; Khandagale, K.; Nadaf, A.; Henry, R. Thirty-three years of 2-acetvl-1-pyrroline, a principal basmati aroma compound in scented rice (*Oryza sation* L.): A status review. *Sci. Food Agric.* **2017**, *97*, 384–395. [CrossRef]
19. Zhou, Y.; Gao, S.; Wei, J.; Chen, X.; Zhu, S.; Zhou, X. Systematical construction of rice flavor types based on HS-SPME-GC–MS and sensory evaluation. *Food Chem.* **2023**, *413*, 135604. [CrossRef]
20. Griglione, A.; Liberto, E.; Cordero, C.; Bressanello, D.; Cagliero, C.; Rubiolo, P.; Bicchi, C.; Sgorbini, B. High-quality ltalian rice cultivars: Chemical indices of ageing and aroma quality. *Food Chem.* **2015**, *172*, 305–313. [CrossRef]
21. Buttery, R.G.; Ling, L.C. 2-Acetyl-1-pyrroline: An important aroma component of cooked rice. *Chem. Ind.* **1982**, 958–959.
22. Routray, W.; Rayaguru, K. 2-Acetyl-1-pyrroline:key aroma component of aromatic rice and other food products. *Food Rev. Int.* **2018**, *34*, 539–565. [CrossRef]
23. Dias, L.G.; Duarte, G.H.B.; Mariutti, L.R.B.; Bragagnolo, N. Aroma profile of rice varieties by a novel SPME method able to maximize 2-acetyl-1-pyrroline and minimize hexanal extraction. *Food Res. Int.* **2019**, *123*, 550–558. [CrossRef]
24. Pal, S.; Bagchi, T.B.; Dhali, K.; Kar, A.; Sanghamitra, P.; Sarkar, S.; Majumder, J. Evaluation of sensory, physicochemical properties and Consumer preferenceof black rice and their products. *J. Food Sci. Technol.* **2019**, *56*, 1484–1494. [CrossRef] [PubMed]
25. Lee, Y.S.; Oh, Y.; Kim, T.H.; Cho, Y.H. Quantitation of 2-acetyl-1-pyrroline inaseptic-packaged cooked fragrant rice by HS-SPME/GC-MS. *Food Sci. Nutr.* **2019**, *7*, 266–272. [CrossRef] [PubMed]

https://doi.org/10.3390/foods15020200

26. Dong, S.Y.; Robert, L.; Shewfelt, K.-S.L.; Stanley, J.; Kays, J. Comparison of odor-active compounds from six distinctly different fice flavor types. *Agric. Food Chem.* **2008**, *56*, 2780–2787. [CrossRef]
27. *GB/T 16291.1-2012;* Sensory Analysis—General Guidance for the Selection, Training and Monitoring of Assessors—Part 1: Selected Assessors. Standards Press of China: Beijing, China, 2012.
28. Brown, M.D.; Chambers, D.H. Sensory characteristics and comparison of commercial plain yogurts and 2 new production sample options. *J. Food Sci.* **2015**, *80*, S2957–S2969. [CrossRef]
29. Guo, F.; Ma, M.J.; Yu, M.; Bian, Q.; Hui, J.; Pan, X.; Su, X.X.; Wu, J.H. Classification of Chinese fragrant rapeseed oil based on sensory evaluation and gas chromatography-olfactometry. *Front. Nutr.* **2022**, *9*, 945144. [CrossRef]
30. Guan, L.; Mi, Y.; Zhang, M.; Li, S.; Ye, G. Construction of aroma association network of cooked rice based on gas chromatography–mass spectrometry (GC–MS) and sensory analysis. *Flavour. Fragr. J.* **2024**, *39*, 45–57. [CrossRef]
31. Zeng, H.; Han, H.; Huang, Y.; Wang, B. Rapid prediction of the aroma type of plain yogurts via electronic nose combined with machine learning approaches. *Food Biosci.* **2023**, *56*, 103269. [CrossRef]
32. Kind, T.; Wohlgemuth, G.; Lee, D.Y. FiehnLib: Mass spectral and retention index libraries for metabolomics based on quadrupole and time-of-flight gas chromatography/mass spectrometry. *Anal Chem.* **2009**, *81*, 10038–10048. [CrossRef]
33. Ye, G.; Guan, L.; Zhang, M.; Li, S.; Mi, Y. Study on the correlation between aroma compounds and texture of cooked rice: A case study of 15 Japonica rice species from Northeast China. *Cereal Chem.* **2025**, *102*, 167–180. [CrossRef]
34. Xiao, Y.; Huang, Y.; Chen, Y.; Xiao, L.; Zhang, X.; Yang, C.; Wang, Y. Discrimination and characterization of the volatile profiles of five Fu brick teas from different manufacturing regions by using HS-SPME/GC–MS and HS-GC–IMS. *Curr. Res. Food Sci.* **2022**, *5*, 1788–1807. [CrossRef]
35. Gemert, L.J.V. *Compilations of Odour Threshold Values in Air, Water and Other Media;* Oliemans Punter: Zeist, The Netherlands, 2003.
36. Fan, Y.; Liu, W.; Xu, F.; Huang, Y.; Zhang, N.; Li, K.; Hu, H.; Zhang, H. Comparative flavor analysis of eight varieties of Xinjiang flatbreads from the Xinjiang Region of China. *Cereal Chem.* **2019**, *96*, 1022–1035. [CrossRef]
37. Liu, D.Y.; Bai, L.; Feng, X.; Chen, Y.P.; Zhang, D.; Yao, W.S. Characterization of *Jinhua ham* aroma profiles in specific to aging time by gas chromatography-ion mobility spectrometry (GC-IMS). *Meat Sci.* **2020**, *168*, 108178. [CrossRef] [PubMed]
38. Sailimuhan, A.; Ren, X.; Zhang, M.; Tuohetisayipu, T.; Li, S.; Wang, Z.; Liang, S.; Wang, Z. Characterization of japonica rice aroma profiles during in vitro mastication by gas chromatography-ion mobility spectrometry (GC-IMS) and electronic nose technology. *Int. J. Food Eng.* **2022**, *18*, 679–688.
39. Zhou, C.G.; Tan, Y.Y.; Gossner, S.; Li, Y.F.; Shu, Q.Y.; Engel, K.H. Impact of Crossing Parent and Environment on the Metabolite Profiles of Progenies Generated from a Low Phytic Acid Rice (*Oryza sativa* L.) Mutant. *J. Agric. Food Chem.* **2019**, *67*, 2396–2407. [CrossRef]
40. Ajarayasiri, J.; Chaiseri, S. Comparative study on aroma-active compounds in Thai, black and white glutinous rice varieties. *Kasetsart J. Nat. Sci.* **2008**, *42*, 715–722.
41. Cho, S.; Kays, S.J. Aroma-active compounds of wild rice (*Zizania palustris* L.). *Food Res. Int.* **2013**, *54*, 1463–1470. [CrossRef]
42. Hinge, V.; Patil, H.; Nadaf, A. Aroma volatile analyses and 2AP characterization at various developmental stages in Basmati and Non-Basmati scented rice (*Oryza sativa* L.) cultivars. *Rice* **2016**, *9*, 38. [CrossRef]
43. Jezussek, M.; Juliano, B.O.; Schieberle, P. Comparison of key aroma compounds in cooked brown rice varieties based on aroma extract dilution analyses. *J. Agric. Food Chem.* **2002**, *50*, 1101–1105. [CrossRef]
44. Limpawattana, M.; Yang, D.S.; Kays, S.J.; Shewfelt, R.L. Relating sensory descriptors to volatile components in flavor of specialty rice types. *J. Food Sci.* **2008**, *73*, 456–461. [CrossRef]
45. Mahattanatawee, K.; Rouseff, R.L. Comparison of aroma active and sulfur volatiles in three fragrant rice cultivars using GC-Olfactometry and GC-PFPD. *Food Chem.* **2014**, *154*, 1–6. [CrossRef]
46. Maraval, I.; Mestres, C.; Pernin, K.; Ribeyre, F.; Boulanger, R.; Guichard, E.; Gunata, Z. Odor-active compounds in cooked rice cultivars from Camargue (France) analyzed by GC-O and GC-MS. *J. Agric. Food Chem.* **2008**, *56*, 5291–5298. [CrossRef]
47. Park, J.S.; Kim, K.Y.; Baek, H.H. Potent aroma-active compounds of cooked Korean non-aromatic rice. *Food Sci. Biotechnol.* **2010**, *19*, 1403–1407. [CrossRef]
48. Zhao, Q.; Xue, Y.; Shen, Q. Changes in the major aroma-active compounds and taste components of Jasmine rice during storage. *Food Res. Int.* **2020**, *133*, 109160. [CrossRef] [PubMed]
49. Dias, L.G.; Hacke, A.; Bergara, S.F.; Villela, O.V.; Mariutti, L.R.B.; Bragagnolo, N. Identification of volatiles and odor-active compounds of aromatic rice by OSME analysis and SPME/GC-MS. *Food Res. Int.* **2021**, *142*, 110206. [CrossRef] [PubMed]
50. Zheng, Z.; Zhang, C.; Liu, K.; Liu, Q. Volatile Organic Compounds, Evaluation Methods and rocessing Properties for Cooked Rice Flavor. *Rice* **2022**, *15*, 53. [CrossRef]
51. Nishant, G.; Mansi, G.; Devansh, B.; Neelansh, G.; Rudraksh, T.; Apuroop, S.; Ganesh, B. FlavorDB2: An Updated Database of Flavor Molecules. *J. Food Sci.* **2024**, *89*, 7076–7082. [CrossRef]
52. Li, W.; Chen, Y.P.; Blank, I.; Li, F.; Li, C.; Liu, Y. GC × GC-ToF-MS and GC-IMS based volatile profile characterization of the Chinese dry-cured hams from different regions. *Food Res. Int.* **2021**, *142*, 110222. [CrossRef]

https://doi.org/10.3390/foods15020200

53. Guo, X.; Schwab, W.; Ho, C.T.; Song, C.; Wan, X. Characterization of the aroma profiles of oolong tea made from three tea cultivars by both GC-MS and GC-IMS. *Food Chem.* **2022**, *376*, 131933. [CrossRef]
54. Feng, H.; Timira, V.; Zhao, J.; Lin, H.; Wang, H.; Li, Z. Insight into the characterization of volatile compounds in smoke-flavored sea bass (lateolabrax maculatus) during processing via HS-SPME-GC-MS and HS-GC-IMS. *Foods* **2022**, *11*, 2614. [CrossRef]
55. Yang, X.; Chen, Q.; Liu, S.; Hong, P.; Zhou, C.; Zhong, S. Characterization of the effect of different cooking methods on volatile compounds in fish cakes using a combination of GC-MS and GC-IMS. *Food Chem.* **2024**, *22*, 101291. [CrossRef]
56. Zhao, Q.; Xi, J.; Xu, D.; Jin, Y.; Wua, F.; Tong, Q.; Yin, Y.; Xu, X. A comparative HS-SPME/GC-MS-based metabolomics approach for discriminating selected japonica rice varieties from different regions of China in raw and cooked form. *Food Chem.* **2022**, *385*, 132701. [CrossRef] [PubMed]
57. Xiong, Y.; Zheng, X.; Tian, X.; Wang, C.; Chen, J.; Zhou, L.; Xu, D.; Wang, J.; Gilard, V.; Wu, M.; et al. Comparative study of volatile organic compound profiles in aromatic and non-aromatic rice cultivars using HS-GC-IMS and their correlation with sensory evaluation. *LWT Food Sci. Technol.* **2024**, *203*, 116321. [CrossRef]
58. Widjaja, R.; Craske, J.D.; Wootton, M. Comparative studies on volatile components of non-fragrant and fragrant rice. *J. Sci. Food Agric.* **1996**, *70*, 151–161. [CrossRef]
59. Bryant, R.J.; Mcclung, A.M. Volatile profiles of aromatic and nonaromatic rice cultivars using SPME/GC-MS. *Food Chem.* **2010**, *124*, 501–513. [CrossRef]
60. Monsoor, M.A.; Proctor, A. Volatile component analysis of commercially milled head and broken rice. *J. Food Sci.* **2004**, *69*, 632–636. [CrossRef]
61. Bi, X.; Asimi, S.; Zhang, M.; Zhou, Q.; Yang, Z.S. Effects of fatty acids on taste quality of cooked rice. *Food Sci.* **2019**, *40*, 8–14.
62. Verma, D.K.; Mahato, D.K.; Srivastav, P.P. Simultaneous distillation extraction (SDE): A traditional method for extraction of aroma chemicals in rice. In *Science and Technology of Aroma, Flavor, and Fragrance in Rice. Science and Technology of Aroma, Flavor, and Fragrance in Rice*, 1st ed.; Apple Academic Press: Palm Bay, FL, USA, 2018.
63. Wang, Z.Y.; Su, H.M.; Bi, X.; Zhang, M. Effect of fragmentation degree on sensory and texture attributes of cooked rice. *J. Food Process Preserv.* **2019**, *43*, e13920. [CrossRef]
64. Matsui, K.; Sasahara, S.; Akakabe, Y.; Kajiwara, T. Linoleic Acid 10-Hydroperoxide as an Intermediate during Formation of 1-Octen-3-ol from Linoleic Acid in Lentinus decadetes. *Biosci. Biotechnol. Biochem.* **2003**, *67*, 2280–2282. [CrossRef]
65. Choi, S.; Seo, H.S.; Lee, K.R.; Lee, S.; Lee, J. Effect of milling degrees on volatile profiles of raw and cooked black rice (*Oryza sativa* L. Cv. Sintoheugmi). *Appl. Biol. Chem.* **2018**, *61*, 91–105. [CrossRef]
66. Bullard, R.W.; Holguin, G. Volatile components of unprocessed rice (*Oryza sativa* L.). *J. Agric. Food Chem.* **1977**, *25*, 99–103. [CrossRef]
67. Cao, Y.; Xie, G.; Wu, C.; Lu, J. A study on characteristic flavor compounds in traditional chinese rice wine-guyue longshan rice wine. *J. Inst. Brewing.* **2010**, *116*, 182–189. [CrossRef]
68. Liu, J.; Tang, X.; Zhang, Y.; Zhao, W. Determination of the volatile composition in brown millet, milled millet and millet bran by gas chromatography/mass spectrometry. *Molecules* **2012**, *17*, 2271–2282. [CrossRef]
69. Izumi, Y.A.; Tetsuya, Y.A.; Mikio, N.; Hidemasa, S.; Tsutomu, H. Volatile Flavor Components of Cooked Rice. *Agric. Biol. Chem.* **1978**, *42*, 1229–1233.
70. Zhao, Q.Y.; Xi, J.Z.; Xu, X.M.; Yin, Y.; Xu, D.; Jin, Y.M.; Wu, F.F. Volatile fingerprints and biomarkers of Chinese fragrant and non-fragrant japonica rice before and after cooking obtained by untargeted GC/MS-based metabolomics. *Food Biosci.* **2022**, *47*, 101764. [CrossRef]
71. *GB/T 15682-2008*; Inspection of Grain and Oils-Method for Sensory Evaluation of Paddy or Rice Cooking and Eating Quality. Standards Press of China: Beijing, China, 2008.

https://doi.org/10.3390/foods15020200

chemosensors

MDPI

Article

A Simplified Multi-Residue Method Using μSPE Clean-Up Combined with Gas Chromatography-High Resolution Mass Spectrometry for the Determination of 250 Pesticide Residues in Cow's Milk

Marta Vargas-Pérez [1], Osmar D. Prestes [2], Roberto Romero-González [1] and Antonia Garrido Frenich [1,*]

[1] Research Group "Analytical Chemistry of Contaminants", Department of Chemistry and Physics, Research Center for Mediterranean Intensive Agrosystems and AgriFood Biotechnology (CIAIMBITAL), University of Almeria, Agrifood Campus of International Excellence, ceiA3, 04120 Almeria, Spain; mvp864@ual.es (M.V.-P.); rromero@ual.es (R.R.-G.)

[2] Laboratory of Pesticide Residue Analysis (LARP), Chemistry Department, Federal University of Santa Maria (UFSM), Santa Maria 97105-900, RS, Brazil; osmar.prestes@ufsm.br

* Correspondence: agarrido@ual.es

Abstract

The complexity of the milk matrix, driven by its lipid-rich composition, complicates pesticide residue analysis. This study developed a simplified and robust analytical procedure for the quantification of 250 pesticides in cow's milk. Sample preparation involved acidified ethyl acetate extraction followed by centrifugation at 0 °C. A subsequent clean-up step was performed using micro solid-phase extraction (μSPE) in a 96-well format with the enhanced matrix removal-lipid (EMR-lipid) sorbent. Final extracts were analyzed by gas chromatography coupled to high-resolution mass spectrometry (GC-Q-Orbitrap-MS). Method validation demonstrated satisfactory linearity within the 5–100 μg/L range, recoveries between 70.6% and 119.8%, and precision, expressed as relative standard deviation (RSD), was acceptable for both intraday (1.8–19.2%) and interday (1.6–18.5%) conditions. The limit of quantification (LOQ) was set at 10 μg/kg for all compounds. The method was applied to 23 commercial cow's milk samples, and no pesticide residues were detected above the current European Union (EU) maximum residue limits (MRLs).

Keywords: milk; Q-Exactive orbitrap; gas chromatography; EMR lipid; pesticide residues

1. Introduction

Milk is a globally produced and consumed food commodity, essential in a balanced diet [1]. Beyond its direct consumption, milk serves as a key ingredient in important products, such as formulated milk for infants. However, fatty food products such as milk can contain relatively high levels of pesticide residues, which can enter the food chain through contaminated animal feed and water, posing a risk of exposure to these toxic substances for both animals and humans [2–4]. To ensure food safety, MRLs have been set by various countries and organizations. The EU has established MRLs for most pesticides in milk and dairy products, with limits as low as 0.8 μg/kg for endrin and 2 μg/kg for chlordane [5]. Therefore, effective analytical methods are necessary to monitor pesticide residues in milk to confirm compliance with these safety limits. Although gas chromatography coupled with mass spectrometry (GC-MS) is highly sensitive and capable of accurately detecting target compounds in complex matrices, it is now less frequently used due to its sensitivity to matrix effects [6]. However, GC coupled to high-resolution mass spectrometry (HRMS), compared to traditional GC-MS, offers full-scan capabilities, high accuracy, and selective screening of target compounds. This allows low levels of pesticides to be detected and quantified, even in complex matrices such as milk. Milk contains organic acids, sugars, fats, proteins, and other impurities that can interfere with chromatographic analysis. Furthermore, the use of full scan HRMS is becoming increasingly popular in the residue analysis of food samples, especially when a large number of analytes must be monitored [7]. Compared to other high-resolution mass analyzers, such as time-of-flight (TOF), the Orbitrap offers superior mass accuracy, resolving power, and dynamic range, along with the ability to operate at different resolution settings depending on analytical needs. These features make it particularly suitable for detecting trace-level compounds in complex matrices and allow for more confident identification of both target analytes and non-target analytes in full-scan acquisition modes.

The currently available methods for the extraction of multiple pesticide residues from fatty milk are still restricted by the laborious and time-consuming sample pretreatment process, limited sensitivity, and selective clean-up [8]. These methods involve the use of solvents such as *n*-hexane [9] and dichloromethane [10], sonication times of 20–30 min, laborious cleaning steps using solid phase extraction (SPE), or combinations of various salts [4], followed by evaporation under nitrogen flow [11].

The extraction of multiple compounds with diverse chemical properties and polarities represents a significant challenge. The QuEChERS (Quick, Easy, Cheap, Effective, Rugged, Safe) method offers advantages, such as rapid processing, high throughput, efficiency, and low cost. However, milk is a complex matrix and has a high matrix effect, characterized by high concentrations of organic acids, sugars, lipids, and proteins. For instance, pesticide residues and other contaminants were analyzed in milk samples [12,13] using QuEChERS extraction. PAHs and pesticide residues were also evaluated in milk samples [14,15]. The clean-up step can be carried out as using dSPE clean-up [12,13], and both methods required manual weighing of small quantities of "enhanced matrix removal-lipid" (EMR-lipid) and other sorbents such as PSA and Z-Sep$^+$ (zirconia-based sorbent) to remove fatty matrix components, whereas other authors used Z-Sep, C18, and PSA during the dSPE clean-up [14,15]. For this work, various clean-up approaches after the extraction of QuEChERS have been proposed, with the aim of reducing the number of co-extractive interferences during pesticide residue analysis, using many steps and sorbents, which are difficult to handle [8,16–19]. An interesting option is the use of the clean-up step with the EMR-lipid material that has been proposed to remove fat from fat-rich food products [4,17]. In most published articles, EMR-lipid has been used as a cleaning sorbent in dispersive SPE (dSPE) in matrices with high fat content, such as pork, fish, smoked foods [20], olive oil or avocado fruit [21], biological matrices such as human plasma [22], or even in milk formulas for analysis of mycotoxins, polycyclic aromatic hydrocarbons or neonicotinoids [15]. The use of well-plate format (μSPE) minimizes weighing errors, improves reproducibility, and facilitates high-throughput workflows, which are key issues for routine laboratories. However, the use of EMR-lipid in a 96-well plate format for μSPE is rather limited, although it has demonstrated clinical applicability in blood [23] and even in infant milk formulas [24]. The present study builds upon the well-established QuEChERS extraction and GC-HRMS principles, but its novelty lies in the first application, to the best of our knowledge, of EMR-lipid in a ready-to-use 96-well plate format for the determination of 250 pesticide residues in whole milk, combining broad analyte coverage with reduced solvent use and simplified sample handling.

In summary, the aim of the study was to develop a simple method that reduces extraction time by using a small amount of salts and solvents for the analysis of pesticide residues in whole milk. The final extracts were analyzed using the analytical methodology already developed [25], based on GC-HRMS, using a single quad-Orbitrap as an analyzer, which was utilized to perform selective target screening for the detection and quantification of low levels of pesticides.

2. Materials and Methods

2.1. Reagents and Chemicals

Standards for 275 pesticides were purchased from Dr. Ehrenstorfer (Augsburg, Germany) or Sigma Aldrich (St. Louis, MO, USA). One isotopically labelled compound, atrazine-d5 from Sigma Aldrich, was used as the surrogate standard. All standards had a purity greater than 95%.

HPLC gradient grade acetonitrile and HPLC grade acetic acid were acquired from Honeywell (Seelze, Germany) and Panreac AppliChem (Barcelona, Spain), respectively. PAR grade ethyl acetate was purchased from Fluka (St. Louis, MO, USA). Formic acid (FA) (Optima LC–MS) was supplied by Fisher Scientific (Geel, Belgium). Magnesium sulfate and sodium acetate were obtained from Fluka, while tri-sodium citrate dehydrate, sodium chloride, and sodium citrate dibasic sesquihydrate were sourced from Panreac AppliChem, Scharlab (Barcelona, Spain), and Sigma Aldrich, respectively. Primary secondary amine (PSA) was obtained from Scharlab, C18 (octadecyl) bonded phase from Supelco (Bellefonte, PA, USA), and Agilent Captiva EMR-lipid in a 96-well plate format from Agilent Technologies (Santa Clara, CA, USA).

Stock standard solutions of the target pesticides were prepared in a suitable solvent (methanol or ethyl acetate) at a concentration of 1000 mg/L. The stock solutions were stored in amber screw-capped glass vials. After that, working solutions containing all pesticides were then prepared at 4.5 mg/L in acetonitrile and at 0.5, 0.2, and 0.1 mg/L in ethyl acetate. The surrogate standard (atrazine-d5) was added at a final concentration of 25 μg/kg prior to extraction. This surrogate was used to check whether sample preparation and analysis were performed correctly. All standard solutions were stored at −20 °C.

2.2. GC-Orbitrap-MS Parameters

The GC system consisted of a TriPlus RSH autosampler, a TRACE 1300 gas chromatograph, and a Q-Exactive Orbitrap mass analyzer (Thermo Fisher Scientific, Bremen, Germany). The injector was equipped with a single taper liner (78.5 mm × 4 mm internal diameter (ID)). Sample introduction was performed using hot splitless injection (1 μL, 280 °C, splitless time 1 min). High-purity helium (99.999%) served as a carrier gas at a constant flow rate of 1 mL/min. Separation was achieved on a VF-5 MS capillary column (30 m × 0.25 mm i.d. and 0.25 μm film thickness from Agilent, Santa Clara, CA, USA). The oven temperature program lasted 29.00 min and was as follows: initial temperature of 50 °C held for 1 min, then ramped to 170 °C at 20 °C/min (6 min), followed by an increase to 310 °C at 10 °C/min (14 min). The final temperature of 310 °C was maintained for 8 min. The Q-Exactive Orbitrap mass analyzer was operated in electron ionization (EI) mode at 70 eV with an emission current of 50 μA. Mass calibration was performed weekly using perflurotributylamine (PFTBA) from Thermo Fisher Scientific.

The filament delay was set for 5 min, and data acquisition was performed from 5 to 29 min. The ion source and transfer line temperatures were maintained at 250 °C. Full-scan (FS) MS acquisition was carried out in profile mode over an m/z range of 40 to 500. The maximum injection time was set to 1 μscan, with a resolution power of 60,000 full width at half maximum (FWHM). The automatic gain control (AGC) target was set to 1×10^6 ions. Instrument control was performed using Xcalibur 4.1 and chromatographic data were processed with TraceFinder 4.1 software (Thermo Fisher Scientific,

Les Ulis, France). Identification criteria included: (i) at least two fragment ions with mass accuracy $\leq$ 5 ppm, and (ii) ion ratio within ($\pm$30% (relative) of the average calibration standard from the same sequence. Detailed GC-MS parameters are provided in Table S1.

2.3. Sample Extraction

Cow's milk samples were processed using a modified version of AOAC QuEChERS, a typical multiresidue extraction method commonly applied in routine laboratories. Prior to extraction, milk samples were homogenized by vortex mixing to ensure a representative distribution of both aqueous and lipid phases. A 5 g aliquot of the sample was weighed into a 50 mL polypropylene centrifuge tube. Ethyl acetate (10 mL) containing 1% acetic acid and 10 µL of the surrogate standard (atrazine-d5) was added, and the mixture was vortexed for 2 min. Subsequently, 4 g of $MgSO_4$ and 1.5 g of sodium acetate were added, followed by another 2 min of vortex mixing. The tube was then centrifuged at 7500 rpm (6200$\times$ *g*) at 0 °C for 8 min.

A 1 mL aliquot of the supernatant was loaded onto a well Agilent Captiva EMR-lipid. After the liquid was passed through the cartridge by gravity, vacuum was applied to remove any remaining liquid. The resulting extract was then transferred to sample vials for GC-Orbitrap-MS analysis, and 1 µL was injected into the system.

2.4. Validation

The performance of the analysis was evaluated in accordance with the analytical quality control and validation guidelines for the analysis of pesticide residues in food and feed, as described in SANTE 11312/2021v2 [26]. The parameters assessed included linearity, limit of quantification (LOQ), recovery, and precision. Additionally, the expanded measurement uncertainty (U) was calculated using intra-laboratory validation data, according to the SANTE guidelines. The whole cow's milk used for the development and validation of the method was also sourced directly from local grocery stores in Almeria (Spain). This represents typical market samples and provides realistic matrix conditions for pesticide residue analysis.

According to the SANTE guideline, an extension of the scope of the method was carried out to other matrices within the milk subgroup (milk and dairy products group). This verification procedure was applied to various types of milk, including semi-skimmed, lactose-free semi-skimmed, skimmed, and evaporated milk, ensuring the reliability of the method in various matrices. This was achieved by analyzing three spiked samples (spiked at the target LOQ, 25 µg/kg, and at 50 µg/kg for all pesticides) and one blank matrix. Quantification was carried out using a matrix-matched calibration prepared in their specific matrix. The method performance criteria remained consistent with the validation parameters for recovery (70–120%) and repeatability ($\leq$20%).

2.5. Applicability of the Method

A total of 23 cow's milk samples with different compositions (whole, semi-skimmed, skimmed, lactose-free, calcium-enriched, and omega-3-enriched) were purchased from local grocery stores in Almería (Spain). The samples, representing different commercial brands, were stored in their original containers under the recommended conditions until analysis. For the development and validation of the method, whole cow's milk was selected as the most challenging matrix in terms of sample preparation. An overview of the samples included in this study is provided in Table S2 (see Supplementary Materials).

3. Results

3.1. Optimization of the Extraction Procedure

3.1.1. Optimization of Extraction Salts

The QuEChERS method was modified and optimized considering the high protein and lipid content of milk, with lipids being one of the most important and significant sources of interferences. As an extraction solvent, acetonitrile was tested because it must meet specific requirements, including the ability to precipitate fats and proteins while efficiently extracting analytes from milk samples. This solvent not only enables the extraction of a wide range of compounds, but also serves as an effective protein precipitant [27], a process that can be enhanced by adding FA [28]. In this study, whole cow's milk was used and three different versions of QuEChERS were compared. Using acetonitrile with 1% FA as the extraction solvent, a comparison of extraction salts was made by spiking whole cow's milk (5 g) with a standard mixture at 100 μg/kg. This spiking level for the preliminary evaluation of extraction salts was selected to ensure that the target compounds could be detected with sufficient sensitivity and to evaluate whether their absence in the chromatogram, if any, was due to poor extraction efficiency rather than limitations of instrumental sensitivity. Three different extraction salts were tested: (A) Original QuEChERS salts [29] containing 4 g of $MgSO_4$ and 1 g of NaCl, (B) the AOAC version [30] containing 6 g of $MgSO_4$ and 1.5 g of sodium acetate, and (C) the EN version [31] consisting of 4 g of $MgSO_4$, 1 g of NaCl, 3.5 g of disodium hydrogen citrate sesquihydrate, and 1 g of trisodium citrate dihydrate. The clean-up step was based on the addition of 150 mg of $MgSO_4$ combined with 25 mg of PSA and C18 as sorbents [32], followed by filtration through a nylon filter. In the last step, an aliquot of extract was evaporated to dryness under a N_2 stream and redissolved with 1 mL of ethyl acetate before chromatographic analysis.

The extraction efficiency was evaluated on recovery values obtained from triplicate experiments, with acceptable recovery defined as 70 to 120%. As shown in Figure 1, a total of 122 compounds achieved suitable recovery using method (A), while 196 and 74 compounds met the recovery criteria with methods (B) and (C), respectively. The improved recoveries obtained using the AOAC QuEChERS version may be attributed to the slightly lower pH provided by the sodium acetate buffering system compared to the citrate-based European version. Based on these results, AOAC salts (B) were selected as the optimal approach for extracting pesticides from whole cow's milk.

3.1.2. Optimization of the Extraction Solvent

Although acetonitrile is compatible with both liquid chromatography coupled to MS (LC-MS) and GC-MS, the potential benefits of using ethyl acetate as the extraction solvent were also tested, taking into account that the evaporation step could be omitted. Having already decided on the use of AOAC salts, a comparison was made using acetonitrile with 1% FA and ethyl acetate with 1% acetic acid as extraction solvents, spiking whole cow's milk (5 g) with a standard mixture at 100 μg/kg.

Since fatty substances have lower melting points than solvents, frozen lipids can be removed by centrifugation, while pesticides remain dissolved in the solvent [10]. To take advantage of this property for the removal of lipids, an additional centrifugation step was introduced at 7500 rpm (6200× *g*) at 0 °C for 8 min. The use of ethyl acetate increased the number of compounds with acceptable recovery (205) compared to acetonitrile (186), while also eliminating the need for a drying step. Consequently, ethyl acetate with 1% acetic acid was used as the extraction solvent.

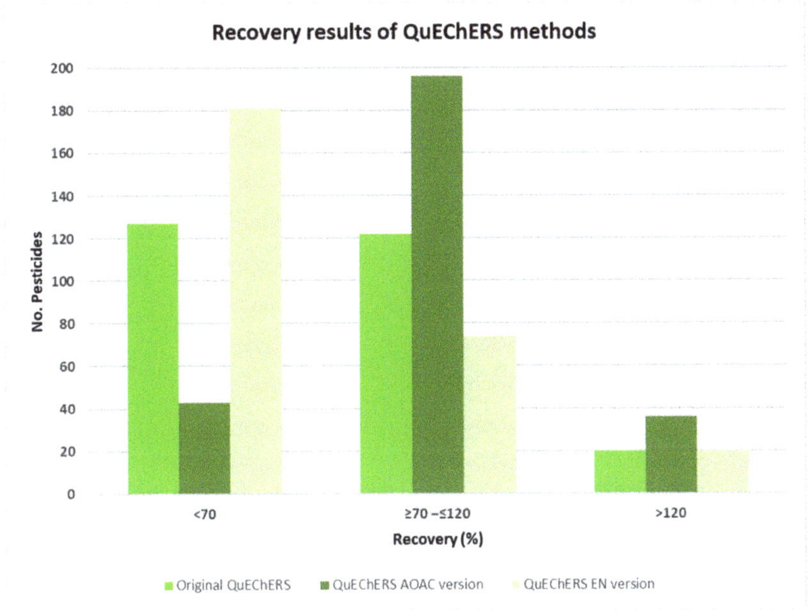

Figure 1. Recovery results of the Original QuEChERS, QuEChERS AOAC, and QuEChERS EN methods.

3.1.3. Optimization of the Clean-Up Step

An efficient clean-up step is critical for removing the maximum possible amount of matrix co-extracts and improving the quality of analytical results. For this purpose, dSPE and μSPE (EMR-lipid 96-well plate) were compared for the cleaning step, evaluating the cleanliness of the extract, the workflow, and the recovery of the pesticides under study. Although EMR-lipid tubes have been used in some previous publications [20,27,33], their application as a 96-well plate format or μSPE is less frequent [24,34]. This format offers the advantage of a streamlined, single-step process, where 1 mL of sample extract is simply passed through the well. Although automation is possible, manual processing is also feasible with the use of a vacuum pump, requiring only a few seconds per sample.

For dSPE clean-up, 1.5 mL of sample extract was transferred to a 15 mL centrifuge tube containing 50 mg of PSA and 25 mg of C18, combined with 150 mg of $MgSO_4$ to remove any moisture traces from the previous step, and then the sample was vortexed for 1 min. The tube was subjected to a centrifugal process for a duration of 5 min at 3700 rpm (3061 × *g*). In contrast, the μSPE cleaning involved directly loading 1 mL of sample extract into a well and placing it under vacuum. Both methods produced comparable pesticide recoveries, with 238 compounds showing acceptable recoveries using dSPE and 242 compounds with μSPE. Despite a similar performance, the μSPE approach aligns with the study's goal of developing a simplified method. The use of EMR-lipid significantly reduced the complexity of the workflow by eliminating the need to weigh multiple salts and minimized the use of disposable materials, such as plasticware. Additionally, μSPE offers a higher sample throughput, which is a key advantage for routine analysis, allowing for simultaneous cleaning of a large number of samples (up to 96). For these reasons, EMR-lipid was selected as the preferred clean-up sorbent over dSPE. However, to demonstrate matrix removal by μSPE in future studies, a comparison of Total Ion Current (TIC) profiles before and after the μSPE process was carried out.

3.2. Method Validation

For method validation, whole cow's milk was selected as the representative matrix according to the SANTE/11312/2021 v2 guidelines [26], which designate it as the standard dairy matrix. Thus, given that it has the highest fat content, it also represents the most challenging matrix, making it appropriate to evaluate the performance of the method in other types of milk.

3.2.1. Linearity and Matrix Effect (ME)

In this study, a calibration curve was prepared using a blank sample extract of whole cow's milk. Linearity was initially evaluated for 275 pesticides within the concentration range of 5 to 100 μg/L, covering five concentration levels (5, 10, 25, 50, and 100 μg/L). A solvent-based calibration curve was also prepared using ethyl acetate at the same concentration levels. The determination coefficients (R^2) were calculated for 265 of the 275 compounds studied, and the R^2 values were consistently above 0.99. Table S3 (see Supplementary Materials) lists the 275 pesticides that were included in the analysis and highlights with an asterisk those that did not meet the established validation criteria. Only compounds with R^2 values greater than 0.99 were considered for the validation study, establishing the linear range of the method between 5–100 μg/L.

The ME was assessed in whole cow's milk using Equation (1):

$$\%ME = \left(\frac{slope\ of\ calibration\ curve\ in\ matrix}{slope\ of\ calibration\ cruve\ in\ solvent} - 1\right) \times 100 \quad (1)$$

As shown in Figure 2, a strong ME (>50%) was observed for 248 compounds. A moderate ME (20% < ME < 50%) was detected in 17 compounds. In particular, none of the compounds exhibited a low or negligible ME (<20%). To mitigate for the pronounced matrix effects, matrix-matched calibration was employed to ensure accurate quantification. The EMR-Lipid clean-up was primarily aimed at enhancing extract cleanliness and improving instrument stability, rather than completely eliminating the matrix effects.

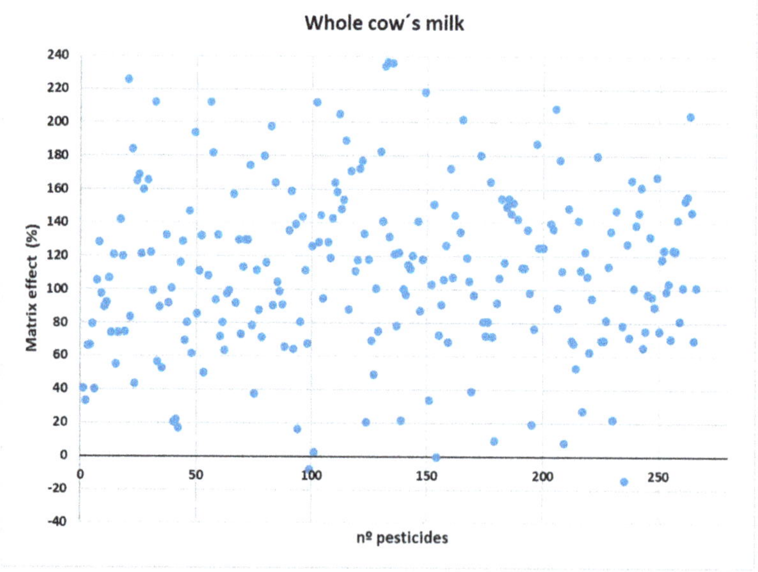

Figure 2. Matrix effect (%) comparing the slope of the curve in solvent and in whole cow's milk.

3.2.2. Recovery, Precision, and Uncertainty

Recovery tests were carried out at 10 μg/kg (MRL), 25 μg/kg (2.5× MRL), and 50 μg/kg (5× MRL), using five replicates at each level. Mean recoveries ranged from 71.1 to 119.7%, with an associated repeatability, expressed as RSD, below 20%, as can be observed in Table 1.

Table 1. Summary of method performance for 250 pesticides in cow's milk.

Parameter	Value		
Linearity (R^2)	≥0.99 (for 96.3% of compounds)		
Matrix Effect (%)	>50% (for 90.1% of compounds)		
LOQ (μg/kg)	10 μg/kg		
	Value (Range or Mean ± RSD)		
	10 μg/kg	25 μg/kg	50 μg/kg
Recovery (%)	70.6–117.0	72.0–117.0	71.8–119.7
Intraday precision (%)	3.2–19.2	3.2–18.5	1.8–18.4
Interday precision (%)	3.0–18.5	2.9–18.3	1.6–18.2
Expanded Uncertainty (%)	13–49		

Intraday precision, expressed as relative standard deviation (% RSD), was calculated from recovery data at these concentrations and ranged from 1.8% to 19.2%. Interday precision, determined over five different days, ranged from 1.6% to 18.5%, all within acceptable limits. The complete dataset is provided in Table S3. Regarding the recovery values, no significant losses were observed during the μSPE step, confirming the efficiency of the extraction procedure.

For most of the 265 compounds that met the linearity criteria, the results were acceptable. However, fifteen compounds showed extraction difficulties and did not achieve acceptable recovery or precision values: 1,2,3,6-tetrahydrophthalimide cis, bifenox, chlordane cis, chlordane trans, difenoconazole, endrin, fenamiphos, fenamiphos sulfone, fenamiphos sulfoxide, fipronil, fipronil sulfone, heptachlor, heptachlor-epoxide-cis, heptachlor-epoxide-trans, and methamidophos. In summary, 250 compounds were successfully validated, meeting the established performance criteria for recovery and precision.

The standard uncertainty was determined using Equation (2). Uncertainty due to bias, u(bias), was estimated from the recovery experiment, while % RSD of inter-day precision was used to estimate the uncertainty arising from precision, u(precision). The expanded uncertainty was calculated for each pesticide by multiplying the standard uncertainty by a coverage factor (k) of 2, which approximately gives a confidence level of 95%. Further details on the calculation procedure of the uncertainty are provided in the Supplementary Materials (Section S1).

$$u = \sqrt{\mathrm{u(bias)}^2 + \mathrm{u(precision)}^2} \quad (2)$$

As can be observed in Table 1, for the 250 validated pesticides, the expanded uncertainty at 10 μg/kg was found to be below 50% in all cases, but they were still in compliance with the requirement (50%) of the SANTE/11312/2021 v2 guidelines [26].

Unlike most previous studies, which involve more complex procedures in terms of sample preparation and time requirements, with lower recovery rates, and face reproducibility challenges, this study achieved broader and more reliable pesticide detection. Although previous research has often focused on specific pesticide families, such as organochlorine pesticides [28,35] or organophosphorus pesticides in milk samples [9,36], this study covered a comprehensive range of 250 compounds, including both long-banned substances, like organochlorine compounds, and, more recently, introduced pesticides for crop protection. Table 1 summarizes the validation results for these compounds.

Finally, the suitability of the method was confirmed by testing other types of milk. Estimated mean recovery and RSD values at 10 and 50 µg/kg for semi-skimmed, lactose-free semi-skimmed, skimmed, and evaporated milk are shown in Table S4. For all pesticides, recoveries ranged from 70.1–119.8%, with RSD values between 0.1% and 19.7%, confirming the applicability of the proposed method to different milk types commonly available on the market.

3.2.3. Limits of Quantification

LOQs were assessed by determining the lowest concentration of spiked samples that produced satisfactory recovery (70–120%) and precision (RSD < 20%). The LOQs were established at 10 µg/kg for 250 pesticides, as these compounds met the established criteria.

The validated LOQ value of 10 µg/kg aligns with the EU MRL for the target pesticides. Analytes that did not meet the validation criteria and were not included in the validation list may still be detected through screening analysis, indicating their potential presence.

3.3. Sample Analysis

The performance of the developed methodologies were evaluated through the analysis of 23 cow's milk samples. An internal quality control procedure was implemented to ensure the reliability of the analytical results. This procedure included: (i) the analysis of one blank matrix sample prior to each batch to confirm the absence of the target pesticides; (ii) the use of matrix-matched calibration, and (iii) the inclusion of two spiked blank samples at 10 and 25 µg/kg with the pesticide mixture within each batch. This approach allowed for the assessment of recovery rates and the verification of potential cross-contamination. In all cases, recoveries at both spiking levels were within the range of 70–120%. Additionally, a matrix-matched calibration curve was used to quantify any detected compound. None of the 250 target pesticides were detected in the analyzed samples. Furthermore, the analysis confirmed that none of the 250 pesticides exceeded the established LOQ of 10 µg/kg in any of the samples.

4. Conclusions

The implementation of µSPE clean-up contributed to reduced reagents and materials, streamlined laboratory workflow, and enhanced sample throughput, key advantages for routine sample analysis. Moreover, this study presents the first GC-HRMS method capable of simultaneously detecting a broad spectrum of 250 pesticides in milk. Nevertheless, a limited number of compounds did not meet the validation criteria, primarily due to low recovery or poor precision values. For these analytes, alternative extraction conditions would be necessary to achieve reliable quantification. Additionally, to further mitigate ME, several strategies could be considered in future studies. These include the use of additional lipid-targeting sorbents during the clean-up step to enhance the removal of co-extracted fats, sample dilution prior to instrumental analysis to reduce the impact of matrix components, and the implementation of advanced instrumental configurations such as online GC–APCI interfaces, which offer improved selectivity and reduced susceptibility to matrix interference. However, given the wide range of physicochemical properties among the 250 pesticides analyzed, the effectiveness of additional sorbents may vary depending on the compound, and thus their use should be carefully optimized to avoid compromising recovery or selectivity.

In summary, this study developed and successfully validated a new QuEChERS extraction method for the analysis of pesticides in milk, which incorporated acidified ethyl acetate extraction and centrifugation at 0 °C followed by a clean-up step with EMR-lipid. This method is both simplified and efficient in workflow. In conclusion, 250 pesticides provided suitable validation parameters with LOQ values lower than the established MRLs for the target compounds, after GC-HRMS analysis. Additionally, the applicability of the method was tested for other types of milk that differed mainly in fat and lactose content. Automating µSPE cleaning could further improve the efficiency, although this would require significant investment in automation equipment. Furthermore, the high sensitivity of the method (LOQ of 10 µg/kg) and the advanced analytical performance through GC-Q-Orbitrap make it a valuable approach for the complete analysis of pesticide residues in complex matrices. The proposed workflow also shows potential for future automation and digital integration, as the 96-well µSPE format could be adapted to LIMS-based environments and batch processing systems, improving efficiency in routine pesticide monitoring.

Supplementary Materials: The following supporting information can be downloaded at: https://www.mdpi.com/article/10.3390/chemosensors13120405/s1, Table S1: GC-MS parameters of the targeted pesticides, Table S2: Milk types and sample numbers included in this study, Table S3: Summary of data obtained for validation tests performed to whole milk matrix, Section S1: Measurement Uncertainty Estimation, Table S4: Summary of data obtained for average recovery and estimated RSD values calculated by quantifying 3 spiked samples of semi-skimmed, lactose-free semi-skimmed, skimmed, and evaporated milk.

Author Contributions: M.V.-P.: conceptualization, formal analysis, methodology, writing—original draft preparation, data curation, validation; O.D.P.: methodology, supervision, formal analysis, writing—review and editing; R.R.-G.: methodology, investigation, data curation, writing—reviewing & editing; A.G.F.: conceptualization, supervision, resources, funding acquisition, methodology, writing—reviewing & editing. All authors have read and agreed to the published version of the manuscript.

Funding: This research received no external funding.

Data Availability Statement: The original contributions presented in this study are included in the article. Further inquiries can be directed to the corresponding author.

Acknowledgments: The authors gratefully acknowledge the financial support to PPIT-UAL, Junta de Andalucía-ERDF 2021–2027. Program: 54.A.

Conflicts of Interest: The authors declare no conflicts of interest.

References

1. Linehan, K.; Patangia, D.V.; Ross, R.P.; Stanton, C. Production, Composition and Nutritional Properties of Organic Milk: A Critical Review. *Foods* **2024**, *13*, 550. [CrossRef]
2. González-Jartín, J.M.; Rodríguez-Cañás, I.; Alvariño, R.; Alfonso, A.; Sainz, M.J.; Vieytes, M.R.; Gomes, A.; Ramos, I.; Botana, L.M. Occurrence of mycotoxins in total mixed ration of dairy farms in Portugal and carry-over to milk. *Food Control* **2024**, *165*, 110682. [CrossRef]
3. Butovskaya, E.; Caprai, E.; Peloso, M.; Gasparini, M.; Borgia, M.; Abdul, M.E.; Candotti, P.; Menotta, S. Plant-based milk alternatives: Assessing the occurrence of chemical and microbiological contaminants in soy, oat, rice and almond beverages from Italian market. *Food Control* **2025**, *169*, 111005. [CrossRef]
4. Madej, K.; Kalenik, T.K.; Piekoszewski, W. Sample preparation and determination of pesticides in fat-containing foods. *Food Chem.* **2018**, *269*, 527–541. [CrossRef]
5. European Commission. Regulation (EU) No 396/2005 on maximum residue levels of pesticides in or on food and feed of plant and animal origin, and amending Council Directive 91/414/EEC. *Off. J. Eur. Union* **2005**, *L70*, 1–16.
6. Wei, Y.; Huang, C.; Chen, L.; Chen, Q.; Hou, J.; Wu, H.; Han, C.; Shen, Y. Determination of chlordimeform and its metabolite residue in milk by gas chromatography–tandem mass spectrometry. *Food Res. Int.* **2024**, *192*, 114754. [CrossRef] [PubMed]
7. Mol, H.G.J.; Tienstra, M.; Zomer, P. Evaluation of gas chromatography–electron ionization–full scan high resolution Orbitrap mass spectrometry for pesticide residue analysis. *Anal. Chim. Acta* **2016**, *935*, 161–172. [CrossRef]

8. Peng, H.; Li, H.; Li, X.; Wei, B.; Du, Z.; Wei, G.; Wang, S. Determination of multi-residue pesticides in dairy products using single-step emulsification/demulsification clean-up strategy combined with low-pressure gas chromatography-tandem mass spectrometry. *Food Chem.* **2024**, *458*, 140246. [CrossRef]
9. Chen, X.; Panuwet, P.; Hunter, R.E.; Riederer, A.M.; Bernoudy, G.C.; Barr, D.B.; Ryan, P.B. Method for the quantification of current use and persistent pesticides in cow milk, human milk and baby formula using gas chromatography tandem mass spectrometry. *J. Chromatogr. B Anal. Technol. Biomed. Life Sci.* **2014**, *970*, 121–130. [CrossRef]
10. Gomes Martins, J.; Amaya Chávez, A.; Waliszewski, S.M.; Colín Cruz, A.; García Fabila, M.M. Extraction and clean-up methods for organochlorine pesticides determination in milk. *Chemosphere* **2013**, *92*, 233–246. [CrossRef] [PubMed]
11. Talari, K.; Ganji, S.K.; Kommu, M.; Tiruveedula, R.R.; Upadhyayula, V. Quantitative determination of targeted and untargeted pesticide residues in coconut milk by liquid chromatography—Atmospheric pressure chemical ionization–high energy collisional dissociation tandem high-resolution mass spectrometry. *J. Chromatogr. A* **2021**, *1659*, 462649. [CrossRef]
12. Guo, Q.; Zhang, J.; Shao, B.; Yin, J.; Yang, Y.; Yang, Y. Rapid Screening for Hazardous Substances with Regulatory Differences in Milk Between Countries Using Ultra-High Performance Liquid Chromatography Ion Mobility Quadrupole Time-of-Flight Mass Spectrometry. *Foods* **2025**, *14*, 967. [CrossRef]
13. Mokhtar, H.I.; Salama, G.M.; El Gindy, A.; Abdel Hameed, E.A. Optimization of the Combined Use of Z-Sep Plus and EMR-Lipid in QuEChERS Procedure for the Analysis of Eight Pesticides in Real Milk Samples. *Food Anal. Methods* **2024**, *18*, 245–263. [CrossRef]
14. Qin, J.; Tong, K.; Chang, Q.; Xie, Y.; Wu, X.; Fan, C.; Zhang, H.; Shi, Z.; Chen, H. Simultaneous determination of PAHs and pesticides in milk based on improved QuEChERS combined with GC-MS/MS. *J. Food Compos. Anal.* **2025**, *141*, 107336. [CrossRef]
15. Galindo, M.V.; Hantao, L.W.; Sampaio, N.M.F.M.; Pessoto, M.A.; Oliveira, W.d.S.; Godoy, H.T. Analysis of polycyclic aromatic hydrocarbons in Brazilian human milk: A simple and effective approach. *Food Control* **2025**, *167*, 110796. [CrossRef]
16. Oyekunle, J.A.O.; Adekunle, A.S.; Adewole, A.M.; Elugoke, S.E.; Durodola, S.S.; Oyebode, B.A. Determination of organochlorine pesticide residues in some evaporated milk samples in Nigeria using gas chromatography-mass spectrometry. *Chem. Africa* **2021**, *4*, 349–366. [CrossRef]
17. Koesukwiwat, U.; Vaclavik, L.; Mastovska, K. Method development and validation for total haloxyfop analysis in infant formulas and related ingredient matrices using liquid chromatography-tandem mass spectrometry. *Anal. Bioanal. Chem.* **2018**, *410*, 5521–5528. [CrossRef]
18. Theurillat, X.; Dubois, M.; Huertas-Pérez, J.F. A multi-residue pesticide determination in fatty food commodities by modified QuEChERS approach and gas chromatography-tandem mass spectrometry. *Food Chem.* **2021**, *353*, 129039. [CrossRef]
19. Sereshti, H.; Jazani, S.S.; Nouri, N.; AliAbadi, M.H.S. Development of a green miniaturized quick, easy, cheap, effective, rugged, and safe approach in tandem with temperature-assisted solidification of floating menthol droplet for analysis of multiclass pesticide residues in milk. *J. Sep. Sci.* **2022**, *45*, 1106–1115. [CrossRef]
20. Castilla-Fernández, D.; Moreno-González, D.; Beneito-Cambra, M.; Molina-Díaz, A. Critical assessment of two sample treatment methods for multiresidue determination of veterinary drugs in milk by UHPLC-MS/MS. *Anal. Bioanal. Chem.* **2019**, *411*, 1433–1442. [CrossRef]
21. Slámová, T.; Sadowska-rociek, A.; Fraňková, A.; Surma, M.; Banout, J. Application of QuEChERS-EMR-Lipid-DLLME method for the determination of polycyclic aromatic hydrocarbons in smoked food of animal origin. *J. Food Compos. Anal.* **2020**, *87*, 103420. [CrossRef]
22. Arce-López, B.; Lizarraga, E.; Flores-Flores, M.; Irigoyen, Á.; González-Peñas, E. Development and validation of a methodology based on Captiva EMR-lipid clean-up and LC-MS / MS analysis for the simultaneous determination of mycotoxins in human plasma. *Talanta* **2020**, *206*, 120193. [CrossRef]
23. Wu, W.; Wang, K.; Liu, J.; So, P.; Leung, T.; Wong, M.; Zhao, D. A High-Throughput Integrated Nontargeted Metabolomics and Lipidomics Workflow Using Microelution Enhanced Matrix Removal- Lipid for Comparative Analysis of Human Maternal and Umbilical Cord Blood Metabolomes. *Anal. Chem.* **2025**, *97*, 2629–2638. [CrossRef]
24. Galindo, M.V.; Vargas-Pérez, M.; Lopez-Ruíz, R.; Oliveira, W.d.S.; Godoy, H.T.; Garrido-Frenich, A.; Romero-González, R. Monitoring of pesticide residues in Brazilian infant formulas using the suspect screening methodology by GC-HRMS. *J. Food Compos. Anal.* **2025**, *139*, 107166. [CrossRef]
25. Vargas-Pérez, M.; Domínguez, I.; González, F.J.E.; Frenich, A.G. Application of full scan gas chromatography high resolution mass spectrometry data to quantify targeted-pesticide residues and to screen for additional substances of concern in fresh-food commodities. *J. Chromatogr. A* **2020**, *1622*, 461118. [CrossRef]
26. Document No. SANTE. 11312/2021. Analytical Quality Control and Method Validation Procedures for Pesticide Residues Analysis in Food and Feed, 1–51. Available online: https://www.eurl-pesticides.eu/userfiles/file/EurlALL/SANTE_11312_2021.pdf (accessed on 19 November 2025).

27. Zhang, L.Q.; Zhang, X.M.; Zhang, H.W.; Wang, H.; Xu, H.; Wang, F.M.; Lin, C.; Xiao, J.; Xu, W.Y. Multiclass and multiresidue screening of veterinary drugs and pesticides in infant formula using Quadrupole-Orbitrap MS with PRM scan mode. *J. Mass Spectrom.* **2020**, *55*, e4497. [CrossRef] [PubMed]
28. Ferronato, G.; Viera, M.S.; Prestes, O.D.; Adaime, M.B.; Zanella, R. Determination of organochlorine pesticides (OCPs) in breast milk from Rio Grande do Sul, Brazil, using a modified QuEChERS method and gas chromatography-negative chemical ionisation-mass spectrometry. *Int. J. Environ. Anal. Chem.* **2018**, *98*, 1005–1016. [CrossRef]
29. Anastassiades, M.; Lehotay, S.J.; Štajnbaher, D.; Schenck, F.J. Fast and easy multiresidue method employing acetonitrile extraction/partitioning and "dispersive solid-phase extraction" for the determination of pesticide residues in produce. *J. AOAC Int.* **2003**, *86*, 412–431. [CrossRef]
30. Anastassiades, M.; Scherbaum, E.; Taşdelen, B.; Štajnbaher, D. *Recent Developments in QuEChERS Methodology for Pesticide Multiresidue Analysis. Pesticide Chemistry: Crop Protection, Public Health, Environmental Safety*; Wiley-VCH Verlag GmbH & Co.: Weinheim, Germany; Berlin, Germany, 2007; pp. 439–458. [CrossRef]
31. Lehotay, S.J.; O'Neil, M.; Tully, J.; Valverde, A.; Contreras, M.; Mol, H.; Heinke, V.; Anspach, T.; Lach, G.; Fussell, R.; et al. Determination of Pesticide Residues in Foods by Acetonitrile Extraction and Partitioning with Magnesium Sulfate: Collaborative Study. *J. AOAC Int.* **2007**, *90*, 485–520. [CrossRef]
32. Tienstra, M.; Mol, H.G.J. Application of Gas Chromatography Coupled to Quadrupole-Orbitrap Mass Spectrometry for Pesticide Residue Analysis in Cereals and Feed Ingredients. *J. AOAC Int.* **2018**, *101*, 342–351. [CrossRef]
33. Sanchez Costa, L.; Rodríguez Martínez, P.; Medina Sala, M. Determination of 23 organochlorine pesticides in animal feeds by GC-MS/MS after QuEChERS with EMR-lipid clean-up. *Anal. Methods* **2018**, *10*, 5171–5180. [CrossRef]
34. Manzano Sánchez, L.; Jesús, F.; Ferrer, C.; Gómez-Ramos, M.M.; Fernández-Alba, A. Evaluation of automated clean-up for large scope pesticide multiresidue analysis by liquid chromatography coupled to mass spectrometry. *J. Chromatogr. A* **2023**, *1694*, 463906. [CrossRef] [PubMed]
35. Lobato, A.; Fernandes, V.C.; Pacheco, J.G.; Delerue-Matos, C.; Gonçalves, L.M. Organochlorine pesticide analysis in milk by gas-diffusion microextraction with gas chromatography-electron capture detection and confirmation by mass spectrometry. *J. Chromatogr. A* **2021**, *1636*, 461797. [CrossRef] [PubMed]
36. Pedersen, T.L.; Smilowitz, J.T.; Winter, C.K.; Emami, S.; Schmidt, R.J.; Bennett, D.H.; Hertz-Picciotto, I.; Taha, A.Y. Quantification of Nonpersistent Pesticides in Small Volumes of Human Breast Milk with Ultrahigh Performance Liquid Chromatography Coupled to Tandem Mass Spectrometry. *J. Agric. Food Chem.* **2021**, *69*, 6676–6689. [CrossRef]

foods

MDPI

Article

Evaluation of Methods to Quantify Sialic Acid on Glycomacropeptide

Madison L. Dirks [1], Joseph Hale [2], Eric Theiste [2] and Owen M. McDougal [3,*]

1 Biomolecular Sciences Graduate Programs, College of Arts and Sciences, Boise State University, Boise, ID 83725, USA
2 Protein Research Center, Agropur, Le Sueur, MN 56058, USA
3 Department of Chemistry and Biochemistry, Food and Dairy Innovation Center, Boise State University, Boise, ID 83725, USA
* Correspondence: owenmcdougal@boisestate.edu

Abstract

Glycomacropeptide (GMP) is isolated from whey and used as an ingredient in phenylketonuria-safe foods because it does not contain phenylalanine. GMP is highly glycosylated and has several sites where *N*-acetylneuraminic acid (NANA) is bound. In the dairy industry, quantification of NANA from dairy proteins is accomplished by colorimetric, fluorometric, enzymatic, and chromatographic procedures; there is no uniformly accepted industry-wide standard method. In this investigation, NANA quantification methods were evaluated using GMP, and a comparison was made based on the length of time to complete the assay, protein-specificity, linearity, precision, and accuracy. From the methods evaluated, the chromatography protocol was determined to have the greatest benefit for use as a dairy industry standard to measure NANA on GMP. The average mass percent of NANA in 10 statistically independent replicates from a commercial GMP product was measured to be 6.18% $\pm$ 0.12%, with a relative standard deviation of 1.94%, which was the lowest of all the methods tested. The accuracy of the chromatographic approach was validated using spike and recovery experiments that provided an average recovery of 90.25%.

Keywords: *N*-acetylneuraminic acid; glycomacropeptide; chromatographic; mass spectrometry; acid hydrolysis; enzymatic; colorimetric; fluorometric

1. Introduction

1.1. Glycomacropeptide (GMP)

Glycomacropeptide (GMP) is obtained during the cheesemaking process when rennet, containing the enzyme chymosin, cleaves the peptide bond on κ-casein between phenylalanine-105 and methionine-106, coagulating casein proteins and leaving whey proteins in solution [1,2]. The resulting 64 amino acid whey protein does not contain phenylalanine (Phe), which makes it a usable source of amino acids for phenylketonuria-safe foods [1,3]. Patients suffering from phenylketonuria (PKU) have a deficiency or mutation in their phenylalanine-4-hydroxlyase enzyme, which is responsible for the breakdown of Phe [4–6]. In addition to being a source of amino acids, GMP also promotes favorable bioactivity including bone remineralization, anti-inflammation, metabolism modulation, and antibacterial, antiviral, and prebiotic effects [1,7]. The mechanism of bioactivity is not known, but current theories attribute the activity of GMP to the peptide, the attached sugars, or a combination of the two [1,8,9]. The antibacterial and antiviral effects from GMP ingestion, as well as the inhibition of cholera toxin, are due to GMP binding to the bacteria, virus, or toxin, preventing them from binding to cells [1,10–12]. GMP binding is proposed to be directly related to the presence of the sugar N-acetylneuraminic acid (NANA) on the peptide [1,10–12]. NANA is the most abundant of a nine-carbon acidic family of sugars, referred to as sialic acids, that are found in humans and mammals. The extent of NANA glycosylation affects the functional properties of GMP, including foaming, emulsification, aggregation, gel formation, and solubility, which are important factors when formulating new food and beverage products [2,13,14].

1.2. NANA on GMP

GMP can be highly glycosylated with O-linked sugars that have five confirmed glycan structures (Figure 1) [15,16]. These carbohydrate groups are post-translationally added to serine (Ser) and threonine (Thr) residues including Thr-121, 131, 133, 136, 142, and 165, and Ser-141 and 142, among other potential sites [1,15,17–19]. The sugars are one to four units in length, composed solely of combinations of NANA, N-acetylgalactosamine (GalNAc), and galactose (Gal) [15,16,19]. With regard to GMP from bovine milk, the sites and level of glycosylation as well as the amino acid sequence vary with cow breed, lactation, and processing techniques [1,20–23]. Knowing the amount of NANA on GMP provides additional information to determine product value. NANA concentration can be used to properly adjust formulations for products requiring foams, gels, emulsification, or a soluble peptide source. This information can also be used to develop new products with health improvement claims, including cell protection via anti-adhesion of viruses, bacteria, or toxins. Measuring NANA on GMP can provide further opportunities and increased value for this glycosylated peptide. Establishing a standard method for NANA quantification is intended to lead to consistency in NANA measurement across the dairy industry.

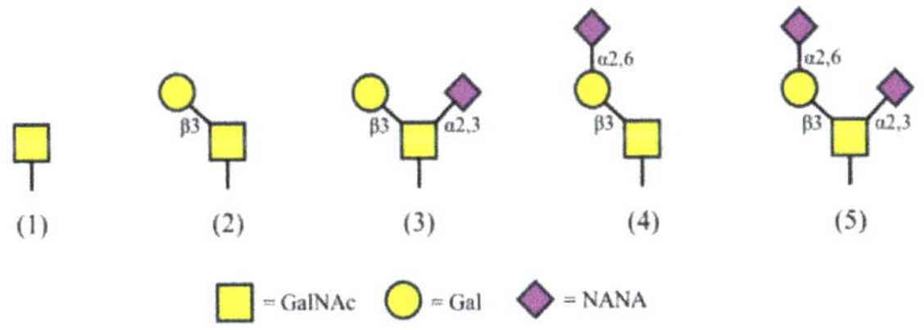

Figure 1. GMP's five glycan structures composed of N-acetylgalactosamine (GalNAc), galactose (Gal), and N-acetylneuraminic acid (NANA).

1.3. Historical Methods for NANA Quantification

In order to measure O-linked glycosylation, the accuracy of the method is dependent on customization for the peptide/protein and sugar units targeted for quantification [23]. Sialic acids are found on the surfaces of both eukaryotic and prokaryotic cells, where they are used by viruses for binding, and they also occur in fungi [24]. The release of NANA from glycoproteins is influenced by the interaction of the physicochemical environment surrounding the sugar-connecting bond [23,25]. Many different methods have been attempted for quantifying NANA, but to date, there is no industry-standard method to accurately measure the amount of NANA on GMP. Methods to hydrolyze NANA from proteins utilize either the enzyme activity, neuraminidase or the chemical effect of acids, including hydrochloric, formic, sulfuric, acetic, or propionic acid [23,25–30]. Incomplete hydrolysis, detection interference from non-NANA sugars, and degradation of NANA due to its relative acid lability are common issues that limit the usefulness of NANA quantification protocols [23,29,31]. Enzymatic assays generally exhibit increased specificity, but occasionally their function can be disrupted by NANA's bonding orientation and variation in O-acetylation and N-glycolyl groups [26]. Often samples require multiple enzymes to achieve complete NANA hydrolysis and transform NANA into detectable products [26]. Enzyme activity can vary from lot to lot, which can change the amount of NANA released from a sample, leading to poor accuracy for quantitation. Even when successfully used, enzymatic assays are expensive and time consuming [26]. When these methods were surveyed at Agropur, enzymatic assays yielded 0.25–4.11% (*w*/*w*) NANA on a single lot of GMP and acid hydrolysis assays reported 7.68–11.98% (*w*/*w*) NANA on a different lot of GMP. The lack of tight tolerance for results and poor accuracy to quantify NANA from GMP inspired the current work to develop a reliable method to measure NANA on GMP.

Once NANA has been hydrolyzed, colorimetric, UV, fluorometric, or chromatographic analysis can be employed for quantitation [26]. Colorimetric methods typically transform hydrolyzed NANA into an intermediate, which then reacts to form a chromophore to be measured spectrophotometrically [26]. These methods are subject to interference from hexoses, pentoses, 2-deoxyribose, and others, which limit their specificity for NANA [26]. Fluorometric methods often have similar, but longer, procedures to colorimetric assays, where NANA is hydrolyzed from a sample and then oxidized before reacting to form a fluorophore; in some cases the method will call for the addition of a fluorescent tag (such as 3,5-diaminobenzoic acid) instead of the fluorophore-forming reaction [26]. Some of the fluorometric assays produce the same toxic waste as the colorimetric methods and also show interference from hexoses, 2-deoxy sugars, aldehydes, and other compounds, but other fluorometric assays have far greater specificity to NANA [26,32]. Gas chromatography (GC) and high-performance liquid chromatography (HPLC) were studied using UV detection, and in some instances GC was paired with mass spectrometry detection (GC-MS) [26]. The GC-MS methods provided greater sensitivity, but also required derivatization and time-consuming purification processes, which lead to the loss of NANA [26,33]. Though many methods have been attempted for NANA quantification, there is still not an industry standard for GMP. The variety of methods employed give a range of results for NANA content on GMP, and some produce toxic waste, lead to degradation or incomplete release of the product, or have poor reproducibility [1,23,34]. The focus of the current investigation was to compare the four main types of experimental methods: colorimetric, fluorometric, enzymatic, and chromatographic, to determine which method is the most reliable, reproducible, and accurate to quantify NANA on GMP.

1.4. Current Methods for NANA Quantification

In 2017, an updated version of the Warren method was developed and packaged as a commercially available kit (Sialic Acid Assay Kit, MAK314, Sigma-Aldrich, Saint Louis, MO, USA) [34,35]. The Sialic Acid Assay Kit can be used for colorimetric or fluorometric detection of NANA, free or bound, within a sample of serum, plasma, saliva, milk, etc. [35]. For the hydrolysis of bound NANA, the sample is mixed with sulfuric acid and heated at 80 °C for 1 h. The freed NANA is oxidized to form formylpyruvic acid, which reacts with thiobarbituric acid, resulting in a chromophore that reflects red light and absorbs blue-green wavelengths at 549 nm using a microplate reader [35]. Fluorometric detection can measure NANA with ten times greater sensitivity at an excitation wavelength of 555 nm and an emission wavelength of 585 nm. The fluorometric assay uses the same workup as the colorimetric method, but the NANA standards used in the calibration curve are diluted 10-fold.

The enzyme α(2→3,6,8,9) neuraminidase from *Arthrobacter ureafaciens* cleaves glycosidic bonds attached to NANA [36,37]. The enzyme is available, along with several other reagents, in a commercially available kit (Sialic Acid Quantitation Kit, SIALICQ, Sigma-Aldrich, Saint Louis, MO, USA) for the measurement of NANA bound to glycoproteins, polysialic acids, or free in solution [36,37]. After enzymatic cleavage, NANA is converted to *N*-acetylmannosamine and pyruvic acid via *N*-acetylneuraminic acid aldolase [36]. The resulting pyruvic acid is reduced to lactic acid with lactic dehydrogenase and *beta*-nicotinamide adnenine dinucleotide (β-NADH) [36]. An indirect method of quantitation is used by measuring the remaining β-NADH post reaction [36]. Detection and quantitation of NANA can be achieved between 1 and 200 nmoles, using a microplate reader in the UV range at 340 nm.

Agropur US (Protein Research Center, Le Sueur, MN, USA) has developed an assay specific to GMP that employs weak acid hydrolysis in combination with high-performance liquid chromatography mass spectrometry (HPLC-MS) as a chromatographic approach to NANA measurement. This chromatographic method originated from non-satisfactory results obtained by experimental methods including the Warren method [34], enzymatic oxidation, and liquid chromatography with and without derivatization. A central composite Design of Experiments (DOE) was performed to determine the best hydrolysis conditions based on time, temperature, and acid concentration to fully remove NANA from GMP. MS detector drift was accounted for using internal NANA standards (C-12 and C-13). The Agropur chromatography method was internally replicated followed by third-party lab validation.

The purpose of the current project was to compare commercially available methods for measurement of NANA on GMP based on the time requirement for the assay, specificity of the hydrolysis method, the acid or enzyme, abilty to cleave NANA from GMP, linearity of calibration curves, precision of quantification, and accuracy of the assay.

2. Materials and Methods

2.1. GMP

Agropur produces a very high quality BiPRO® GMP 9000 product (Agropur, Appleton, WI, USA) that is used in medicinal foods for individuals with PKU and as a functional ingredient in foods and beverages. The GMP tested in these experiments was lot number JE 0020-20-440. The certificate of analysis stated that it was 96.1% protein, 97.0% of which was GMP.

2.2. Colorimetric Kit

A commercial kit (MAK314, Sigma-Aldrich, Saint Louis, MO, USA) was used to measure NANA content from GMP obtained from Agropur US, by colorimetry, following the manufacturer's technical bulletin and with additional guidance provided by Sigma-Aldrich technical support [35,38]. The NANA standard (10 mM) provided in the kit was serially diluted to provide concentrations of 1000, 600, 300, and 0 μM. The standards (20 μL) were combined with a 10% trichloroacetic acid (TCA) solution (5 μL) in microcentrifuge tubes in duplicate. For the hydrolysis reaction, a 20 μL aliquot of 600 μg/mL GMP was diluted with 40 μL nanopure (18.2 Mohm resistivity) water and mixed with 40 μL of hydrolysis reagent. Each sample was prepared in duplicate. The tubes were placed in a heat block set to 80 °C for one hr. The samples were allowed to cool to room temperature for 5–10 min before they were centrifuged at 14,000 rpm for 3 min. The 10% TCA solution (25 μL) was added to each sample and vortexed for 30 s. The tubes were centrifuged again at 14,000 rpm for 10 min. A working reagent was prepared in bulk for all samples and standards by combining the hydrolysis reagent (15 μL/tube), nanopure water (50 μL/tube), and oxidation reagent (65 μL/tube). Each sample (25 μL in a new centrifuge tube) and standard was combined with working reagent (125 μL) and allowed to sit at room temperature for one hr. For the color reaction, dye reagent containing 2-thiobarbituric acid (50 μL) was mixed into each tube of sample and standard. All tubes were placed in the heat block that was set to 100 °C, and heated for 10 min. The tubes were removed and allowed to cool for 5–10 min before being combined with dimethyl sulfoxide (DMSO) (100 μL), followed by centrifugation at 14,000 rpm for 5 min. For detection, samples and standards in aliquots of 250 μL were transferred to a clear, Greiner 96-well plate (Sigma-Aldrich, Saint Louis, MO, USA) and analyzed by Agilent BioTek Epoch 2 microplate spectrophotometer (Agilent, Santa Clara, CA, USA). Optical density (OD) was recorded at 549 nm. The OD values were averaged for all duplicates of standards and samples. The OD for the blank (0 μM NANA standard) was subtracted from the OD values of all other standards and samples. A calibration curve was generated with these corrected OD values and the reported concentrations. The following equation (Equation (1)) was used to calculate the concentration (μM) of NANA in each sample:

$$[Sialic\ acid] = \frac{R_{sample} - R_{blank}}{Slope\left(\mu M^{-1}\right)} \times n(\mu M) \quad (1)$$

where R_{sample} and R_{blank} are the OD or fluorescence intensity values of the sample and the blank, respectively, Slope is the slope of the calibration curve, and n is the dilution factor (5).

The concentration values (μM) were converted to NANA content (μg/mL) using the molecular mass of NANA (309 g/mol) and then to the mass percent (%*w*/*w*) by dividing by the mass of each GMP sample.

2.3. Fluorometric Kit

The same commercial kit (MAK 314, Sigma-Aldrich, Saint Louis, MO, USA) was used to measure NANA concentration by fluorimetry. The fluorimetry procedure was consistent with the colorimetric assay with a few exceptions. The NANA standards were prepared at 10-fold dilute concentrations of 100, 60, 30, and 0 μM. For detection, the samples were added to a black, clear bottom, Nunc_96-well plate and analyzed on a BioTek Synergy H1 Microplate Reader (Agilent, Santa Clara, CA, USA). Fluorescence intensity was measured using an excitation wavelength of 555 nm and an emission wavelength of 585 nm. In addition to subtracting the fluorescence intensity of the blank from all samples and standards, the fluorescence intensity of the empty plate was also subtracted from all wells.

2.4. Enzymatic Kit

A neuraminidase assay was performed using a kit (SIALICQ, Sigma-Aldrich, Saint Louis, MO, USA) by following the included instructions, and with additional assistance from the manufacturer [36,37]. First, a time-course experiment was conducted. The buffers and reagents were prepared from the stock solutions provided in the kit. The tris(hydroxymethyl)aminomethane hydrochloride (Tris-HCl) reaction buffer was prepared by diluting the 1 M Tris-HCl to 25 mM with deionized (DI) water. The β-NADH solution was generated by adding Tris reaction buffer (640 μL) to one vial of β-NADH (5 mg). A reaction mix was prepared with Tris reaction buffer (6510 μL), β-NADH solution (140 μL), *N*-acetylneuraminic acid aldolase (7 μL), and L-lactic dehydrogenase (7 μL). The reaction mix (1000 μL) was added to six microcentrifuge tubes, and what remained was reserved for a blank. For the initial time (T_0 min), the GMP sample (40 μL of 600 μg/mL GMP in nanopure water) and sialidase buffer (10 μL) were added to one of the microcentrifuge tubes containing the reaction mix and incubated for one hr. in a water bath at 37 °C. For continuous digestion, the GMP sample (200 μL of 600 μg/mL GMP in nanopure water), sialidase buffer (50 μL), and α-(2→3,6,8,9)-neuraminidase (5 μL) were combined in a microcentrifuge tube and incubated in a water bath at 37 °C. Every hr. (for a total of five hr.), the sample (51 μL) was removed from the microcentrifuge tube and added to one of the microcentrifuge tubes containing the reaction mix. The reaction mix tube was returned to the water bath for an additional hr. before detection. The T_0 sample, the blank, and each digested sample (200 μL in triplicate) were added to a clear, Costar 48-well cell culture plate and analyzed by absorbance at 340 nm using an Agilent BioTek Epoch 2 microplate spectrophotometer (Agilent, Santa Clara, CA, USA).

The enzymatic reaction began with mixing the GMP sample (40 μL of 600 μg/mL GMP in nanopure water), sialidase buffer (10 μL), and α-(2→3,6,8,9)-neuraminidase (1 μL). This mixture was prepared in duplicate and incubated for 4 h. in a water bath at 37 °C. Meanwhile standards of 1000, 600, 300, and 0 μM NANA were generated from the provided NANA standard (0.01 M). The Tris reaction buffer, β-NADH solution, and reaction mix were prepared in the same way as the time-course experiment with the volumes corrected for the number of samples being analyzed. After 4 h. of incubation, each sample (12 μL) was combined with reaction mix (238 μL) in new microcentrifuge tubes. Three blanks (250 μL reaction mix) were also added to microcentrifuge tubes. All samples, standards, and blanks were incubated for one hr. in a water bath at 37 °C. For detection, the blanks and samples (200 μL) were added to a clear, Greiner 96-well plate (Sigma-Aldrich, Saint Louis, MO, USA), and absorbance was measured at 340 nm on an Agilent BioTek Epoch 2 microplate spectrophotometer (Agilent, Santa Clara, CA, USA). Data analysis was performed in the same way as for the colorimetric method, but the dilution factor used in Equation (1) was 1.275 instead of 5.

2.5. *Chromatographic*

The chromatographic method was developed by scientists at Agropur US (Protein Research Center, Le Sueur, MN, USA). A central composite DOE was used to build a response surface model system that would predict NANA hydrolysis reaction outcomes for experimental parameters: acetic acid concentration (1.18 M–2.82 M), reaction temperature (70–90 °C), reaction time (22–218 min), and GMP sample concentration. In each experiment of the design, 200 μL of acid and 200 μL of GMP sample solution were combined and analyzed for the liberation of NANA from GMP. The measured NANA responses were used to build the response surface model optimizing for maximum measured NANA. The design and data analysis were performed using Minitab® (Version 17, Minitab LLC, State College, PA, USA). Parameters were adjusted further using 1.1 M acetic acid at 67.5 °C for a range of 30–360 min with differing concentrations of GMP (1000 μg/mL, 800 μg/mL, 600 μg/mL, and 500 μg/mL). These concentration values were plotted to find where NANA liberation stabilized. An extended time reaction was also assessed to measure the loss of NANA over 20 h.

A modified version of the Agropur US chromatographic method was performed at the Food and Dairy Innovation Center, Boise State University, as follows. The standard addition calibration matrix (0.1200 ± 0.0050 g GMP) and sample (0.0600 ± 0.0010 g GMP) were dissolved in nanopure water on stir plates and brought to a volume of 100.0 mL in volumetric flasks. An NANA (97%, Thermo Scientific, Waltham, MA, USA) standard solution (1 mg/mL) was prepared in nanopure water and stored in the freezer when not in use. From this stock solution (spiking stock #4), a serial dilution (1:4) was used to generate spiking stocks #3, #2, and #1. Five different calibration standards were prepared in microcentrifuge tubes by combining the calibration matrix (500.0 μL) with either one of the spiking stocks or nanopure water (500.0 μL) for the low calibration point. Each calibration standard and duplicate GMP test samples (200.0 μL) were added to 2 mL polypropylene microtubes (Thermo Scientific, Waltham, MA, USA) and aspirated via pipet tip with 2.2 M acetic acid (200.0 μL). The tubes were incubated overnight at 60 °C to hydrolyze NANA from GMP. The following day, samples and standards were evaporated to dryness using a SpeedVac with the temperature set to 65 °C. The samples and standards were reconstituted in a surrogate internal standard (200.0 μL) of $^{13}C_3$-NANA (Sigma-Aldrich, Saint Louis, MO,

USA) in 50/50 acetonitrile (>99% purity, Fisher Scientific, Pittsburgh, PA, USA)/nanopure water (100 μg/mL), which was stored in the freezer when not in use. The solutions were vortexed until the solid dissolved, and were then sonicated for 10 min. A 50/50 acetonitrile (>99% purity, Fisher Scientific, Pittsburgh, PA, USA)/nanopure water solution (800.0 μL) was added to each tube and vortexed for 10 s. The solutions were syringe-filtered with 0.2 μm nylon filters into HPLC vials.

NANA separation was performed using an Agilent 1260 Infinity II HPLC system (Agilent, Santa Clara, CA, USA) with a quaternary pump, thermostatted column compartment, and split loop autosampler using a Waters Corp XBridge BEH Amide Column, 3.5 μm, 2.1 mm × 100 mm (Waters, Milford, MA, USA). The HPLC injection volume was 10 μL and the flow rate was 1 mL/min with the column temperature set at 30.0 °C. Mobile phase A was 200 mM ammonium formate (Sigma-Aldrich, Saint Louis, MO, USA) acidified to pH 3.3 using formic acid (Sigma-Aldrich, Saint Louis, MO, USA) (A); mobile phase B was acetonitrile (>99% purity, Fisher Scientific, Pittsburgh, PA, USA) (B); and mobile phase C was nanopure water (C). Mobile phase A was held at 5.0% for the entirety of the method. From 0.00 to 1.50 min: (B) = 90.0%, (C) = 5.0%. From 1.50 to 5.10 min: (B) 90.0% to 67.1%, (C) 5.0% to 27.9%. From 5.10 to 5.70 min: (B) 67.1 to 50.0%, (C) 27.9% to 45.0%. From 5.70 to 6.70 min: (B) = 50.0%, (C) = 45.0%. From 6.70 to 6.80 min: (B) 50.0% to 90.0%, (C) 45.0% to 5.0%. From 6.80 to 18.50 min: (B) = 90.0%, (C) = 5.0%. Analyte detection was accomplished with an Agilent G6125C single quadrupole liquid chromatography–mass spectrometry detector system in positive ion mode set to 310.1 m/z and 313.1 m/z, capillary voltage positive setpoint: 4000, negative setpoint: 4000, nebulizing gas flow of 7.0 L/min, nebulizer set to 15 psi, gas temperature set to 300 °C, and electrospray ionization as the ion source. MS peaks eluted between 4 and 7 min for both the 310.1 m/z and 313.1 m/z. The areas under each curve were integrated, and their ratios were plotted vs. standard concentrations to generate a linear standard addition calibration curve for NANA. The curve was corrected by subtracting the y-intercept from the equation of the line. The equation of the line from the calibration curve was used to calculate the concentration of NANA in each sample (μg/mL). The density of water (1 mg/mL), the dilution factor, and the mass of the GMP sample were used to compute the percentage of NANA per total sample mass (%w/w).

2.6. Method Validation

Five different criteria were used to evaluate each method: time, protein-specificity, linearity, precision, and accuracy. For the time evaluation, the time to complete the entire assay, from sample preparation to data analysis, was determined. This was broken down into active and inactive time for the analyst. Protein-specificity was explored based on the level of specificity the hydrolysis agent (acid or enzyme) had for removing NANA from GMP. Linearity of each assay was derived from the calibration curves generated from NANA standards. The slope, intercept, and coefficient of determination for each regression equation were calculated. The limit of detection (LOD) and limit of quantification (LOQ) were generated from the standard error of the intercept and the slope of each regression equation. The precision of each method was assessed by calculating the average NANA content (%w/w) on GMP, as well as the standard deviation (SD) and the relative standard deviation (RSD) based on the number of GMP replicates tested. Finally, accuracy was measured by comparing to a reference material (bovine fetuin) for the enzymatic assay or by spike and recovery for the chromatographic approach. The fetuin sample was treated in the same way as the GMP test solution preparation, and the resulting NANA content (%w/w) was compared to the predicted NANA content. Spike and recovery experiments were performed in triplicate for the chromatographic method. A volume of 50 μL of

1 mg/mL NANA standard was spiked into 1.0 mL filtered GMP test solution. The spiked samples went through the same procedures as the normal GMP test solution preparation.

2.7. Statistical Analysis

All statistical analysis was performed using Microsoft 365 Excel (Microsoft, Redmond, WA, USA).

3. Results

The measurement parameters used to assess each NANA quantitation method were time, protein-specificity, linearity, precision, and accuracy. These evaluation strategies have been used to review similar methods in the literature [39,40].

3.1. Time

The first point of interest is the length of time required to perform each method, including sample preparation, experimentation, waiting periods, and data analysis. The protocol to use the colorimetric/fluorometric kit has two steps with 1 h waiting periods, complemented by multiple timed centrifugation steps and cooling periods. When paired with sample and standard preparation and data analysis, the kit requires a minimum of 4.5 h to complete. The enzymatic kit suggests a hydrolysis time-course experiment prior to running the complete assay. During the time-course experiment, the optimum time for hydrolysis was determined to be 4 h (Figure 2). With the 4 h hydrolysis, the required incubation period of 1 h, the sample and standard preparation required, and analysis of the resulting data, the entire experiment takes approximately 8 h to complete. The chromatographic approach has an overnight hydrolysis period and a drying period that takes at least 2 h to achieve. With the sample preparation and LC-MS quantitation, the chromatographic method requires a total of 31 h, inclusive of 2 h sample preparation, an overnight incubation period, 2 h drying periods, 4 h for LC-MS sample preparation, and 1 h for manual peak integration and data analysis. A summary of protocol times is provided in Table 1.

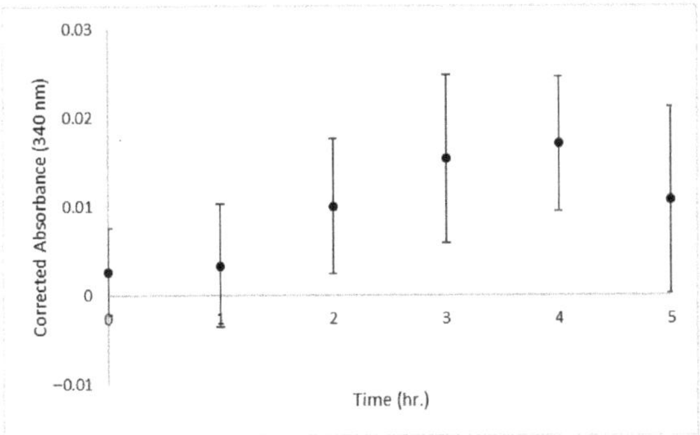

Figure 2. The time-course experiment for enzymatic hydrolysis of *N*-acetylneuraminic acid (NANA) from glycomacropeptide (GMP). UV absorbance at 340 nm over time showed an optimum incubation time of 4 h for maximum hydrolysis of NANA. Standard deviation error bars for n = 3 samples.

The fastest method to perform is the colorimetric/fluorometric kit, that takes 4.5 h, followed by the enzymatic protocol at 8 h, and lastly, the chromatographic approach is the most time consuming at 31 h. The majority of the time for each method comes from the hydrolysis reaction, drying, and/or separation times. Each experiment has comparable times where the analyst is in the laboratory, actively conducting the study or analyzing data. The colorimetric/fluorometric method involves 2.5 h of laboratory work and data analysis, the enzymatic method requires 3 h, and the chromatographic approach involves 4 h of scientist effort. These times are based on the current study, where an analyst performed each method three or more times.

Table 1. Approximate time required to complete each method for quantitation of *N*-acetylneuraminic acid (NANA) in glycomacropeptide (GMP). Active time is the duration an analyst spends performing a test method step excluding incubations and waiting.

Method	* Colorimetric	* Fluorometric	* Enzymatic	* Chromatographic
Sample Preparation	**2 h**	**2 h**	**2.5 h**	**3 h**
Reaction or Drying	2 h	2 h	5 h	24 h
Detection	30 s	30 s	30 s	3 h
Data Analysis	**0.5 h**	**0.5 h**	**0.5 h**	**1 h**
Total	4.5 h	4.5 h	8 h	31 h
Total Active Time	**2.5 h**	**2.5 h**	**3 h**	**4 h**

* Active time for the analyst is in bold.

3.2. *Protein-Specificity*

NANA hydrolysis requires the appropriate acid strength or enzymatic reaction, proper temperature control, and sufficient time. The characteristics of these parameters are protein dependent, which can make methods tedious and expensive when they are very specific, accurate, and reliable [41]. The colorimetric/fluorometric kit is marketed for quantification of free and total NANA in matrices including serum, saliva, and milk [35]. Similarly, the enzymatic assay is described as determining total NANA content free in solution or bound to glycoproteins or polysaccharides regardless of bond type [36,37]. The enzymatic time-course experiment allowed for maximum release of NANA from GMP to be explored before proceeding with the entirety of the method. The chromatographic approach utilized central composite DOE to model reaction conditions for liberation of NANA from GMP based on empirical measurements. This technique allowed for the maximum amount of NANA detection before degradation by testing hydrolysis time, acid concentration, oven temperature, and GMP sample concentration. The results showed that the maximum amount of NANA was released from GMP when the temperature was held at 67.5 °C for 6 h for a sample consisting of 600 μg/mL GMP and the hydrolysis acid, acetic acid, at a concentration of 1.1 M.

The colorimetric/fluorometric and the enzymatic assays are not sample specific, making their use generic to detect free or bound NANA. The kits can be used for a wide range of applications, but their lack of specificity for NANA detection on a glycosylated protein, like GMP, equates to non-optimal quantitation. Widely variable results and poor quantitation likely originate from the rapid, intense protocol that inconsistently hydrolyzes NANA from GMP. In contrast, the chromatographic approach was specifically designed to cleave and measure NANA on GMP. The central composite DOE was used to determine the exact specifications to achieve maximum NANA hydrolysis and measurement from GMP. The DOE result called for a longer incubation period at a lower temperature with a less concentrated weak acid than the other methods. The chromatographic method was further extended to overnight hydrolysis because it provided a greater level of convenience for the analyst than a 6 h hydrolysis period, but the result was within 0.5% NANA content.

3.3. *Linearity*

Linearity ensures a method is fit for its intended purpose by relating analyte concentration to instrument response. Linearity of each calibration curve was measured via regression analysis. The number of NANA standards used to generate the curve, the concentration, slopes, intercepts, coefficients of determination, LOD, and LOQ, and regression equations for each method are provided in the supplemental materials (Tables S1–S5). A summary of calibration results is listed in Table 2. LOD and LOQ were determined using Equations (2) and (3), respectively.

$$\text{LOD} = 3.3 \times \frac{\sigma}{\text{s}} \quad (2)$$

where σ is the standard error of the intercept, and s is the slope.

$$\text{LOQ} = 10 \times \frac{\sigma}{\text{s}} \quad (3)$$

where σ is the standard error of the intercept, and s is the slope.

Table 2. Summary of method linearity including number of standards, range of concentration, number of calibration curves, and the average ± standard deviation for slope, intercept, coefficient of determination, LOD, and LOQ.

Method	Colorimetric	Fluorometric	Enzymatic	Chromatographic
Number of standards (n)	4	4	4	5
Range of concentration (μg/mL)	0–309	0–30.9	0–309	0–485.0
Number of calibration curves	3	3	2	10
The slope of the regression equation	0.0017 ± 0.0001	39.747 ± 6.741	0.0003 ± 0.0000	9.6162 ± 0.4943
The intercept of the regression equation	−0.0056 ± 0.0106	71.532 ± 44.610	−0.0029 ± 0.0113	0.387 ± 0.030
Coefficient of determination	0.9979 ± 0.0024	0.9969 ± 0.0032	0.9858 ± 0.0113	0.9992 ± 0.0006
Limit of detection (μg/mL)	16.5 ± 11.5	2.13 ± 1.20	52.1 ± 3.2	10.7 ± 3.3
Limit of quantification (μg/mL)	50.1 ± 34.7	6.44 ± 3.63	158 ± 10	32.5 ± 10.0

The LOD represents the lowest concentration of NANA the instrument could detect with a certainty of ≥95%. The lowest average LOD of the four methods was 2.13 ± 1.20 μg/mL for the fluorometric, followed by the chromatographic at 10.7 ± 3.3 μg/mL, and the colorimetric at 16.5 ± 11.5 μg/mL, and finally the enzymatic at 52.1 ± 3.2 μg/mL. The LOQ is the lowest value at which a quantified measurement of NANA can be considered accurate based on Equation (3). The lowest quantifiable value was for the fluorometric method at 6.44 ± 3.63 μg/mL, the next lowest was the chromatographic at 32.5 ± 10.0 μg/mL, then the colorimetric at 50.1 ± 34.7 μg/mL, and the highest was the enzymatic at 158 ± 10 μg/mL. The values reported for the enzymatic method are problematic considering none of the five reported sample concentrations are greater than the LOQ. Out of the 12 samples analyzed using the colorimetric method, only 42% of them measured NANA above the LOQ. For the chromatographic method, 80% of samples were above the LOQ. Only the fluorometric method had 100% of samples calculated above the LOQ.

3.4. Precision

Precision was assessed to determine the inter-day reproducibility of each measurement. The amount of NANA on GMP was measured as the mass of NANA over the mass of GMP (%*w*/*w*). To report precision, the average, standard deviation (SD), and relative standard deviation (RSD) were calculated for all samples and for those above the LOQ (Table 3).

Table 3. Inter-day reproducibility of NANA measurement of the mass of NANA over the mass of GMP in %*w*/*w* based on number of measurements, average, standard deviation (SD), and relative standard deviation (RSD).

Method	* Colorimetric	* Fluorometric	* Enzymatic	* Chromatographic
Number of observations (n)	12	11	5	10
Average	**5.25** (5.11)	**7.26**	(14.09)	**6.18** (6.17)
SD	**0.24** (0.27)	**1.37**	(3.25)	**0.12** (0.15)
RSD (%)	**4.48** (5.19)	**18.90**	(23.09)	**1.94** (2.44)

* Bold values represent numbers within the LOQ, and numbers in parentheses represent the average of all samples.

There was variability in the number of samples included in this study due to the quantity of reagents available in each kit and assay. The colorimetric/fluorometric kit was prepared with sufficient reagents to measure 100 samples and was determined to contain leftover reagents after 100 assays had been run. Three different kits were purchased for the colorimetric/fluorometric assays, but only one kit was viable for use. Therefore, the data included is solely from the second kit. The 12 colorimetric samples were measured across three different days, and the 11 fluorometric samples were also quantified on three different days. Two enzymatic kits were purchased, and each was intended to be used for 25 samples. Neither kit had sufficient reagents to conduct 25 assays, and, of the experiments completed, not all the calibration curves were usable for quantifying NANA content. The usable data was collected on two different days. The chromatographic approach did not have the reagent constraints of the other methods; thus, it was run 10 times for statistical significance. Each experiment was performed on a different day to demonstrate inter-day reproducibility.

The average, standard deviation (SD), and relative standard deviation (RSD) were calculated for each method. The enzymatic method provided the least consistent results with the largest RSD at 23.09%. The fluorometric method was next with an RSD of 18.90%. The colorimetric and chromatographic assays had much lower RSD values of 4.48% and 1.94%, respectively. The chromatographic method had the lowest relative standard deviation among the methods evaluated, providing the most precise measurements.

3.5. Accuracy

Each method was evaluated for accuracy to determine which analysis aligned most closely with the actual value of NANA on GMP. Accuracy is typically determined by comparing to a reference method or by using a standard reference material. Neither of these currently exist for the measurement of NANA content. Bovine fetuin (0.5 mg) was provided as a standard reference material for the enzymatic assay and should have yielded approximately 48 nmoles of NANA [38]. When following the procedures for digestion and calculation of nmoles for fetuin, the result was 61.24 nmoles. This value was determined using a calibration curve with an R^2 value of 0.8168, which was ruled unfit for analysis.

When there is not a standard method or reference material for comparison, accuracy can be assessed by performing a spike and recovery analysis. This approach was used for the chromatographic method. Three different samples were spiked with a known amount of NANA standard, and the percent recovery was measured. The acceptable recovery for this type of validation is within the range from 80 to 120%. Each of the three samples were within the accepted range, and the average recovery was 90.25% (Table 4).

Spike and recovery experiments were successfully performed for the chromatographic approach and were attempted for the colorimetric method. A valid calibration curve could not be produced for the colorimetric method, so the spike and recovery results were discarded. The spike and recovery results for the chromatographic method suggest the accuracy of this approach.

Table 4. *N*-acetylneuraminic acid (NANA) spike and recovery for chromatographic method.

Sample	Measured NANA (%*w*/*w*)	Recovery (%)
1	6.25	96.32
2	5.94	81.18
3	6.44	93.24
Average	6.21	90.25

After performing the previously described methods, three distinct hybrid approaches were attempted by combining the acid hydrolysis portion of the chromatographic method with colorimetric detection. The most promising results were from one of the colorimetric kits, but when it was attempted again with another colorimetric kit, it was unsuccessful. Variations on Skoza's [42] method were trialed, but a spike and recovery analysis performed to validate the results provided values outside of the acceptable range of 80–120%. The hybrid approach was abandoned in favor of the more robust and reliable chromatographic method.

4. Discussion

The NANA measurement methods that are available generally employ a combination of hydrolysis and detection of NANA through colorimetric methods. As far back as 1959, Warren measured NANA free in solution and in tissues and fluids obtained from bovine, human, mouse, rat, guinea pig, and chicken sources [34]. This and other similar methods are not well-designed for quantitation of NANA due to limited specificity, poor sensitivity, and their propensity to produce toxic waste [29,34,43]. The commercially available kit used for this work is an improvement from these earlier methods, but it still has drawbacks. This method is rapid to perform, only requiring a total of 4.5 h, where the analyst is actively working for 2.5 h. Another benefit of this method is its precision. For all the samples that were above the LOQ, there was a 4.48% RSD. This was one of the lowest RSD values out of the four methods tested, second only to the chromatographic approach. This kit is not protein-specific, which means that its hydrolysis conditions do not work equally for every protein. While the hydrolysis time is short, the conditions are harsh, which led to variability in the reported NANA content. The LOD for this method is 16.5 ± 11.5 μg/mL, and the LOQ is 50.1 ± 34.7 μg/mL. Over 50% of the samples tested using this method were not within the LOQ. This alone renders this method highly untrustworthy. Based on the issues with linearity and protein-specificity it was deemed unnecessary to proceed with detailed studies on accuracy for this assay. If the goal is to measure NANA on GMP, the benefits of this method do not outweigh the drawbacks.

4.1. Fluorometric

The only difference between the fluorometric and colorimetric procedures was that all the standards were diluted tenfold. Therefore, the time required to complete the experiment was still 4.5 h with 2.5 h of active work for the analyst and the lack of protein-specificity was the same. There were a few differences in the results of these two assays. The linearity of the fluorometric assay was its most notable positive feature. It had an LOD of 2.13 ± 1.20 μg/mL and an LOQ of 6.44 ± 3.63 μg/mL. This assay had the lowest LOD and LOQ and is the only method tested where all the sample concentrations of NANA were above the LOQ. This did not translate to the results on precision. The RSD was nearly 20%, which showed that the results were not clustered near the mean of 7.26% NANA. This variation indicates that this method does not provide consistent results for the same reference material, and thus cannot be considered reliable. Spike and recovery studies to test accuracy were not attempted due to this level of unreliability.

4.2. Enzymatic

A variety of enzymatic assays were explored in the 1960s and 1970s [26,28]. These often involved the conversion of NANA into a series of products including hydrogen peroxide or pyruvate that were able to be detected and measured using UV-Vis spectroscopy or biosensors [26]. The method used here utilized a neuraminidase enzyme, which can cleave $\alpha(2\rightarrow3,6,8,9)$ bonds. This should allow for the cleavage of any bonded NANA. The time-course experiment was performed to determine the best hydrolysis parameters for GMP, which demonstrated a level of protein-specificity. Having a 4 h hydrolysis period along with sample preparation and other procedures resulted in this method requiring about eight hr., where for three hr. the analyst is actively working. This can be completed in a standard day of work, but does not have the speed of the colorimetric and fluorometric assays. Drawbacks of this method began with the linearity results, which were by far the highest of the four methods at an LOD (52.1 ± 3.2 µg/mL) and LOQ (158 ± 10 µg/mL). None of the values reported for the five samples tested on this method were within the LOQ, which causes this method to be unusable for this application. In addition, this method had the worst precision with an RSD of nearly 25%. The accuracy was tested using the provided bovine fetuin standard, and the reported NANA value was over 125% of the expected value. This method may have had increased protein-specificity due to the initial time-course experiment, but it does not have sufficient linearity, precision, or accuracy to be recommended for use.

4.3. Chromatographic

Chromatographic assays for measuring NANA content became prevalent in the literature reports in the early 1980s [26]. One of the key features of the chromatographic method used for this work was the DOE for the hydrolysis parameters. The protein-specificity of the chromatographic method is unmatched compared to the other three methods tested. Hydrolysis temperatures and times were evaluated here, and previous work performed by Joseph Hale at Agropur's Protein Research Center determined the type of acid, concentration of acid, and concentration of sample. One of the highlights of this method is its precision, which was the most superior of any of the methods. The RSD was 1.94% for the samples that were above the LOQ, meaning the sample values were tightly grouped around the mean. Measured NANA content was reproducible. This method was also tested for accuracy via spike and recovery experiments. Each of the triplicate runs was within the accepted tolerance of 80–120%, and the average recovery was 90.25%. Having an accurate and reliable method is necessary for standardization across the industry. The chromatographic approach had the second best LOD (10.7 ± 3.3 µg/mL) and LOQ (32.5 ± 10.0 µg/mL) after the fluorometric method. Out of the 10 samples tested, 8 of them were within the LOQ. This is a limitataion of the assay, though it still performed sufficiently 80% of the time. The biggest drawback is the length of time required to perform this experiment. With the overnight method, a total of 31 h are needed with 4 of those being active time for the analyst. Overall, the benefits of this assay far outweigh the drawbacks, making it the best option for a standard method.

5. Conclusions

The results of this work compared facets of four different methods used to hydrolyze and quantify NANA on GMP. When considering the length of time, protein-specificity, linearity, precision, and accuracy of each method, the chromatographic approach was unequivocally the best for this purpose. Scientists at Agropur US developed the chromatography method as a robust and accurate means to quantify NANA on GMP. The protocol was altered slightly to require a total of 31 h. to complete, during which the analyst is actively involved for 4 h. Out of the 10 GMP samples examined, 8 of the measured values were above the 32.5 ± 10 µg/mL LOQ threshold for NANA. The chromatography method is the only one specifically tailored by DOE to the GMP matrix, and it also provided the lowest RSD across all methods at 1.94%. The chromatography method did not have the issue of inconsistency between kits, and when testing for accuracy by spike recovery experiments, the average NANA recovery was 90.25%. Although the chromatographic method requires more time to perform than other methods, and the LOQ was determined to be satisfied for 80% of test cases, the specificity of this method for NANA on GMP demonstrated superior precision and accuracy, setting it apart from the other methods. The ability to measure NANA on GMP is critical for consistent formulation of existing and new products within the dairy industry. At this time, the functional and bioactive properties of NANA on GMP are largely underutilized in GMP's limited application as a protein source for PKU-safe foods. With implementation of this chromatographic method, the functionality and bioactivity could be harnessed for the development of novel products.

Supplementary Materials: The following supporting information can be downloaded at: https://www.mdpi.com/article/10.3390/foods14223939/s1, Table S1: Linearity of colorimetric method; Table S2: Linearity of fluorometric method; Table S3: Linearity of enzymatic method; Table S4: Linearity of chromatographic method; Table S5: Regression equations for each method.

Author Contributions: Conceptualization, M.L.D., J.H., E.T. and O.M.M.; methodology, M.L.D., J.H., E.T. and O.M.M.; software, M.L.D., J.H. and E.T.; validation, M.L.D., J.H. and E.T.; formal analysis, M.L.D., J.H. and E.T.; investigation, M.L.D., J.H. and E.T.; resources, O.M.M.; data curation, M.L.D. and O.M.M.; writing—original draft preparation, M.L.D.; writing—review and editing, M.L.D., J.H., E.T. and O.M.M.; visualization, M.L.D.; supervision, O.M.M.; project administration, O.M.M.; funding acquisition, O.M.M. All authors have read and agreed to the published version of the manuscript.

Funding: This research was funded by the National Science Foundation Convergence Accelerator, grant number 2345069 and the BUILD Dairy program with co-sponsorship from Agropur.

Institutional Review Board Statement: Not applicable.

Informed Consent Statement: Not applicable.

Data Availability Statement: The original contributions presented in this study are included in the article/Supplementary Material. Further inquiries can be directed to the corresponding author.

Acknowledgments: We would like to recognize the generous administrative, technical, material, and financial support of Anand Rao and the Protein Research Center at Agropur, Le Sueur, MN. Minitab® and all other trademarks and logos for the Company's products and services are the exclusive property of Minitab, LLC. All other marks referenced remain the property of their respective owners. See minitab.com for more information.

Conflicts of Interest: Authors Joseph Hale and Eric Theiste were employed by the company Agropur. The remaining authors declare that the research was conducted in the absence of any commercial or financial relationships that could be construed as a potential conflict of interest. The funders had no role in the design of the study; in the collection, analyses, or interpretation of data; in the writing of the manuscript; or in the decision to publish the results.

Abbreviations

The following abbreviations are used in this manuscript:

beta-NADH	*beta*-nicotinamide adenine dinucleotide
DI	Deionized
DMSO	Dimethyl sulfoxide
DOE	Design of Experiments
Gal	Galactose
GalNAc	*N*-acetylgalactosamine
GC	Gas chromatography

GC-MS	Gas chromatography-mass spectrometry
GMP	Glycomacropeptide
HPLC	High-performance liquid chromatography
HPLC-MS	High-performance liquid chromatography–mass spectrometry
LOD	Limit of detection
LOQ	Limit of quantification
NANA	*N*-acetylneuraminic acid
OD	Optical density
Phe	Phenylalanine
PKU	Phenylketonuria
RSD	Relative standard deviation
SD	Standard deviation
Ser	Serine
TCA	Trichloroacetic acid
Thr	Threonine
Tris-HCl	Tris(hydroxymethyl)aminomethane hydrochloride

References

1. Córdova-Dávalos, L.E.; Jiménez, M.; Salinas, E. Glycomacropeptide Bioactivity and Health: A Review Highlighting Action Mechanisms and Signaling Pathways. *Nutrients* **2019**, *11*, 598. [CrossRef]
2. O'Riordan, N.; Kane, M.; Joshi, L.; Hickey, R.M. Structural and Functional Characteristics of Bovine Milk Protein Glycosylation. *Glycobiology* **2014**, *24*, 220–236. [CrossRef]
3. Lim, K.; van Calcar, S.C.; Nelson, K.L.; Gleason, S.T.; Ney, D.M. Acceptable Low-Phenylalanine Foods and Beverages Can Be Made with Glycomacropeptide from Cheese Whey for Individuals with PKU. *Mol. Genet. Metab.* **2007**, *92*, 176–178. [CrossRef]
4. Sirtori, L.R.; Dutra-Filho, C.S.; Fitarelli, D.; Sitta, A.; Haeser, A.; Barschak, A.G.; Wajner, M.; Coelho, D.M.; Llesuy, S.; Belló-Klein, A.; et al. Oxidative Stress in Patients with Phenylketonuria. *Biochim. Biophys. Acta* **2005**, *1740*, 68–73. [CrossRef]
5. Hendriksz, C.J.; Walter, J.H. Update on Phenylketonuria. *Curr. Paediatr.* **2004**, *14*, 400–406. [CrossRef]
6. Proserpio, C.; Pagliarini, E.; Zuvadelli, J.; Paci, S.; Re Dionigi, A.; Banderali, G.; Cattaneo, C.; Verduci, E. Exploring Drivers of Liking of Low-Phenylalanine Products in Subjects with Phenyilketonuria Using Check-All-That-Apply Method. *Nutrients* **2018**, *10*, 1179. [CrossRef]
7. Morrin, S.T.; Buck, R.H.; Farrow, M.; Hickey, R.M. Milk-Derived Anti-Infectives and Their Potential to Combat Bacterial and Viral Infection. *J. Funct. Foods* **2021**, *81*, 104442. [CrossRef]
8. Fukudome, H.; Yamaguchi, T.; Higuchi, J.; Ogawa, A.; Taguchi, Y.; Li, J.; Kabuki, T.; Ito, K.; Sakai, F. Large-Scale Preparation and Glycan Characterization of Sialylglycopeptide from Bovine Milk Glycomacropeptide and Its Bifidogenic Properties. *J. Dairy. Sci.* **2021**, *104*, 1433–1444. [CrossRef]
9. Idota, T.; Kawakami, H.; Nakajima, I. Growth-Promoting Effects of *N*-Acetylneuraminic Acid-Containing Substances on Bifidobacteria. *Biosci. Biotechnol. Biochem.* **1994**, *58*, 1720–1722. [CrossRef]
10. Nakajima, K.; Tamura, N.; Kobayashi-Hattori, K.; Yoshida, T.; Hara-Kudo, Y.; Ikedo, M.; Sugita-Konishi, Y.; Hattori, M. Prevention of Intestinal Infection by Glycomacropeptide. *Biosci. Biotechnol. Biochem.* **2005**, *69*, 2294–2301. [CrossRef]
11. Kawasaki, Y.; Isoda, H.; Tanimoto, M.; Dosako, S.; Idota, T.; Ahiko, K. Inhibition by Lactoferrin and κ-Casein Glycomacropeptide of Binding of Cholera Toxin to Its Receptor. *Biosci. Biotechnol. Biochem.* **1992**, *56*, 195–198. [CrossRef]
12. Kawasaki, Y.; Isoda, H.; Shinmoto, H.; Tanimoto, M. Inhibition by K-Casein Glycomacropeptide and Lactoferrin of Influenza Virus Hemagglutination. *Biosci. Biotechnol. Biochem.* **1993**, *57*, 1214–1215. [CrossRef] [PubMed]
13. Neelima; Sharma, R.; Singh Rajput, Y.; Mann, B. Chemical and Functional Properties of Glycomacropeptide (GMP) and Its Role in the Detection of Cheese Whey Adulteration in Milk: A Review. *Dairy. Sci. Technol.* **2013**, *93*, 21–43. [CrossRef]
14. Kreuß, M.; Strixner, T.; Kulozik, U. The Effect of Glycosylation on the Interfacial Properties of Bovine Caseinomacropeptide. *Food Hydrocoll.* **2009**, *23*, 1818–1826. [CrossRef]
15. González-Morelo, K.J.; Vega-Sagardía, M.; Garrido, D. Molecular Insights Into *O*-Linked Glycan Utilization by Gut Microbes. *Front. Microbiol.* **2020**, *11*, 591568. [CrossRef]
16. Saito, T.; Itoh, T. Variations and Distributions of O-Glycosidically Linked Sugar Chains in Bovine K-Casein. *J. Dairy. Sci.* **1992**, *75*, 1768–1774. [CrossRef]
17. Huang, L.J.; Lin, J.H.; Tsai, J.H.; Chu, Y.Y.; Chen, Y.W.; Chen, S.L.; Chen, S.H. Identification of Protein O-Glycosylation Site and Corresponding Glycans Using Liquid Chromatography–Tandem Mass Spectrometry via Mapping Accurate Mass and Retention Time Shift. *J. Chromatogr. A* **2014**, *1371*, 136–145. [CrossRef]

18. Minkiewicz, P.; Slangen, C.J.; Lagerwerf, F.M.; Haverkamp, J.; Rollema, H.S.; Visser, S. Reversed-Phase High-Performance Liquid Chromatographic Separation of Bovine κ-Casein Macropeptide and Characterization of Isolated Fractions. *J. Chromatogr. A* **1996**, *743*, 123–135. [CrossRef] [PubMed]
19. Thomä-Worringer, C.; Sørensen, J.; López-Fandiño, R. Health Effects and Technological Features of Caseinomacropeptide. *Int. Dairy. J.* **2006**, *16*, 1324–1333. [CrossRef]
20. Qu, Y.; Kim, B.J.; Koh, J.; Dallas, D.C. Analysis of Bovine Kappa-Casein Glycomacropeptide by Liquid Chromatography–Tandem Mass Spectrometry. *Foods* **2021**, *10*, 2028. [CrossRef]
21. Poulsen, N.; Jensen, H.; Larsen, L. Factors Influencing Degree of Glycosylation and Phosphorylation of Caseins in Individual Cow Milk Samples. *J. Dairy. Sci.* **2016**, *99*, 3325–3333. [CrossRef]
22. Taylor, C.M.; Woonton, B.W. Quantity and Carbohydrate Content of Glycomacropeptide Fractions Isolated from Raw and Heat-Treated Milk. *Int. Dairy. J.* **2009**, *19*, 709–714. [CrossRef]
23. Hubl, U.; Sun, C.Q. Comparison of Conditions for the Hydrolytic Release of *N*-Acetylneuraminic Acid from Bovine Glycomacropeptide. *Food New Zealand* **2022**, *22*, 22–29.
24. Ghosh, S. Sialic Acid and Biology of Life: An Introduction. In *Sialic Acids and Sialoglycoconjugates in the Biology of Life, Health and Disease*; Elsevier: Amsterdam, The Netherlands, 2020; pp. 1–61. ISBN 9780128161265.
25. Karamanos, N.K.; Wikstrom, B.; Antonopoulos, C.A.; Hjerpe, A. Determination of *N*-Acetyl- and *N*-Glycolylneuraminic Acids in Glycoconjugates by Reversed-Phase High-Performance Liquid Chromatography with Ultraviolet Detection. *J. Chromatogr.* **1990**, *503*, 421–429. [CrossRef]
26. Cheeseman, J.; Kuhnle, G.; Spencer, D.I.R.; Osborn, H.M.I. Assays for the Identification and Quantification of Sialic Acids: Challenges, Opportunities and Future Perspectives. *Bioorg. Med. Chem.* **2021**, *30*, 115882. [CrossRef] [PubMed]
27. Spichtig, V.; Michaud, J.; Austin, S. Determination of Sialic Acids in Milks and Milk-Based Products. *Anal. Biochem.* **2010**, *405*, 28–40. [CrossRef]
28. Schauer, R. Characterization of Sialic Acids. *Methods Enzym.* **1978**, *50*, 64–89. [CrossRef]
29. Hara, S.; Takemori, Y.; Yamaguchi, M.; Nakamura, M.; Ohkura, Y. Fluorometric High-Performance Liquid Chromatography of N-Acetyl- and N-Glycolylneuraminic Acids and Its Application to Their Microdetermination in Human and Animal Sera, Glycoproteins, and Glycolipids. *Anal. Biochem.* **1987**, *164*, 138–145. [CrossRef]
30. Varki, A.; Diaz, S. The Release and Purification of Sialic Acids from Glycoconjugates: Methods to Minimize the Loss and Migration of O-Acetyl Groups. *Anal. Biochem.* **1984**, *137*, 236–247. [CrossRef]
31. Chen, X.; Varki, A. Advances in the Biology and Chemistry of Sialic Acids. *ACS Chem. Biol.* **2010**, *5*, 163–176. [CrossRef] [PubMed]
32. Hammond, K.S.; Papermaster, D.S. Fluorometric Assay of Sialic Acid in the Picomole Range: A Modification of the Thiobarbituric Acid Assay. *Anal. Biochem.* **1976**, *74*, 292–297. [CrossRef]
33. Shukla, A.K.; Schauer, R. Analysis Of N,O-Acylated Neuraminic Acids By High-Performance Liquid Anion-Exchange Chromatography. *J. Chromatogr.* **1982**, *244*, 81–89. [CrossRef]
34. Warren, L. The Thiobarbituric Acid Assay of Sialic Acids*. *J. Biol. Chem.* **1959**, *234*, 1971–1975. [CrossRef]
35. Sigma-Aldrich. Sialic Acid Assay Kit. Available online: https://www.sigmaaldrich.com/US/en/product/sigma/mak314 (accessed on 17 March 2024).
36. Sigma-Aldrich. Sialic Acid Quantitation Kit Technical Bulletin. 2019. Available online: https://www.sigmaaldrich.com/deepweb/assets/sigmaaldrich/product/documents/228/218/sialicqbul.pdf (accessed on 2 November 2025).
37. Sigma-Aldrich. Sialic Acid Quantitation Kit Sufficient for 25 Reactions. Available online: https://www.sigmaaldrich.com/US/en/product/sigma/sialicq (accessed on 18 March 2024).
38. Sigma-Aldrich. Sialic Acid Assay Kit Technical Bulletin. 2017. Available online: https://www.sigmaaldrich.com/deepweb/assets/sigmaaldrich/product/documents/185/226/mak314bul.pdf (accessed on 2 November 2025).
39. Demir, I.; Bulduk, I.; Darwish, I.A.; Enginar, H. A Green Approach for Metoclopramide Quantification in Pharmaceutical Products: New HPLC and Spectrophotometric Methods. *Sci. Rep.* **2024**, *14*, 8765. [CrossRef] [PubMed]
40. Bilskey, S.R.; Olendorff, S.A.; Chmielewska, K.; Tucker, K.R. A Comparative Analysis of Methods for Quantitation of Sugars during the Corn-to-Ethanol Fermentation Process. *SLAS Technol.* **2020**, *25*, 494–504. [CrossRef] [PubMed]
41. Schauer, R.; Kamerling, J.P. Chemistry, Biochemistry and Biology of Sialic Acids. In *Glycoproteins II*; Montreuil, J., Vliegenthart, J.F.G., Schachter, H., Eds.; Elsevier: Amsterdam, The Netherlands, 1997; pp. 243–402.
42. Skoza, L.; Mohos, S. Stable Thiobarbituric Acid Chromophore with Dimethyl Sulphoxide. *Biochem. J.* **1976**, *159*, 457–462. [CrossRef]
43. Fernando, S.F.; Woonton, B.W. Quantitation of *N*-Acetylneuraminic (Sialic) Acid in Bovine Glycomacropeptide (GMP). *J. Food Compos. Anal.* **2010**, *23*, 359–366. [CrossRef]

foods

MDPI

Article

Variety Identification of Corn Seeds Based on Hyperspectral Imaging and Residual Mamba 1D CNN

Guoqing Qi [1], Chengzhong Liu [1,*], Junying Han [1], Yuqian Zhou [2], Yongsheng Li [2] and Yalong Wang [1]

[1] College of Information Sciences and Technology, Gansu Agricultural University, Lanzhou 730070, China
[2] Crop Research Institute of Gansu Academy of Agricultural Sciences, Lanzhou 730070, China
* Correspondence: liucz@gsau.edu.cn; Tel.: +86-138-9321-8815

Abstract

Maize is a globally important crop, and reliable identification of seed varieties is vital for breeding and quality assurance. To overcome the limitations of conventional methods, this study developed a non-destructive approach integrating hyperspectral imaging (HSI) and deep learning. A Residual Mamba One-Dimensional Convolutional Neural Network (RM1DNet) is proposed, which integrates residual and Mamba modules to enhance feature learning. RM1DNet achieved 94.85% accuracy in classifying 20 maize seed varieties, outperforming traditional classifiers and baseline deep models. These results demonstrate the robustness and efficiency of RM1DNet, highlighting its potential to advance intelligent seed variety identification using hyperspectral data.

Keywords: seed classification; non-destructive testing; spectral analysis; residual neural networks; crop phenotyping

1. Introduction

Maize (*Zea mays* L.) is one of the world's major staple crops, widely cultivated for food, feed, and industrial applications. Accurate variety identification is essential for breeding, cultivation optimization, and quality control [1]. However, conventional approaches such as morphological inspection and molecular marker analysis are often time-consuming, labor-intensive, and costly, making them unsuitable for the modern agricultural demand for rapid, non-destructive, and high-throughput identification [2–4].

Hyperspectral imaging (HSI) has emerged as a promising alternative, offering the capability to capture both spatial and spectral information across continuous wavelength bands [5,6]. Its rich phenotypic representation makes it particularly suitable for seed variety identification [7]. Recent studies have confirmed the effectiveness of HSI when combined with machine learning and deep learning techniques [8–11]. For example, Huang et al. applied an LS-SVM model to classify maize seeds, achieving 94.4% accuracy [12], while Fu et al. developed an SSAE-CS-SVM approach with 95.81% accuracy [13]. More recently, Zhang et al. employed a five-layer CNN with mapped hyperspectral features, reaching 96.65% accuracy [14], and Li et al. introduced the ERNet model, which achieved 98.36% accuracy [6]. These findings highlight the rapid advancement of HSI-based methods in crop variety recognition.

Despite these advances, most existing approaches remain dependent on spectral band selection followed by classifier construction. Although techniques such as Successive Projection Algorithm (SPA) and Competitive Adaptive Reweighted Sampling (CARS)

improve computational efficiency, they often discard valuable spectral information, resulting in incomplete feature representation. Furthermore, traditional classifiers and even many deep models struggle to capture the complex nonlinear relationships and long-range dependencies inherent in hyperspectral data, thereby limiting their robustness and generalization ability.

To address these limitations, we propose Residual Mamba One-Dimensional Convolutional Neural Network (RM1DNet), an improved framework that integrates residual modules for deep local spectral feature extraction with Mamba modules for efficient global sequence modeling. The main contributions of this study are as follows: (1) designing a novel 1D CNN architecture that combines residual and Mamba mechanisms; (2) systematically comparing RM1DNet with traditional classifiers and baseline deep models on 20 maize varieties; and (3) evaluating the influence of SPA and CARS feature selection on recognition accuracy. In addition, future work will focus on validating the model under different sampling and environmental conditions to further assess its robustness and practical applicability. The results demonstrate that RM1DNet effectively leverages both local and global spectral information, achieving superior accuracy and robustness for maize variety identification based on hyperspectral imaging.

2. Materials and Methods

2.1. Experimental Materials

All maize seed samples used in this study were provided by the Gansu Academy of Agricultural Sciences. Twenty widely cultivated maize varieties from Northwest China, including R1831, Longsun 632, Longsun 655 (backcross), Sianyu 1483, Ripu 909, and DF899, were selected and labeled sequentially from 0 to 19. All samples were sealed in kraft paper bags and stored in a cool, dry environment to minimize moisture absorption and prevent deterioration. For each variety, 150 intact and undamaged seeds were selected, resulting in a total of 3000 samples for experimentation. The detailed list of maize varieties is provided in Table 1.

Table 1. Maize varieties.

No.	Name	No.	Name
0	R1831	10	Xianyu335
1	Longdan632	11	Longdan636
2	Longdan633	12	Longdan2463
3	Longdan635	13	Longdan24159
4	Longdan655	14	Longdan634
5	Longdan655fanjiao	15	Ruipu909
6	Xianyu1483	16	DF899
7	Longdan656	17	Xianyu698
8	Longdan657	18	Xianyu1620
9	Longdan659	19	Xianyu1516

2.2. Hyperspectral Image Acquisition

2.2.1. Hyperspectral Imaging Systems

A GaiaField-V10E portable hyperspectral imaging system (Sichuan Dualix Spectral Imaging Technology Co., Ltd., Chengdu, China) was used in this study (Figure 1). The camera covered a spectral range of 380–1018 nm with 320 bands at a spectral resolution of 2.8 nm. The imaging system was equipped with a standardized light source and a dark box to ensure stable acquisition. The push-broom scanning mode was adopted to generate hyperspectral cubes, while system control and calibration were performed using SpecView-50 software. Detailed specifications are provided in the Supplementary Material.

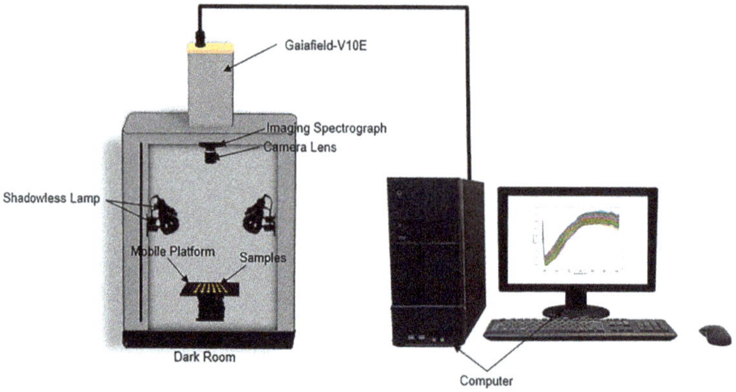

Figure 1. Hyperspectral image acquisition system.

2.2.2. Image Acquisition and Calibration

Prior to image acquisition, the hyperspectral system was preheated for 30 min to minimize baseline drift and then calibrated using SpecView software. Images were captured using the GaiaField-V10E camera under fixed parameters (exposure time: 49 ms; gain: 2; frame rate: 18 Hz; scanning speed: 0.0064 cm/s). For each of the 20 maize varieties, five independent acquisitions were performed, with 30 randomly selected seeds (embryo side up) scanned per acquisition, resulting in 150 seeds per variety and a total of 3000 samples (100 images) [15]. After image acquisition, the raw hyperspectral images were corrected using a black-and-white reference to eliminate dark current noise introduced by the camera. The black-and-white correction equation is shown in Equation (1):

$$I_c = \frac{I_{raw} - I_{dark}}{I_{white} - I_{dark}} \quad (1)$$

Here $Iraw$ represents the raw image, $Iwhite$ represents the white reference image, $Idark$ represents the all-black calibrated image, and Ic represents the corrected image.

2.3. Spectral Extraction and Preprocessing

After black-and-white correction of the original hyperspectral images, the largest possible rectangular region of interest (ROI) was selected on the germinal side of each maize seed using ENVI 5.6 software to minimize the influence of chemical composition heterogeneity. The spectral extraction process is illustrated in Figure 2. The spectral reflectance values of all pixels within the ROI were averaged to obtain the mean spectrum for each seed sample [16].

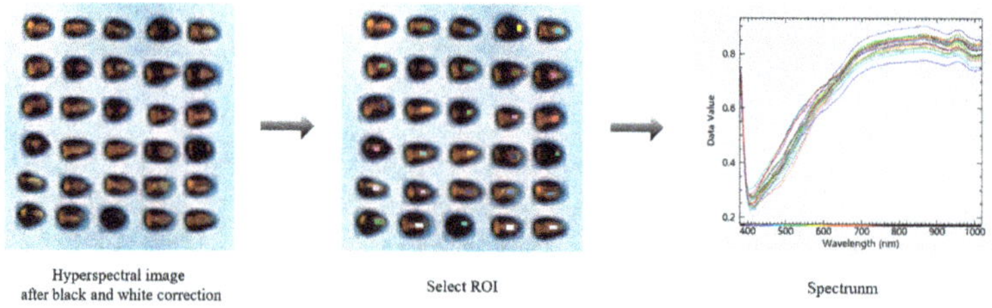

Figure 2. Spectral extraction process.

During spectral acquisition, external interferences such as background noise can compromise data reliability and adversely affect subsequent spectral analysis and model performance. To improve the accuracy and robustness of the model, preprocessing of the raw spectral data is essential to mitigate the impact of these irrelevant factors. In this study, the Savitzky–Golay (SG) smoothing algorithm was employed to preprocess the spectral data. A three-point window was used, including one point on each side of the central point, to form a symmetric kernel.

After spectral feature extraction, all samples were randomly divided into a training set and a test set with a ratio of 8:2 to ensure a fair evaluation of the model's generalization ability. The partitioning was conducted in a stratified manner based on maize variety labels to maintain class balance. During training, the training subset was used for model optimization, while the test subset was reserved for independent performance evaluation.

2.4. Classification Modeling

In this study, a novel convolutional neural network (CNN) architecture, termed RM1DNet, is proposed for the identification of maize varieties. This model incorporates both a residual module and a Mamba module to enhance classification performance. The detailed structure and functional contributions of each component are described below.

2.4.1. ResidualBlock and Mamba Module

In recent years, convolutional neural networks (CNNs) with residual structures have been widely applied in spectral recognition and image processing, as they enable deeper architectures while alleviating gradient vanishing and network degradation [6,17]. Building on this principle, we designed RM1DNet by integrating three residual modules and a Mamba module to achieve efficient local feature extraction and global dependency modeling (Figure 3).

During convolutional encoding, Residual Blocks are employed to improve feature extraction efficiency and ensure stable gradient flow [18]. Each block consists of two 1D convolution layers, ReLU activation, layer normalization, and a skip connection. The skip connection allows cross-layer feature transfer, while a 1×1 convolution is applied if channel dimensions differ. This design accelerates convergence and enhances representational capacity (Figure 3a).

To capture long-range dependencies at a lower computational cost than Transformer-based attention mechanisms, the Mamba module was introduced. Based on a State Space Model (SSM), it incorporates depthwise convolution for local feature extraction, an SSM unit for global modeling, and a gating mechanism to filter informative features selectively. The final output is combined with the input through a residual connection, improving training stability and representational power (Figure 3b).

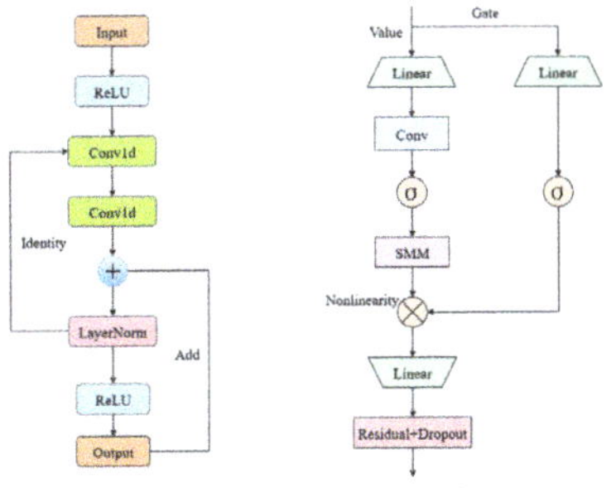

(**a**) (**b**)

Figure 3. Module structure diagram: (**a**) ResidualBlock, (**b**) Mamba module.

2.4.2. RM1DNet

RM1DNet is a novel hybrid architecture that integrates convolutional feature extraction with state space sequence modeling, aiming to fully exploit both spatial patterns and temporal dependencies within hyperspectral data to enhance the modeling capacity and generalization performance in spectral classification tasks [19]. The model architecture comprises a one-dimensional convolutional neural network (1DCNN), residual modules, Mamba modules, a positional encoder, and a fully connected classification head, as shown in Figure 4.

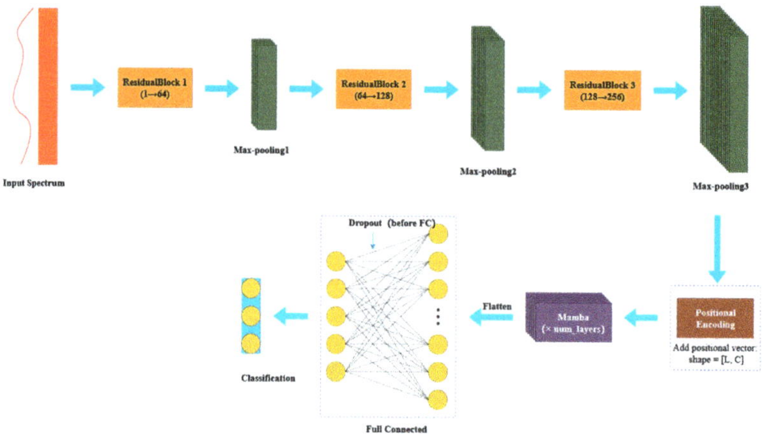

Figure 4. RM1DNet structure diagram.

The front end consists of three Residual Blocks, each containing two 1D convolutional layers and a skip connection, followed by MaxPooling layers for dimensionality reduction. The output channels of the convolutional layers are set to 64, 128, and 256, respectively, facilitating hierarchical feature extraction while mitigating gradient degradation in deep networks. After convolution, the feature maps are reshaped into a [L, B, C] format and augmented with learnable positional encodings before being input to the stacked Mamba modules. Leveraging state space models (SSMs), the Mamba modules efficiently capture long-range dependencies with linear computational complexity, offering improved scalability over conventional Transformer-based designs. The extracted contextual features are subsequently flattened, regularized via dropout, and mapped to class probabilities through a fully connected layer. ReLU activation is applied throughout the network to accelerate convergence and alleviate vanishing gradient issues.

Given the multi-class nature of maize variety classification, this study employs the Label Smoothing (LS) cross-entropy loss function to evaluate model performance. The specific formulation is presented in Equation (2):

$$\text{Loss} = -\sum_{i=1}^{C}\left[(1-\varepsilon)\cdot y_i + \frac{\varepsilon}{C}\right]\cdot \log(\hat{y}_i) \tag{2}$$

where C is the number of categories; $\varepsilon \in [0,1]$ is the smoothing factor (usually taken as 0.1); y_i is the true label of the i th category; and $\hat{y}_i$ is the predicted probability of the i th category as output by the model. The smoothed label distribution retains high confidence in the correct class while allocating small uniform weights to the remaining classes, thereby mitigating overfitting and improving generalization.

The model was trained using the Adam optimizer to minimize the loss and maximize classification accuracy. Training was further stabilized and accelerated through a ReduceLROnPlateau scheduler, which reduced the learning rate by half if validation accuracy did not improve over five consecutive epochs, and by applying an early stopping strategy that terminated training if no improvement was observed within ten consecutive epochs.

2.5. Baseline Models for Comparison

To rigorously evaluate the performance of RM1DNet, six classical models were selected as baselines, covering both traditional machine learning and deep learning approaches.

Support Vector Machine (SVM): A supervised learning method that constructs optimal hyperplanes in high-dimensional space to maximize inter-class margins, widely used in hyperspectral image classification due to its robustness in small-sample and high-dimensional scenarios.

Extreme Learning Machine (ELM): A feedforward neural network with a single hidden layer, characterized by fast training speed and simple structure, making it suitable for hyperspectral classification [20].

Backpropagation Neural Network (BP): A multi-layer feedforward neural network trained via error backpropagation, providing stronger feature representation than ELM [21].

Long Short-Term Memory (LSTM): A recurrent neural network specialized for modeling sequential data and capturing long-term dependencies in spectral sequences [22].

1D Convolutional Neural Network (1DCNN): Applies convolutional filters along one-dimensional spectral inputs to extract local features, showing strong capability in hyperspectral classification [23].

Residual 1DCNN (Res1DCNN): Enhances 1DCNN with residual connections to mitigate gradient degradation and improve deep spectral feature learning.

Mamba1DCNN: A hybrid variant that combines 1DCNN with Mamba modules to capture both local spectral patterns and global dependencies.

2.6. Feature Selection Algorithms

In hyperspectral modeling tasks, feature selection plays a crucial role in reducing dimensionality, eliminating redundancy, and improving the interpretability and stability of classification models. While traditional machine learning methods such as SVM, ELM, and BP generally rely on feature selection to enhance accuracy, its impact on deep learning models remains insufficiently explored [24,25]. To systematically investigate this aspect, two representative algorithms—Successive Projection Algorithm (SPA) and Competitive Adaptive Reweighted Sampling (CARS)—were employed in this study.

SPA is a forward-selection-based feature extraction algorithm that incrementally constructs a subset of variables by minimizing linear correlation, thus reducing redundancy [26,27]. The feature selection process using SPA is illustrated in Figure 5. As shown in Figure 5a, the root mean square error (RMSE) of the model decreases sharply with the increasing number of selected variables, with an inflection point observed when four variables are selected. Beyond this point, the reduction in RMSE slows, suggesting diminishing returns. Therefore, a final subset of four wavelengths is chosen to balance model accuracy and simplicity. Figure 5b visualizes the distribution of selected wavelengths along the mean reflectance spectrum.

The CARS algorithm integrates Monte Carlo sampling with partial least squares (PLS) regression to dynamically select spectral bands that most significantly affect model performance [27,28]. The process of CARS feature selection is depicted in Figure 6. As seen in Figure 6a, the number of retained wavelengths drops rapidly as the iteration progresses, representing a shift from coarse to fine selection. Figure 6b displays the cross-validation root mean square error (RMSECV) for each sampling iteration, with the lowest RMSECV occurring at iteration 33, indicating the optimal variable subset. Figure 6c presents the corresponding PLS regression coefficients, while Figure 6d shows the location of the final selected bands on the average spectral reflectance curve, thereby illustrating their spectral response characteristics.

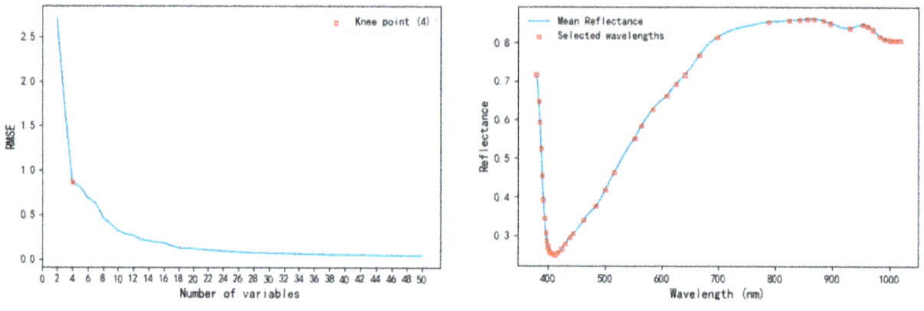

(a) (b)

Figure 5. SPA extraction of characteristic wavelengths. (**a**) Number of variables. (**b**) Position of variables.

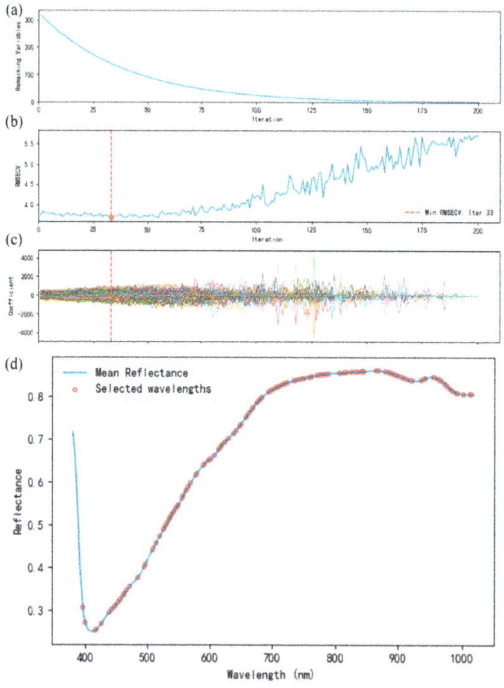

Figure 6. The process of extracting characteristic wavelengths by CARS. (**a**) Preferred number of characteristic wavelength variables. (**b**) Root-mean-square error of cross-validated variances. (**c**) Path diagram of regression coefficients. (**d**) Location of variables.

Furthermore, the discriminative wavelengths selected by SPA and CARS correspond to specific absorption features of key biochemical constituents in maize seeds. For instance, SPA selected characteristic wavelengths around 450 nm, 550 nm, 680 nm, and 970 nm. The 450 nm region corresponds to carotenoids and flavonoids associated with seed pigments; the 550–680 nm range is related to protein and chlorophyll absorption; and the near-infrared region around 950–1000 nm reflects C–H and O–H combination bands linked to starch and lipid molecular vibrations. These findings suggest that the selected spectral bands reflect intrinsic biochemical variations—particularly in pigments, proteins, and carbohydrate composition—among different maize varieties. Even without direct biochemical quantification, these associations provide a plausible physiological basis for the model's discriminative capability [12,14,18,27].

2.7. Software Tools

The operating system used in this experiment was Windows 11 64-bit Professional. The CPU was a 13th generation Intel® Core™ i7-13700@2.10 GHz, the GPU was an NVIDIA GeForce RTX 4090 equipped with 24 GB of video memory, and the system memory was 64 GB. Hyperspectral image preprocessing and spectral extraction were performed using ENVI 5.6, where mean spectra from the embryo face were obtained as regions of interest. Model implementation, including RM1DNet, feature selection algorithms (SPA and CARS), traditional classifiers (SVM, ELM, and BP), and the LSTM model, was carried out in PyCharm 2023.3.

3. Results

3.1. Spectral Characterization

Figure 7a presents the raw spectral curves of all samples from the 20 maize varieties. Figure 7b displays the spectral curves after SG smoothing, which effectively eliminated most of the noise in the original spectral data. Figure 7c illustrates the average reflectance spectra of the 20 maize varieties across the 380–1018 nm wavelength range. It can be observed that the overall shape of the spectral curves was largely consistent among all varieties, showing a general trend of increasing reflectance from the visible to the near-infrared (NIR) region. This spectral pattern reflects the characteristic absorption and reflection behavior of maize seeds at different wavelengths. In the 380–550 nm range, reflectance values are relatively low for all varieties, indicating higher light absorption in this region. Between 550 and 750 nm, the reflectance increases rapidly, then stabilizes within the 750–900 nm range, forming a distinct plateau. This plateau region exhibited slight inter-varietal differences, with some varieties showing higher reflectance values while others were slightly lower. Moreover, noticeable variations in reflectance distributions were observed in the 600–750 nm and 900–1018 nm wavelength intervals. These differences suggest that although the overall spectral profiles were similar, specific spectral bands contained discriminative information that differentiated maize varieties.

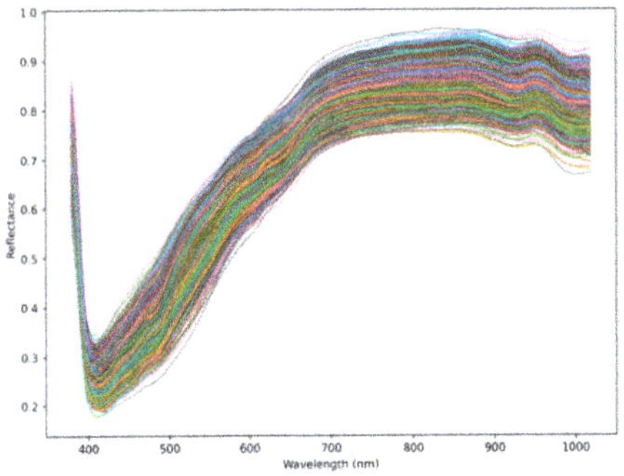

(a)

Figure 7. *Cont.*

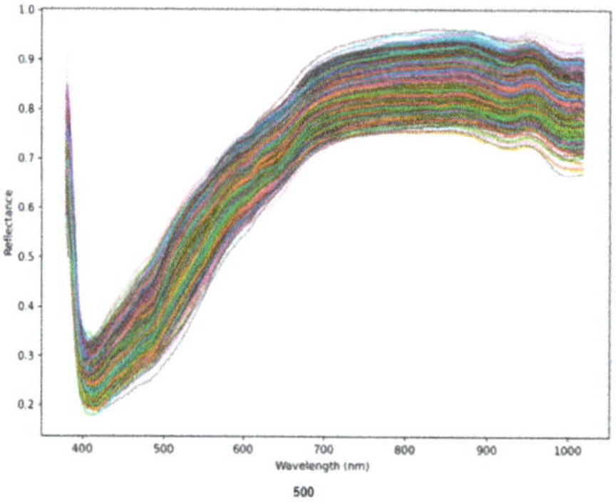

(**b**)

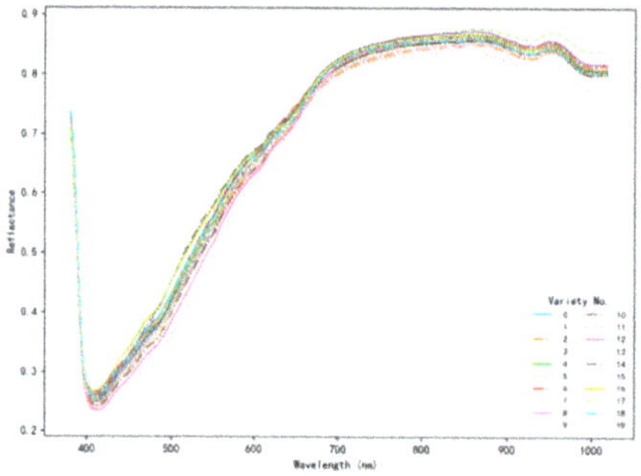

(**c**)

Figure 7. Spectral data: (**a**) all raw spectra, (**b**) all NIR spectra pre-processed by SG, (**c**) average spectra of 20 maize varieties.

3.2. Exploratory Analysis of Multi-Species Clustering

Before establishing the classification model, an exploratory clustering analysis was conducted on the hyperspectral data to evaluate the intrinsic separability among maize varieties. The original high-dimensional spectral signals were projected into a three-dimensional space using two dimensionality-reduction techniques—Principal Component Analysis (PCA) and Linear Discriminant Analysis (LDA)—to enhance the effectiveness of cluster visualization.

Following PCA, the first three principal components accounted for approximately 40.41% of the total variance, effectively preserving the major variance trends in the original spectral data. In contrast, LDA, as a supervised dimensionality-reduction method, was designed to maximize inter-class variance while minimizing intra-class variance by incorporating category labels. As a result, it provides superior enhancement of spectral separability.

Figure 8 illustrates the clustering results after dimensionality reduction. As shown in Figure 8a, the PCA-based 3D scatter plot reveals substantial overlap among data points, with blurred cluster boundaries and ambiguous separations. This indicates that the original spectral feature space contains overlapping distributions across some maize varieties, making it difficult to distinguish them using unsupervised approaches. Conversely, Figure 8b presents the LDA-based clustering visualization, which exhibits clearly separated and well-defined clusters under the guidance of class labels. The enhanced class aggregation and discrimination confirmed the existence of meaningful spectral distinctions among maize varieties.

These exploratory results provided important theoretical support for the subsequent model development. On one hand, PCA serves as a valuable tool for data preprocessing and visual inspection. On the other hand, the pronounced class separability achieved by LDA underscores the feasibility and necessity of incorporating supervised learning strategies—such as deep learning models or ensemble classifiers—during the modeling process. Therefore, this study proposed a novel hybrid architecture integrating RM1DNet. The model's performance is systematically compared against several representative classification algorithms (e.g., SVM, LSTM) to comprehensively validate its effectiveness and robustness in hyperspectral maize variety identification.

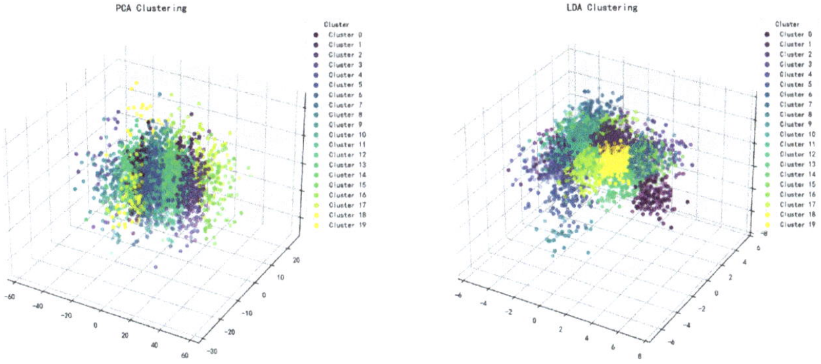

(**a**) (**b**)

Figure 8. PCA (**a**) and LDA (**b**) 3D clustering diagrams.

3.3. Full-Spectrum Based Classification Model Results

This study systematically evaluated the effectiveness of full-spectrum hyperspectral data in maize variety recognition by comparing traditional machine learning methods, recurrent and convolutional neural networks, and the newly developed RM1DNet model. The tested models included SVM, ELM, BP, LSTM, 1DCNN, Res1DCNN, Mamba1DCNN, and RM1DNet, with results summarized in Table 2.

Table 2. Results of full-spectrum based classification of maize varieties.

Models	Average Accuracy (%)	Highest Accuracy (%)	Precision (%)	Recall (%)	F1 (%)
SVM	70.57	72.21	73.10	72.21	72.15
ELM	59.12	60.50	61.20	60.50	60.40
BP	57.21	58.50	58.80	58.50	58.55
LSTM	85.25	86.42	86.70	86.42	86.45
1DCNN	94.14	94.58	94.82	94.58	94.61
Res1DCNN	94.39	94.83	95.04	94.83	94.84
Mamba1DCNN	94.57	94.95	95.12	94.95	94.96
RM1DNet	94.85	95.25	95.40	95.25	95.25

Among traditional machine learning approaches, SVM demonstrated moderate performance (accuracy 70.57%), while ELM and BP showed limited capability in capturing complex spectral features (accuracies 59.12% and 57.21%, respectively). The LSTM model achieved improved recognition accuracy (85.25%) by leveraging sequence modeling; however, its performance was constrained by spectral redundancy and inter-band correlations.

CNN-based methods exhibited more robust classification. The 1DCNN and Res1DCNN achieved accuracies above 94%, highlighting the benefits of convolutional feature extraction and residual connections for mitigating gradient degradation. Incorporating the Mamba module further enhanced sequence modeling, as demonstrated by Mamba1DCNN (accuracy 94.57%).

The proposed RM1DNet, integrating residual and Mamba modules, achieved the highest performance across all metrics, with an accuracy of 94.85%. This indicates that combining local convolutional feature extraction with global sequence modeling effectively captures subtle spectral variations among maize varieties, improving both accuracy and robustness.

Figure 9 presents the confusion matrix and training curve of RM1DNet. The normalized confusion matrix (Figure 9a) indicates that most maize varieties were correctly classified, with classification accuracies exceeding 95%. Minor misclassifications mainly occurred between spectrally similar varieties, such as Longdan656 and Longdan657, which share comparable biochemical compositions. The learning curve (Figure 9b) illustrates a smooth decline in loss and stable convergence, confirming that the model training process was effective and showed no signs of overfitting.

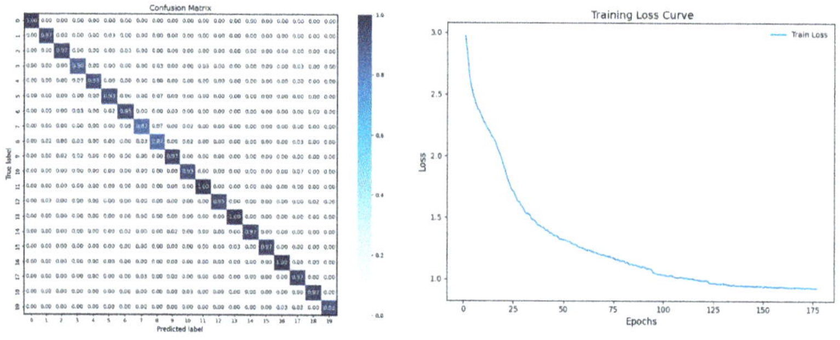

(**a**) (**b**)

Figure 9. Confusion matrix and loss function. The confusion matrix and loss function of RM1DNet are shown separately. Figure (**a**) shows the confusion matrix and Figure (**b**) shows the loss function, which together demonstrate the classification performance of the model.

Table 3 details the per-class performance metrics, including precision, recall, F1-score, true positives (TP), false positives (FP), and false negatives (FN). Most classes achieved precision and recall values above 95%, demonstrating that RM1DNet not only achieved high overall accuracy but also performed robustly across all maize varieties.

Overall, these results demonstrate the potential of RM1DNet for precise and robust maize variety classification using hyperspectral data. Nevertheless, future work should consider external validation datasets to further assess generalization, as well as explore the biological basis of spectral differences to better interpret model predictions and guide practical applications.

Table 3. Classification performance (Precision, Recall, and F1) for 20 maize varieties using the RM1DNet model.

Corn Kernel Varieties	Precision (%)	Recall (%)	F1 (%)	TP	FP	FN
R1831	100.00	100.00	100.00	150	0	0
Longdan632	98.70	97.30	98.00	146	2	4
Longdan633	98.00	98.70	98.30	148	1	2

Table 3. *Cont.*

Corn Kernel Varieties	Precision (%)	Recall (%)	F1 (%)	TP	FP	FN
Longdan635	97.30	96.00	96.60	144	5	6
Longdan655	98.70	97.30	98.00	146	2	4
Longdan655fanjiao	97.30	98.00	97.60	147	1	3
Xianyu1483	98.70	98.00	98.30	147	3	3
Longdan656	91.50	88.70	90.10	133	14	17
Longdan657	89.30	90.70	90.00	136	12	14
Longdan659	97.30	97.30	97.30	146	2	4
Xianyu335	98.00	98.70	98.30	148	1	2
Longdan636	100.00	100.00	100.00	150	0	0
Longdan2463	96.70	98.00	97.30	147	1	3
Longdan24159	100.00	100.00	100.00	150	0	0
Longdan634	98.00	98.70	98.30	148	2	2
Ruipu909	97.30	96.00	96.60	144	5	6
DF899	100.00	100.00	100.00	150	0	0
Xianyu698	97.30	98.00	97.60	147	1	3
Xianyu1620	96.70	97.30	97.00	146	2	4
Xianyu1516	97.30	98.00	97.60	147	1	3

3.4. Classification Model Results Based on Feature Selection

Full-spectrum hyperspectral data often contain substantial redundancy, which can limit the performance of traditional machine learning models that rely heavily on input feature quality. Feature wavelength selection can improve classification accuracy and reduce computational complexity [27]. Compared with conventional algorithms such as SVM, ELM, and BP, 1DCNN exhibits superior automatic feature-extraction capability, thereby reducing the need for manual feature engineering [23].

To evaluate the impact of feature selection, two representative methods—CARS and SPA—were applied to extract informative spectral bands from the original hyperspectral data, yielding 141 and 50 wavelengths, respectively. The classification results based on these selected features are summarized in Table 4. Most models showed decreased or comparable accuracy after feature selection, except for the SPA-LSTM model, which displayed modest improvement. Notably, RM1DNet maintained high performance, achieving test accuracies of 94.14% and 93.27% with CARS- and SPA-selected features, respectively—slightly lower than the 94.85% obtained using the full spectrum.

These results suggest that variations in model performance are influenced by both the intrinsic feature extraction capabilities of each model and the representativeness of the selected wavelengths. RM1DNet consistently demonstrated robust classification accuracy even without feature selection, thanks to its integration of residual convolutional modules with the Mamba state-space mechanism. This architecture enables effective learning of discriminative spectral patterns directly from high-dimensional raw data, reducing the reliance on manual dimensionality reduction.

Overall, RM1DNet provides an efficient and reliable approach for maize variety classification, demonstrating resilience to variations in input feature dimensionality while minimizing the need for explicit feature selection. Future work could further explore the biological basis of spectral differences to enhance model interpretability and practical applicability.

Table 4. Classification results of maize varieties based on feature selection methods.

Models	Methods	Variables	Average Accuracy (%)	Highest Accuracy (%)	Precision (%)	Recall (%)	F1 (%)
SVM	CARS	141	70.14	71.58	71.80	71.58	71.60
	SPA	50	63.53	64.75	64.90	64.75	64.78
ELM	CARS	141	59.54	60.50	60.70	60.50	60.55
	SPA	50	59.26	60.17	60.40	60.17	60.20
BP	CARS	141	51.57	52.67	52.80	52.67	52.70
	SPA	50	48.55	49.33	49.50	49.33	49.35
LSTM	CARS	141	84.86	85.83	86.00	85.83	85.85
	SPA	50	88.27	89.17	89.40	89.17	89.20
1DCNN	CARS	141	91.89	92.42	92.63	92.42	92.45
	SPA	50	91.38	91.92	92.02	91.92	91.90
Res1DCNN	CARS	141	93.25	93.67	93.89	93.67	93.67
	SPA	50	92.73	93.17	93.44	93.17	93.16
Mamba 1DCNN	CARS	141	93.64	94.10	94.30	94.10	94.15
	SPA	50	92.93	93.30	93.50	93.30	93.35
RM1DNet	CARS	141	94.14	94.58	94.79	94.58	94.61
	SPA	50	93.27	93.67	93.98	93.67	93.69

4. Discussion

This study focused on 20 maize varieties (a total of 3000 seeds), utilizing hyperspectral imaging (380–1018 nm) and developing RM1DNet to investigate the classification performance of different models and the impact of feature-selection strategies on identification accuracy.

In terms of spectral preprocessing, the Savitzky–Golay (SG) smoothing algorithm with a three-point window was employed to mitigate background noise and enhance the reliability of the raw spectral data. Comparative analysis of the preprocessed (Figure 7b) and raw spectra (Figure 7a) showed that SG smoothing effectively preserved the intrinsic spectral characteristics of maize seeds—such as the rapid increase in reflectance between 550 and 750 nm and the stable plateau between 750 and 900 nm—while removing random noise interference. This provided a solid foundation for subsequent model training, consistent with the findings of Chivasa et al. [29], who emphasized that appropriate spectral smoothing enhances the discriminative power of hyperspectral data for crop variety classification.

Regarding model performance, the proposed RM1DNet achieved a test accuracy of 94.85%, significantly outperforming traditional machine learning models (SVM: 70.57%; ELM: 59.12%; BP: 57.21%) and other deep learning architectures (LSTM: 85.25%; 1DCNN: 94.14%; Res1DCNN: 94.39%; Mamba1DCNN: 94.57%). This superior performance arises from RM1DNet's integrated design: the residual modules mitigate gradient-vanishing issues in deep networks, enabling efficient extraction of local spectral features (e.g., subtle reflectance differences in the 600–750 nm and 900–1018 nm ranges), while the Mamba state-space module captures long-range spectral dependencies with linear computational complexity—thereby overcoming the limitations of conventional CNNs in modeling global sequence information and Transformers in computational efficiency. The confusion matrix (Figure 9a) further confirmed the robustness of RM1DNet, showing 100% classification accuracy for varieties such as R1831 (No. 0) and DF899 (No. 16), although minor misclassifications occurred between spectrally similar varieties (e.g., Longdan656 (No. 7) and Longdan657 (No. 8)), likely due to overlapping biochemical compositions [12].

Compared with existing studies on maize variety classification, this work presents three key advancements. First, whereas Fu et al. [13] achieved 95.81% accuracy for four maize varieties using the SSAE-CS-SVM model, this study successfully classified 20 varieties—reflecting a broader diversity representative of real agricultural production—while maintaining high accuracy. Second, compared with Li et al. [6], who relied on feature engineering for ERNet (98.36% accuracy), RM1DNet requires no manual feature selection, thereby reducing subjective bias and information loss caused by heuristic filtering [17]. Feature-selection experiments (SPA, CARS) showed that traditional models (e.g., SVM) suffered significant accuracy declines when using selected features (SPA: 63.53% vs. full spectrum: 70.57%), whereas RM1DNet retained 93.27–94.14% accuracy, demonstrating stronger adaptability to high-dimensional data and reduced reliance on preprocessing workflows [30].

This study also has certain limitations. First, the hyperspectral data were collected using a single system, and the model's performance across broader spectral ranges remains to be validated. Second, all samples were collected from Northwest China and stored under uniform conditions; therefore, model generalization to seeds from other regions or under varying storage environments (e.g., high humidity) requires further verification [31]. Only spectral data were used; integrating hyperspectral information with morphological or molecular markers could further reduce the misclassification of spectrally similar varieties [19]. Future studies will address these limitations by extending spectral ranges, diversifying sample sources, and exploring multimodal data fusion.

5. Conclusions

This study established a non-destructive framework for maize variety identification based on hyperspectral imaging (380–1018 nm) and deep learning. Using 20 widely cultivated maize varieties (a total of 3000 seeds) from Northwest China, the framework directly processed full-spectrum data (320 bands), thereby minimizing information loss due to dimensionality reduction and providing a more comprehensive evaluation of spectral diversity.

To overcome the limitations of traditional machine learning methods (e.g., SVM, BP) in handling high-dimensional spectral data, a novel model, RM1DNet, was developed by integrating residual modules with the Mamba state-space mechanism. Experimental results demonstrated that RM1DNet achieved a test accuracy of 94.85%, outperforming all seven baseline models. These findings indicate that combining local convolutional feature extraction with global sequence modeling effectively captures subtle spectral differences among maize varieties.

Feature-selection experiments (SPA and CARS) further confirmed the robustness of RM1DNet, which maintained high accuracy (93.27–94.14%) even when trained on reduced feature subsets. This suggests that the proposed architecture is capable of learning discriminative spectral features without relying heavily on manual wavelength selection.

Future research will extend this work by (1) applying hyperspectral systems with broader spectral ranges to assess model adaptability; (2) incorporating maize samples collected under diverse geographical and storage conditions to improve generalization; and (3) exploring multimodal fusion (hyperspectral + RGB imaging) to address challenges in distinguishing spectrally similar varieties.

Supplementary Materials: The following supporting information can be downloaded at: https://www.mdpi.com/article/10.3390/foods14203558/s1, Table S1: Hyperspectral Imaging System Specifications; Table S2: Image Acquisition Parameters; Figure S3: Example Spectra; Table S4: Model Implementation Details.

Author Contributions: Conceptualization, G.Q. and C.L.; data curation, G.Q. and Y.W.; formal analysis, G.Q.; funding acquisition, C.L.; investigation, C.L.; methodology, G.Q.; project administration, J.H.; resources, C.L., Y.Z. and Y.L.; software, C.L.; supervision, C.L.; validation, C.L., J.H., Y.Z. and Y.L.; visualization, G.Q.; writing—original draft, G.Q.; writing—review and editing, C.L. All authors have read and agreed to the published version of the manuscript.

Funding: Financial support was received for the research, authorship, and publication of this article. National Natural Science Foundation of China (Grant No. 32360437).

Institutional Review Board Statement: Not applicable.

Informed Consent Statement: Not applicable.Please add the corresponding content of this part.

Data Availability Statement: The data presented in this study are available on request from the corresponding author due to the data being collected and obtained by individuals.

Conflicts of Interest: The authors declare no conflicts of interest.

References

1. Feng, R.J.; Chen, Z.G.; Yi, S.J. Identification of Corn Varieties Based on Bayesian Optimization SVM. *Spectrosc. Spectral Anal.* **2022**, *42*, 1698–1703. [CrossRef]
2. Wang, L.G.; Wang, L.F. Variety Identification Model for Maize Seeds Using Hyperspectral Pixel-Level Information Combined with Convolutional Neural Network. *Natl. Remote Sens. Bull.* **2021**, *25*, 2234–2244. [CrossRef]
3. Wang, D.; Wang, K.; Wu, J.; Han, P. Progress in Research on Rapid and Non-Destructive Detection of Seed Quality Based on Spectroscopy and Imaging Technology. *Spectrosc. Spectral Anal.* **2021**, *41*, 52–59. [CrossRef]
4. Huang, J.; Liu, C.; Song, S.; Tan, Y.; Han, Y.; Xu, L.; Liu, J. Research Progress of Spectroscopic Techniques in Quality Detection of Maize Varieties. *Agric. Technol.* **2022**, *42*, 5–8. [CrossRef]
5. Zhang, L.; Liu, C.; Han, J.; Yang, Y. Variety Identification of Corn Seeds Based on Hyperspectral Imaging and Convolutional Neural Network. *Foods* **2025**, *14*, 3052. [CrossRef]
6. Li, X.; Zhai, M.; Zheng, L.; Zhou, L.; Xie, X.; Zhao, W.; Zhang, W. Efficient Residual Network Using Hyperspectral Images for Corn Variety Identification. *Front. Plant Sci.* **2024**, *15*, 1376915. [CrossRef]
7. Ambrose, A.; Kandpal, L.M.; Kim, M.S.; Lee, W.; Cho, B. High-Speed Measurement of Corn Seed Viability Using Hyperspectral Imaging. *Infrared Phys. Technol.* **2016**, *75*, 173–179. [CrossRef]
8. Song, S.; Dong, Z.; Wu, Y. Classification and Identification of Corn Varieties Based on RF Algorithm and Hyperspectral Information of Glutinous Corn Leaves. *Agric. Technol.* **2024**, *44*, 51–55. [CrossRef]
9. Huang, Y.; Yang, Y.; Sun, Y.; Zhou, H.; Chen, K. Identification of Apple Varieties Using a Multichannel Hyperspectral Imaging System. *Sensors* **2020**, *20*, 5120. [CrossRef]
10. Saha, D.; Manickavasagan, A. Machine Learning Techniques for Analysis of Hyperspectral Images to Determine Quality of Food Products: A Review. *Curr. Res. Food Sci.* **2021**, *4*, 28–44. [CrossRef]
11. Ji, Y.; Sun, L.; Li, Y.; Li, J.; Liu, S.; Xie, X.; Xu, Y. Non-Destructive Classification of Defective Potatoes Based on Hyperspectral Imaging and Support Vector Machine. *Infrared Phys. Technol.* **2019**, *99*, 71–79. [CrossRef]
12. Huang, M.; Tang, J.; Yang, B.; Zhu, Q. Classification of Maize Seeds of Different Years Based on Hyperspectral Imaging and Model Updating. *Comput. Electron. Agric.* **2016**, *122*, 139–145. [CrossRef]
13. Fu, L.; Sun, J.; Wang, S.; Xu, M.; Yao, K.; Cao, Y.; Tang, N. Identification of Maize Seed Varieties Based on Stacked Sparse Autoencoder and Near-Infrared Hyperspectral Imaging Technology. *J. Food Process. Eng.* **2022**, *45*, e14120. [CrossRef]
14. Zhang, F.; Zhang, F.; Wang, S.; Li, L.; Lv, Q.; Fu, S.; Zhang, Y. Hyperspectral Imaging Combined with CNN for Maize Variety Identification. *Front. Plant Sci.* **2023**, *14*, 1254548. [CrossRef]
15. Wang, W.; Guo, L. Effects of Storage Temperature on Physiological Changes of Corn Germ and Endosperm. *Cereals Oils* **2019**, *32*, 68–71. [CrossRef]
16. Feng, Z.; Zhu, Q.; Huang, M. Maize Variety Recognition Using Hyperspectral Image. *J. Jiangnan Univ. Nat. Sci. Ed.* **2012**, *11*, 149–153. [CrossRef]
17. Jin, S.; Zhang, F.; Zheng, Y.; Zhou, L.; Zuo, X.; Zhang, Z.; Zhao, W.; Zhang, W.; Pan, X. CSKNN: Cost-Sensitive K-Nearest Neighbor Using Hyperspectral Imaging for Identification of Wheat Varieties. *Comput. Electr. Eng.* **2023**, *111*, 108896. [CrossRef]
18. Huang, H.; Liu, Y.; Zhu, S.; Feng, C.; Zhang, S.; Shi, L.; Sun, T.; Liu, C. Detection of Mechanical Damage in Corn Seeds Using Hyperspectral Imaging and the ResNeSt_E Deep Learning Network. *Agriculture* **2024**, *14*, 1780. [CrossRef]
19. Yang, X.; Hong, H.; You, Z.; Cheng, F. Spectral and Image Integrated Analysis of Hyperspectral Data for Waxy Corn Seed Variety Classification. *Sensors* **2015**, *15*, 15578–15594. [CrossRef]
20. Yang, R.; Kan, J. Classification of Tree Species at the Leaf Level Based on Hyperspectral Imaging Technology. *J. Appl. Spectrosc.* **2020**, *87*, 184–193. [CrossRef]
21. Wang, Y.; Song, S. Variety Identification of Sweet Maize Seeds Based on Hyperspectral Imaging Combined with Deep Learning. *Infrared Phys. Technol.* **2023**, *130*, 104611. [CrossRef]
22. Yin, J.; Qi, C.; Chen, Q.; Qu, J. Spatial-Spectral Network for Hyperspectral Image Classification: A 3-D CNN and Bi-LSTM Framework. *Remote Sens.* **2021**, *13*, 2353. [CrossRef]

23. Hu, H.; Mei, Y.; Wei, Y.; Xu, Z.; Zhao, Y.; Xu, H.; Huang, L. Chemical Composition Prediction in Goji (*Lycium barbarum*) Using Hyperspectral Imaging and Multi-Task 1DCNN with Attention Mechanism. *LWT* **2024**, *204*, 116436. [CrossRef]
24. Bi, C.; Zhang, S.; Chen, H.; Bi, X.; Liu, J.; Xie, H.; Yu, H.; Song, S.; Shi, L. Non-Destructive Classification of Maize Seeds Based on RGB and Hyperspectral Data with Improved Grey Wolf Optimization Algorithms. *Agronomy* **2024**, *14*, 645. [CrossRef]
25. Zhang, F.; Wang, M.; Zhang, F.; Xiong, Y.; Wang, X.; Ali, S.; Zhang, Y.; Fu, S. Hyperspectral Imaging Combined with GA-SVM for Maize Variety Identification. *Food Sci. Nutr.* **2024**, *12*, 3177–3187. [CrossRef]
26. Xuan, G.; Jia, H.; Shao, Y.; Shi, C. Protein Content Prediction of Rice Grains Based on Hyperspectral Imaging. *Spectrochim. Acta A Mol. Biomol. Spectrosc.* **2024**, *320*, 124589. [CrossRef]
27. Zhu, D.; Han, J.; Liu, C.; Zhang, J.; Qi, Y. Vis-NIR and NIR Hyperspectral Imaging Combined with Convolutional Neural Network with Attention Module for Flaxseed Varieties Identification. *J. Food Compos. Anal.* **2025**, *137*, 106880. [CrossRef]
28. Zhu, Q.; Liu, Q.; Ma, D.; Zhu, Y.; Zhang, L.; Wang, A.; Fan, S. Maize Seed Variety Classification Based on Hyperspectral Imaging and a CNN-LSTM Learning Framework. *Agronomy* **2025**, *15*, 1585. [CrossRef]
29. Chivasa, W.; Mutanga, O.; Biradar, C. Phenology-Based Discrimination of Maize (*Zea mays* L.) Varieties Using Multitemporal Hyperspectral Data. *J. Appl. Remote Sens.* **2019**, *13*, 017504. [CrossRef]
30. Sharma, A.; Singh, T.; Garg, N. Nondestructive Identification of Wheat Seed Variety and Geographical Origin Using Near-Infrared Hyperspectral Imagery and Deep Learning. *J. Chemom.* **2024**, *38*, e3585. [CrossRef]
31. Polder, G.; Blok, P.; Hendrik, A.; Jan, M.; Kamp, J. Potato Virus Y Detection in Seed Potatoes Using Deep Learning on Hyperspectral Images. *Front. Plant Sci.* **2019**, *10*, 209. [CrossRef]

analytica

MDPI

Article

Development of a Sensitive HILIC-MS/MS Method for Quantification of Melamine, Derivatives, and Potential Precursors in Various Water Matrices

Merle Käberich *, Lisann Nemetz and Frank Sacher

TZW: DVGW-Technologiezentrum Wasser (German Water Centre), 76139 Karlsruhe, Germany; lisann.nemetz@tzw.de (L.N.); frank.sacher@tzw.de (F.S.)

* Correspondence: merle.kaeberich@tzw.de; Tel.: +49-721-9678-1279

Abstract

Melamine (MEL) has broad applications and can be released to the aquatic environment from various sources, including industry, agriculture, traffic, and household articles. In addition, MEL derivatives ammeline (AMN), ammelide (AMD), and cyanuric acid (CYA) as well as potential precursors cyromazine (CYRO) and hexa(methoxymethyl)melamine (HMMM) are relevant related substances. However, occurrence and transformation in water resources has not yet been thoroughly investigated. Here, we developed a sensitive analytical method for quantification of these analytes by hydrophilic interaction liquid chromatography (HILIC) coupled to tandem mass spectrometry (MS/MS). Direct injection achieved limits of quantification (LOQs) of 0.1 μg/L (AMN 0.2 μg/L; CYA 1 μg/L), while LOQs could be improved to 0.01 μg/L (CYA 0.05 μg/L) by applying evaporation for analyte pre-concentration. The method was extensively validated, showing good recovery, repeatability, and linearity. The evaluation of the matrix effects revealed method applicability for various water matrices, including surface water and wastewater. During proof-of-concept measurements, HMMM in combination with MEL and its derivatives was found in multiple samples, emphasizing the importance of including precursors. In the future, the developed method with its novelty of covering both MEL derivatives and precursors can be applied for comprehensive monitoring programs elucidating MEL sources and transformation in water resources.

Keywords: melamine; hydrophilic interaction liquid chromatography (HILIC); tandem mass spectrometry; water analysis; ammeline; ammelide; cyanuric acid; hexa(methoxymethyl)melamine; cyromazine

1. Introduction

Melamine (MEL; CAS number 108-78-1) and melamine-based resins are used for a broad variety of applications in industry (e.g., manufacture of plastics, laminates, adhesives, surface coatings, tires, and textiles). Therefore, MEL is a high-production-volume chemical. Although acute toxicity of MEL is low [1], it is suspected to be carcinogenic [2]. Melamine is under assessment as endocrine-disrupting by the European Chemicals Agency (ECHA) [2]. Recent publications, however, have stated that the substance does not meet the criteria for endocrine disruption [3,4].

The relevant derivatives of MEL are ammeline (AMN), ammelide (AMD), and cyanuric acid (CYA), which are all synthesis by-products and transformation products of MEL [5,6], formed by hydrolysis under elimination of ammonia (Figure 1). Especially the simultaneous occurrence of MEL and CYA has been shown to elevate substance toxicity substantially by formation of crystal complexes resulting in kidney stones [1,7]. In addition, MEL may be formed by degradation of precursors. MEL was shown to be the main metabolite of cyromazine (CYRO), an insecticide which can be transformed to MEL by dealkylation [8] (Figure 1). Also, hexa(methoxymethyl)melamine (HMMM), used as a cross-linking agent in coatings and vulcanizer in tires, is suspected to degrade to MEL [9,10].

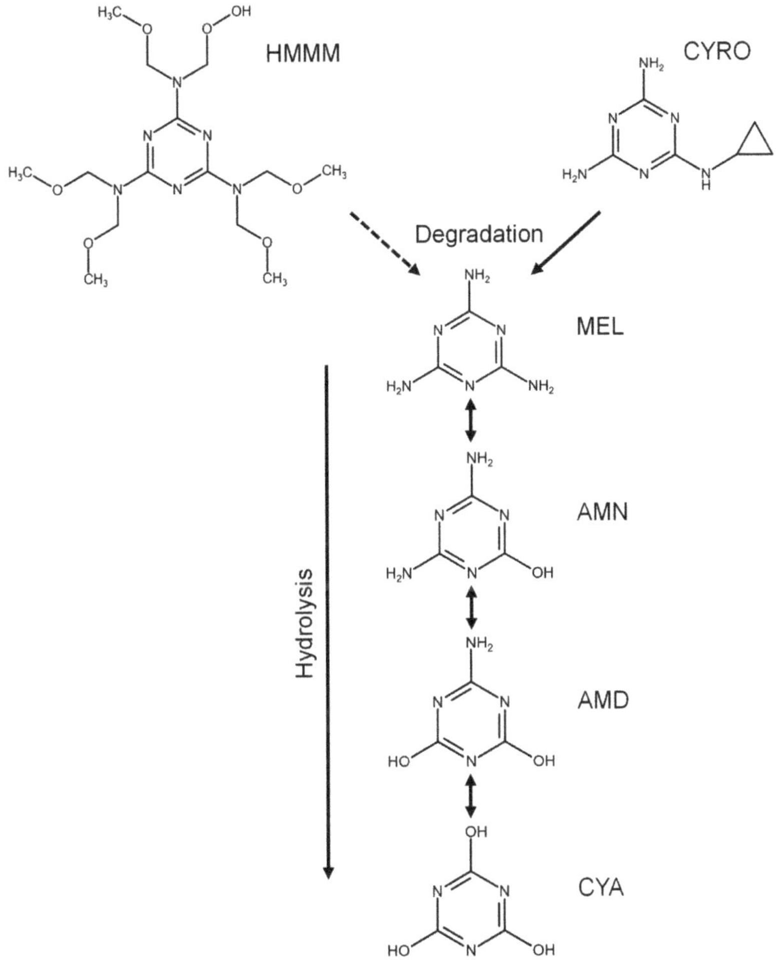

Figure 1. Structures and possible transformations of melamine (MEL), ammeline (AMN), ammelide (AMD), cyanuric acid (CYA), hexa(methoxymethyl)melamine (HMMM), and cyromazine(CYRO).

Melamine has received substantial political and scientific attention since contamination incidents in 2007 and 2008. Milk products and gluten in pet food were found to be illegally contaminated with MEL and its derivatives, resulting in severe renal failure in both humans and animals [11–13]. This led to the development and publication of numerous analytical methods for detection and quantification of MEL and its derivatives in the following years. During a comprehensive literature review, 84 relevant publications describing analytical methods for MEL and at least one of its derivatives were evaluated. Different analytical methods have been applied, mostly based on gas chromatography (GC) or (high-performance) liquid chromatography ((HP)LC). Gas chromatography can be coupled with mass spectrometry (MS) [14,15] or MS/MS detection [16,17], while (HP)LC has been applied in combination with ultraviolet (UV) detection [18], diode array detection (DAD) [19,20], MS [21,22], or MS/MS [23,24].

Although a multitude of applications have been published, the majority is targeting foodstuffs and pet food (43 of 84 relevant publications), with special emphasis on milk and infant formula, e.g., [16,17,22,25,26]. However, MEL and its related substances can also be released to the aquatic environment by discharge and transformation processes which are not yet well understood. Industrial discharges as well as release from the use phase of household products are possible entry pathways [27], defining MEL as an indicator for anthropogenic activity [28]. Monitoring data for MEL occurrence in the aquatic environment is scarce [27] and only a few studies on the analysis and occurrence of MEL and derivatives in water matrices have been published so far [19,23,29–31]. To our knowledge, only Zhu and Kannan [29] measured all four derivatives with limits of quantification (LOQs) below 0.5 μg/L in aquatic matrices. Most published methods either resulted in higher LOQ values or did not include all derivatives. In most publications, solid-phase extraction (SPE) with both cation and anion exchange has been applied for sample preparation [29,30,32]. Moreover, analysis of MEL precursors in water matrices in addition to its transformation products has received very limited scientific attention. Cyromazine has occasionally been included in analytical methods [19,23,30,33]. Hexa(methoxymethyl)melamine has been analyzed in combination with MEL [10,34] but not, however, with its transformation products.

Thus, to gain information about distribution of MEL and its related substances in water resources as well as to elucidate transformation pathways, a sensitive, robust, and efficient quantitative analytical method covering MEL and its derivatives AMN, AMD, and CYA, as well as precursors CYRO and HMMM is needed. Here, we developed a sensitive multi-method for quantification of these six target analytes in various water matrices, including drinking water, surface water, and wastewater. Analysis was conducted by HPLC-MS/MS with chromatographic separation by hydrophilic interaction liquid chromatography (HILIC). The method was validated and showed good recovery, linearity, and repeatability. While already sensitive with direct injection, method sensitivity can be further enhanced by applying evaporation during sample preparation.

2. Materials and Methods

2.1. Reagents and Stock and Standard Solutions

Certified reference materials of analytical standards and stable isotope-labeled internal standards (ISs) were obtained from Dr. Ehrenstorfer (Augsburg, Germany; MEL, CYA, CYRO, and MEL-^{13}C3), Cambridge Isotope Laboratories (Tewksbury, MA, USA; AMN and AMD), and Toronto Research Chemicals Inc. (Toronto, ON, Canada; HMMM, AMN-^{13}C3, AMD-^{13}C3, CYA-^{13}C3, and CYRO-d4). For HMMM, no isotope-labeled IS was commercially available. Ultrapure water from a Sartorius AriumPro system (Goettingen, Germany); methanol from Honeywell (Charlotte, NC, USA); acetonitrile from Supelco (Bellefonte, PA, USA); diethylamine, ammonium acetate, and ammonium formate from Sigma-Aldrich (Saint Louis, MO, USA); acetic acid from Roth (Karlsruhe, Germany); and formic acid from VWR (Darmstadt, Germany) were used for preparation of HPLC eluents and stock and standard solutions (all except for ammonium acetate LC-MS grade; ammonium acetate > 98% purity).

The standards and ISs of MEL, CYA, CYRO, and HMMM were dissolved in methanol. Ammeline and AMD were dissolved in a mixture of ultrapure water:diethylamine (80:20, *v:v*) due to low solubility in methanol. From these stock solutions, standard solutions with a concentration of 1 mg/L were prepared in methanol for all standards and ISs.

These standard solutions were determined to be stable for one year. Combined calibration standards of all the substances were prepared daily in acetonitrile:water (90:10, *v*:*v*). As the water matrix, the local drinking water was used, which was free of the analytes under investigation.

2.2. Instrumentation

Sample evaporation was performed with a vacuum pre-concentrator (RVC 2-33 CD-plus, Martin Christ Gefriertrocknungsanlagen GmbH, Osterode, Germany). LC-MS/MS analysis was implemented on an HPLC Infinity 1290 system (Agilent Technologies, Santa Clara, CA, USA) coupled to an API5500 Triple Quad system with electrospray ionization (AB Sciex, Framingham, MA, USA). For the final method, a Luna HILIC chromatography column (150 × 2 mm, 3 µm; Phenomenex, Torrance, CA, USA) with a HILIC security guard cartridge (4 × 2 mm; Phenomenex) was applied.

2.3. Method Optimization

For sample preparation, both evaporation and SPE were assessed as possible methods for analyte pre-concentration. The SPE protocol was adapted from previous publications [24,29] and varied, using cation exchange (Waters Oasis MCX, 150 mg, 6 mL, 30 µm; Milford, MA, USA) and anion exchange cartridges (Waters Oasis MAX, 150 mg, 6 mL, 30 µm). The details of the tested SPE protocol are given in Table S1.

Aside from the chosen Luna HILIC column, another HILIC column (Sequant ZIC-HILIC, 20 × 2.1 mm, 5 µm; Merck, Darmstadt, Germany) and two mixed-mode columns (Obelisc R, 150 × 2.1 mm, 5 µm; SIELC, Wheeling, IL, USA, and Raptor Polar, 100 × 2.1 mm, 2.7 µm; Restek, Centre County, PA, USA) were tested. However, the mixed-mode columns were excluded from further method development due to insufficient separation of analytes and poor signal intensities. Chromatographic separation of the analytes was also achieved on the ZIC-HILIC column, but peak shapes and sensitivity were slightly better on the chosen Luna HILIC column. Different eluent compositions with an acetonitrile content between 60 and 95 vol% during isocratic elution and a gradient with variation between 100 and 80 vol% were tried out on the HILIC columns to achieve optimum sensitivity and analyte peak shapes. In addition, ammonium acetate, ammonium formate, acetic acid, and formic acid were evaluated as eluent modifiers to increase method sensitivity. The injection volume was varied between 5 and 40 µL to optimize peak shapes.

2.4. Determination of Limits of Detection and Quantification

After optimization, limits of detection (LODs) and LOQs were determined for both the direct injection method and the evaporation method in various matrices, i.e., drinking water, surface water, and wastewater (influent and effluent of wastewater treatment plant (WWTP)), based on signal-to-noise ratios (S/N). The LOD was defined at S/N = 3 and LOQ at S/N = 10.

2.5. Method Validation

The method was validated including both direct injection and analyte pre-concentration by evaporation. The validation protocol included the determination of recovery during analyte pre-concentration, repeatability, calibration linearity, and matrix effects of the method. Recovery of pre-concentration compared to direct injection in matrix drinking water was considered acceptable between 75 and 125% [35]. Calibration linearity was determined for each analyte in a concentration range between 0.5 and 10 × LOQ and the resulting coefficient of determination r^2 was required to exceed 0.99. The matrix effects of surface water and wastewater (influent and effluent of WWTP) were examined and required to remain below a threshold of 25% compared to drinking water as the reference value. Method repeatability was determined by analyzing five replicate samples with a concentration of 5 × LOQ, whereas standard deviation had to remain below 25%.

2.6. Stability Test

Stability of analytes was assessed over 28 days in sample containers consisting of glass, polyethylene (PE), and polystyrene (PS) stored at room temperature (20 °C), refrigerated (4 °C), and frozen (−18 °C). To this end, separate analyte solutions with 100 µg/L were prepared in drinking water, and samples were taken on days 0, 1, 2, 5, 7, 14, 21, and 28 for MEL, AMN, AMD, and CYA. Additionally, samples for CYRO and HMMM were taken on days 0 and 28.

2.7. Proof of Concept

To show the applicability of the developed method and obtain the first results, 10 samples exhibiting different water matrices were analyzed. These included drinking water, groundwater, surface water, road run-off, and wastewater (detailed information on the samples is given in Table S2). Drinking water and groundwater were measured with evaporation as pre-concentration, while all the other samples were measured with direct injection.

3. Results

3.1. Sample Preparation

Both evaporation and SPE proved successful for pre-concentration of the analytes under investigation. During SPE, MEL, AMN, CYRO, and HMMM were pre-concentrated by cation exchange and AMD and CYA by anion exchange (Table S1). However, pre-concentration by SPE was by far more labor-intensive and material-consuming due to the use of two separate cartridges and several conditioning and elution steps. The LOQ values achieved with SPE after optimization (Table S1) were similar to the values after evaporation, thereby not enhancing method sensitivity compared to simple evaporation of the water. Therefore, evaporation was selected as the preferable method for analyte pre-concentration, with an initial sample volume of 10 mL. After evaporation, the dry residue was reconstituted in acetonitrile:ultrapure water (90:10, *v:v*) to match the eluent composition of the chromatographic method. Alternatively, samples can be measured by direct injection if maximizing sensitivity is not crucial. In this case, water samples need to be diluted with acetonitrile to achieve the matrix of acetonitrile:water (90:10, *v:v*; Table 1).

Table 1. Sample preparation steps with direct injection and pre-concentration by evaporation.

	Direct injection
Sample volume	1 mL
Sample pH	Original
Internal standards	MEL-^{13}C3, AMN-^{13}C3, AMD-^{13}C3, CYA-^{13}C3, CYRO-d4, 20 µL of a 0.1 mg/L solution in methanol
Sample dilution	Dilution of 100 µL with 900 µL acetonitrile
	Pre-concentration by evaporation
Sample volume	10 mL
Sample pH	Original
Internal standards	MEL-^{13}C3, AMN-^{13}C3, AMD-^{13}C3, CYA-^{13}C3, CYRO-d4, 5 µL of a 0.1 mg/L solution in methanol

Table 1. *Cont.*

		Start	1	2	3	4	5	6
Pre-concentration program	Time (h:m)		00:03	04:00	00:05	04:00	00:05	04:00
	Temperature (°C)	30	40	40	35	35	30	30
	Vacuum (mbar)		10	10	10	10	10	10
	Rotational speed (min^{-1})		900	900	1200	1200	1200	1200
Reconstitution of the dry residue	In 225 µL acetonitrile + 25 µL ultrapure water							

3.2. *Optimum Chromatographic and MS Conditions*

While multiple chromatographic parameters were optimized, exemplary chromatograms are shown in Figure 2, depicting the impact of isocratic eluent variation on CYA detection. Here, isocratic elution with acetonitrile:ultrapure water (90:10, *v:v*) was determined as the best eluent composition, with addition of 10 mM ammonium acetate and 0.1% formic acid to the water phase. The sensitivity and peak shape improved with increasing acetonitrile content. From acetonitrile:ultrapure water 90:10 (*v:v*; Figure 2C) to 95:5 (*v:v*; Figure 2D), only minimal improvement was achieved, and the higher acetonitrile content would also increase the necessary sample dilution during direct injection to match the eluent composition and sample matrix; therefore, a ratio of 90:10 (*v:v*) was determined as optimum. Optimum peak shapes were achieved with an injection volume of 10 µL. The flow was set to 0.4 mL/min and the column temperature to 30 °C. The method resulted in retention times between 1.2 and 3.4 min (Figure 3) and was continued until 10 min to avoid contamination. The analytical standards and isotope-labeled ISs of the respective analytes showed simultaneous elution. The MS parameters for detection of target analytes and ISs are listed in Table 2.

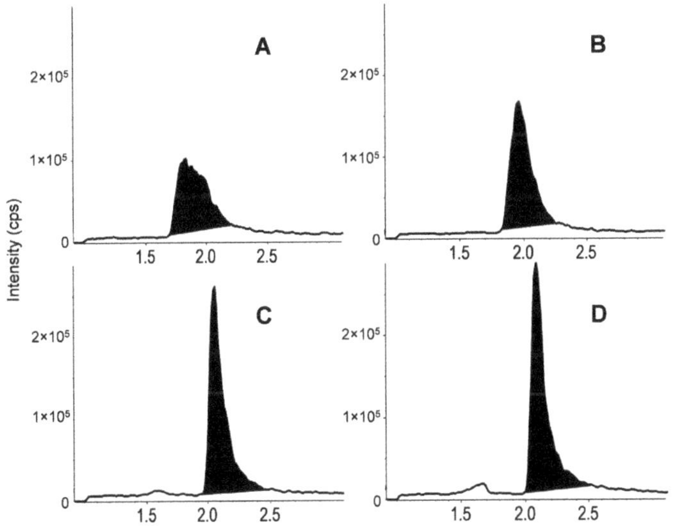

Figure 2. Comparison of the chromatographic signal of CYA as the most challenging analyte during method development. Signals with eluent ratios acetonitrile:ultrapure water with 10 mM ammonium acetate and 0.1% formic acid of (**A**) 70:30 *v:v*, (**B**) 80:20 *v:v*, (**C**) 90:10 *v:v*, and (**D**) 95:5 *v:v* are shown, with all other parameters kept constant. (**C**) was chosen for all further optimization steps.

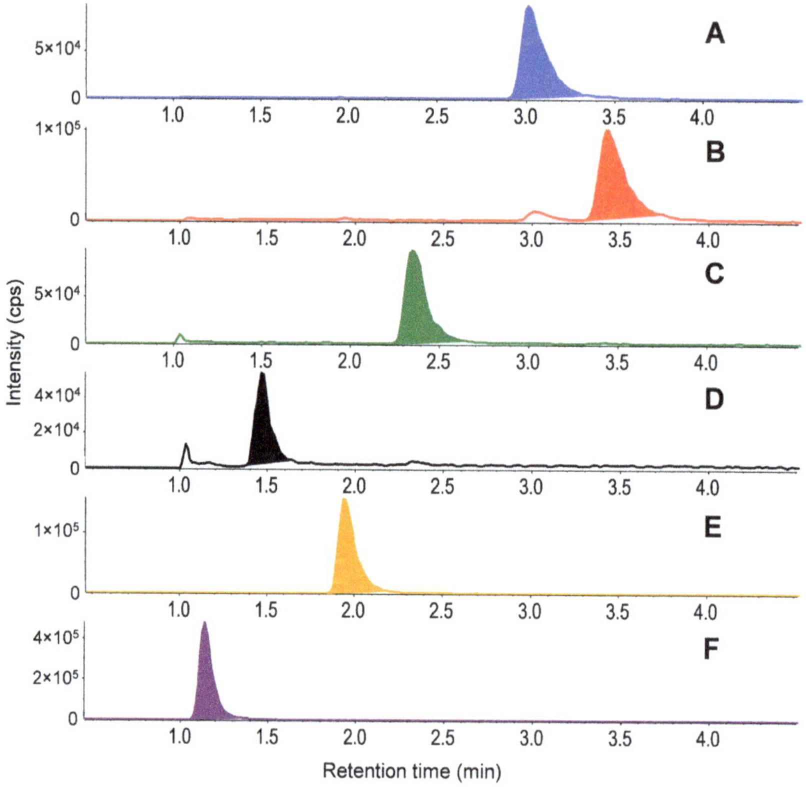

Figure 3. Extracted ion chromatograms of quantifier ions for (**A**) MEL, (**B**) AMN, (**C**) AMD, (**D**) CYA, (**E**) CYRO, and (**F**) HMMM with the final method. Standards with concentrations of 10 µg/L in drinking water were measured with direct injection. Chromatographic separation was achieved on a Luna HILIC column (150 × 2 mm, 3 µm; Phenomenex, Torrance, CA, USA) with a HILIC security guard cartridge (4 × 2 mm; Phenomenex) and isocratic elution with acetonitrile:ultrapure water with 10 mM ammonium acetate and 0.1% formic acid (90:10, *v*:*v*). The flow was set to 0.4 mL/min and the column temperature to 30 °C. An injection volume of 40 µL was applied.

Table 2. Retention times and MS parameters (m/z at quadrupole 1 (Q1), m/z at quadrupole 3 (Q3), declustering potential (DP), entrance potential (EP), collision energy (CE), and collision cell exit potential (CXP)) for detection of target analytes and available ISs.

Analyte	Ionization Mode	Retention Time (min)	Q1	Q3	DP (V)	EP (V)	CE (V)	CXP (V)
MEL	Positive	3.1	127.0	68.0 * 85.0	106	10	37 25	10 12
AMN	Positive	3.4	128.0	86.0 * 43.1	96	10	23 55	12 6
AMD	Negative	2.4	127.0	83.9 * 42.0	−50	−10	−16 −28	−7 −5
CYA	Negative	1.5	127.9	42.1 * 85.0	−45	−10	−26 −14	−5 −7
CYRO	Positive	1.9	167.0	68.0 * 85.0	91	10	43 25	10 12

Table 2. *Cont.*

Analyte	Ionization Mode	Retention Time (min)	Q1	Q3	DP (V)	EP (V)	CE (V)	CXP (V)
HMMM	Positive	1.2	391.1	177.1 * 359.2	31	10	39 11	12 26
IS-MEL-^{13}C3	Positive	3.1	130.0	87.1 *	120	10	25	6
IS-AMN-^{13}C3	Positive	3.4	131.0	88.0 * 71.0	131	10	23 37	10 12
IS-AMD-^{13}C3	Negative	2.4	130.0	86 * 43.0	−55	−10	−16 −36	−5 −5
IS-CYA-^{13}C3	Negative	1.5	130.9	43.1 *	−45	−10	−38	−9
IS-CYRO-d4	Positive	1.9	171.0	86.0 *	71	10	27	10

* Quantifier ion.

3.3. Limits of Detection and Quantification

With direct injection, LOQ values of 0.1 µg/L were achieved for MEL, AMD, CYRO, and HMMM. For AMN and CYA, direct injection resulted in LOQs of 0.2 µg/L and 1.0 µg/L (2.0 µg/L for CYA in wastewater), respectively. By applying evaporation for analyte pre-concentration, LOQ values in the drinking and surface water were lowered to 0.01 µg/L for all analytes but for CYA, where the LOQ was 0.05 µg/L. In the wastewater samples, which had to be diluted in ultrapure water prior to analysis with evaporation (by factor 5 for WWTP effluents and factor 10 for WWTP influents), LOQ differences between direct injection and evaporation were not as significant (Table 3; LOD values in Table S3).

Table 3. Limits of quantification for target analytes in different water matrices with direct injection and evaporation.

	Direct Injection (µg/L)				Evaporation (µg/L)			
Analyte	Drinking Water	Surface Water	WWTP Effluent	WWTP Influent	Drinking Water	Surface Water	WWTP Effluent	WWTP Influent
MEL	0.10	0.10	0.10	0.10	0.01	0.01	0.05	0.10
AMN	0.20	0.20	0.20	0.20	0.01	0.01	0.05	0.10
AMD	0.10	0.10	0.10	0.10	0.01	0.01	0.05	0.10
CYA	1.00	1.00	2.00	2.00	0.05	0.05	0.25	0.50
CYRO	0.10	0.10	0.10	0.10	0.01	0.01	0.05	0.10
HMMM	0.10	0.10	0.10	0.10	0.01	0.01	0.05	0.10

3.4. Method Validation

3.4.1. Recovery of Evaporation

Recoveries of target analytes after pre-concentration by evaporation ranged between 76.1% (HMMM) and 103.2% (MEL; Table 4). Consequently, all the analytes showed sufficient recovery during sample preparation.

Table 4. Recoveries of target analytes after pre-concentration by evaporation (corrected by internal standards, except for HMMM).

Analyte	Recovery of Evaporation Procedure (%)
MEL	103.2
AMN	100.9
AMD	91.8
CYA	81.5
CYRO	96.9
HMMM	76.1

3.4.2. Calibration Linearity

All the analytes showed linear calibrations with $r^2 \geq 0.99$ for both direct injection and evaporation (Table S4). Linearity was therefore ascertained in the concentration range of 0.5 to 10 × LOQ.

3.4.3. Matrix Effects

All the analytes except for HMMM remained well below the threshold of 25% for acceptable matrix effects in surface water and wastewater with both direct injection and evaporation (Table S5). For HMMM, considerably higher matrix effects were observed with evaporation for WWTP effluents (76.6%) and WWTP influents (68.7%). These matrix effects could not be corrected because no IS for HMMM was commercially available and all other applied ISs were unsuitable for HMMM matrix correction. Therefore, it is not recommended to apply the evaporation method for HMMM measurements in wastewater samples. These samples should be measured for HMMM only with direct injection. Also, the application of SPE might be suitable for analysis of HMMM in wastewater samples, leading to the removal of interfering compounds.

3.4.4. Repeatability

Standard deviations of five replicate samples with concentrations of 5 × LOQ remained within 25% of the spiked value for all target analytes (Table S6). Consequently, the method showed sufficient repeatability.

3.5. Stability Test

All the target analytes were stable in water over the test period of 28 days in all the tested sample containers (glass, PE, and PS) and under all the tested temperature conditions (room temperature, 4 °C, and −18 °C). Analytes were considered stable if the concentrations remained within a deviation of ±25% from the initial concentration (Figures S1–S3). These deviations can be attributed to measurement uncertainties. Also, no decreasing trends over time, which would indicate degradation or transformation, were observed. As further confirmation, samples were analyzed for all six target analytes even though only one standard was added. Since some of the analytes are transformation products of other analytes, occurrence and increasing concentrations of such other analytes would suggest degradation. However, no other target analytes were detected, supporting the findings of analyte stability.

3.6. Proof of Concept

Of the 10 analyzed proof-of-concept samples, all except two contained MEL, with concentrations between 0.015 µg/L (groundwater) and 6.9 µg/L (WWTP influent). The MEL concentrations were below the LOQ only in the drinking water and one groundwater sample. Ammeline was not detected in any sample, but AMD and CYA were found in road run-off samples and CYA additionally in both WWTP effluents, with concentrations up to 2.1 µg/L (AMD) and 12 µg/L (CYA). Furthermore, HMMM was detected in road run-off, wastewater, and one surface water sample, with concentrations up to 1.6 µg/L (Table 5).

Table 5. Concentrations of target analytes in samples measured as proof of concept.

Sample	Matrix	MEL (µg/L)	AMN (µg/L)	AMD (µg/L)	CYA (µg/L)	CYRO (µg/L)	HMMM (µg/L)
1	Drinking water	<LOQ	<LOQ	<LOQ	<LOQ	<LOQ	<LOQ
2	Groundwater	<LOQ	<LOQ	<LOQ	<LOQ	<LOQ	<LOQ
3	Groundwater	0.015	<LOQ	<LOQ	<LOQ	<LOQ	<LOQ

Table 5. *Cont.*

Sample	Matrix	MEL (µg/L)	AMN (µg/L)	AMD (µg/L)	CYA (µg/L)	CYRO (µg/L)	HMMM (µg/L)
4	Surface water	0.69	<LOQ	<LOQ	<LOQ	<LOQ	<LOQ
5	Surface water	1.2	<LOQ	<LOQ	<LOQ	<LOQ	1.6
6	Road run-off	0.71	<LOQ	1.3	12.0	<LOQ	0.74
7	Road run-off	0.49	<LOQ	2.1	5.9	<LOQ	0.77
8	WWTP effluent	3.0	<LOQ	<LOQ	1.0	<LOQ	0.29
9	WWTP effluent	1.3	<LOQ	<LOQ	1.1	<LOQ	0.18
10	WWTP influent	6.9	<LOQ	<LOQ	<LOQ	<LOQ	0.23

4. Discussion

In this study, a sensitive and robust method for analysis of MEL; its derivatives AMN, AMD, and CYA; and its potential precursors CYRO and HMMM was developed and validated for various water matrices. Depending on the required sensitivity and expected concentration ranges, samples can either be measured with direct injection or after evaporation as analyte pre-concentration. Only HMMM, for which no isotope-labeled IS is commercially available, should not be measured in wastewater with evaporation due to the possibility of severe matrix effects, such as signal suppression.

Even with evaporation, sample preparation is more efficient in terms of required time and material when compared to methods applying SPE. In addition, with a total run-time of 10 min per sample for separation of all analytes, the chromatographic method is fast while still achieving good sensitivity. Furthermore, the required volumes (1 mL for direct injection; 10 mL for evaporation) are low compared to methods including SPE as analyte pre-concentration, where usually between 20 and 100 mL is necessary to perform both cation and anion exchange [29,32]. However, it is still recommended to collect at least 50 mL of sample, if possible, to allow for repeated measurements including analyte pre-concentration.

Several implications can be drawn from the conducted stability tests, guiding sampling and sample transport strategies. The observed stability of all the analytes under the tested conditions renders addition of stabilizers and sample cooling unnecessary. Still, samples should not be exposed to temperatures above 25 °C and shielded from sunlight. Addition of pH modifiers should also be avoided because both acidic and basic conditions can facilitate transformation between MEL and its derivatives [24,36]. With all the tested sample container materials (glass, PE, and PS) determined as suitable, sampling in PE or PS containers is recommended to avoid glass breakage during sample transport. Samples should be analyzed within 28 days after sampling to remain within the timeframe of ascertained analyte stability.

During proof-of-concept measurements, MEL was found in all the surface water, road run-off, and wastewater samples and additionally in one groundwater sample. These results are in accordance with recent findings, showing almost ubiquitous occurrence of MEL throughout the water cycle [28]. Releases to the aquatic environment may stem from a variety of anthropogenic activities, including, among others, industrial discharges [28], agricultural applications [30], traffic due to tire wear [37–39], laundry (dissolution of wrinkle-proof substances in textiles [40]), concrete additives (superplasticizers [41,42]), and use of MEL-containing household items [27]. In our study, the highest MEL concentration was observed in the WWTP influent, with a concentration decrease towards the effluent. Reduction in MEL concentrations by wastewater treatment has been previously reported [29,43,44], with adsorption on activated sludge determined as a main removal mechanism while microbial degradation was negligible [44]. However, various strains of bacteria are capable of degrading MEL finally into CO_2 and ammonia [6,45,46]. In contrast, other studies reported an increase in MEL concentrations in WWTPs [10,30]. These opposing results suggest the simultaneous occurrence of several degradation, formation, or release processes, which have not yet been thoroughly investigated.

Of the three MEL derivatives, CYA occurred most abundantly during the proof-of-concept measurements, which was also observed by Zhu and Kannan during measurement of various water matrices in the United States [29]. This may either be due to MEL transformation with CYA as final hydrolytic product or to a discharge by separate sources, e.g., presence of CYA as an impurity in dishwashing detergents [26,29] or as an additive in swimming pools [47]. Cyromazine was likely not detected in the measured samples because of seasonal applications in agriculture [8]. The observed occurrence of HMMM in wastewater, road run-off, and surface water has been previously reported [10,34,43]. In the proof-of-concept study, all the samples with HMMM contained MEL and partially AMD and CYA as well. While the number of samples measured as proof of concept during this study is not sufficient to establish statistical correlations between the measured compounds, these findings still suggest some connections. Consequently, in addition to contributing to available monitoring data on MEL, the application of the developed method to various water samples can also elucidate pathways and favoring conditions for transformation of precursors to MEL or its hydrolysis products, which so far remain unclear. Especially HMMM is of interest since it is frequently detected in the aquatic environment and has usually not been included in analytical methods in addition to MEL and its derivatives.

In conclusion, the developed method with efficient sample preparation and sensitive measurement of MEL, derivatives, and precursors is applicable to conduct large-scale studies and monitoring programs. The method adds novelty by covering both derivatives and several potential precursors in addition to MEL, which has not been previously reported. Thus, the method contributes to an improved understanding of MEL distribution in the aquatic environment and its transformation in water resources and therefore to securing water quality.

Supplementary Materials: The following supporting information can be downloaded at https://www.mdpi.com/article/10.3390/analytica6030027/s1. Table S1: Details of tested SPE protocol (adapted from [29]); Table S2: Sample information on proof-of-concept samples; Table S3: Limits of detection for target analytes in different water matrices with direct injection and evaporation; Table S4: Linearity (described by correlation coefficient r^2 and linear equation) of target analytes for direct injection and evaporation in the concentration range 0.5–10 × LOQ (corrected by IS, except for HMMM); Table S5: Matrix effects of analytes at a concentration of 5 × LOQ for direct injection and evaporation in matrices, surface water, WWTP effluent, and WWTP influent, compared to drinking water (corrected by IS, except for HMMM); Table S6: Standard deviation of five replicate measurements of analytes with concentrations of 5 × LOQ for direct injection and evaporation (corrected by IS, except for HMMM); Figure S1: Results of stability tests (corrected by IS) of MEL (top) and AMD (bottom) in the tested container materials glass (left), PE (middle), and PS (right) at tested temperature conditions, room temperature (red), refrigerated (black), and frozen (blue). The green area depicts a deviation of ±25% from the concentration measured at day 0; Figure S2: Results of stability tests (corrected by IS) of AMN (top) and CYA (bottom) in the tested container materials glass (left), PE (middle), and PS (right) at tested temperature conditions, room temperature (red), refrigerated (black), and frozen (blue). The green area depicts a deviation of ±25% from the concentration measured at day 0; Figure S3: Results of stability tests of CYRO (top; corrected by IS) and HMMM (bottom) in the tested container materials glass (left), PE (middle), and PS (right) at tested temperature conditions, room temperature (red), refrigerated (black), and frozen (blue). The green area depicts a deviation of ±25% from the concentration measured at day 0.

Author Contributions: Conceptualization, F.S. and M.K.; methodology, M.K. and L.N.; validation, L.N., M.K. and F.S.; formal analysis, M.K.; investigation, L.N. and M.K.; writing—original draft preparation, M.K.; writing—review and editing, F.S. and L.N.; visualization, M.K.; supervision, F.S.; project administration, F.S.; funding acquisition, F.S. All authors have read and agreed to the published version of the manuscript.

Funding: This research was funded by the European Melamine Producers Association (EMPA) during the project "Development of a standardised analytical method for the detection and quantitation of melamine and related substances in aqueous media." The APC was also funded by EMPA.

Data Availability Statement: The original contributions presented in this study are included in the article/Supplementary Material. Further inquiries can be directed to the corresponding author.

Conflicts of Interest: The authors declare no conflicts of interest. The funders had no role in the design of the study; in the collection, analyses, or interpretation of data; or in the writing of the manuscript.

Abbreviations

The following abbreviations are used in this manuscript:

AMD	Ammeline
AMN	Ammelide
CE	Collision energy
CXP	Collision cell exit potential
CYA	Cyanuric acid
CYRO	Cyromazine
DAD	Diode array detector
DP	Declustering potential
ECHA	European Chemicals Agency
GC	Gas chromatography
EP	Entrance potential
HMMM	Hexa(methoxymethyl)melamine
HILIC	Hydrophilic interaction liquid chromatography
HPLC	High-performance liquid chromatography
IS	Isotope-labeled internal standards
LOD	Limit of detection
LOQ	Limit of quantification
MEL	Melamine
MS	Mass spectrometry
MS/MS	Tandem mass spectrometry
PE	Polyethylene
PS	Polystyrene
REACH	Registration, evaluation, authorization, and restriction of chemicals
S/N	Signal-to-noise ratio
SPE	Solid-phase extraction
UV	Ultraviolet detector
WWTP	Wastewater treatment plant

References

1. Wu, Y.; Zhang, Y. Analytical chemistry, toxicology, epidemiology and health impact assessment of melamine in infant formula: Recent progress and developments. *Food. Chem. Toxicol.* **2013**, *56*, 325–335. [CrossRef] [PubMed]
2. ECHA (European Chemicals Agency). Substance Infocard Melamine. Available online: https://echa.europa.eu/substance-information/-/substanceinfo/100.003.288 (accessed on 15 April 2025).
3. Charron, I.; Maguresse-Battistoni, B.L.; Habert, R.; Canivenc-Lavier, M.C.; Mhaouty-Kodja, S.; Michel-Caillet, C. Melamine regulatory assessment for endocrine disruption. *Environ. Int.* **2024**, *194*, 109188. [CrossRef] [PubMed]

4. Ministère de la Transition Écologique et de la Cohésion des Territoires. Regulatory Management Option Analysis Conclusion Document. 2024. Available online: https://echa.europa.eu/documents/10162/1f5fde3d-96ae-340a-78bd-61a650ab3fb4 (accessed on 12 June 2025).
5. Ono, S.; Funato, T.; Inoue, Y.; Munechika, T.; Yoshimura, T.; Morita, H.; Rengakuji, S.-I.; Shimasaki, C. Determination of melamine derivatives, melame, meleme, ammeline and ammelide by high-performance cation-exchange chromatography. *J. Chromatogr. A* **1998**, *815*, 197–204. [CrossRef]
6. Shelton, D.R.; Karns, J.S.; McCarty, G.W.; Durham, D.R. Metabolism of melamine by Klebsiella terragena. *Appl. Environ. Microbiol.* **1997**, *63*, 2832–2835. [CrossRef]
7. Dobson, R.L.M.; Motlagh, S.; Quijano, M.; Cambron, R.T.; Baker, T.R.; Pullen, A.M.; Regg, B.T.; Bigalow-Kern, A.S.; Vennard, T.; Fix, A.; et al. Identification and characterization of toxicity of contaminants in pet food leading to an outbreak of renal toxicity in cats and dogs. *Toxicol. Sci.* **2008**, *106*, 251–262. [CrossRef]
8. Pote, D.H.; Daniel, T.C.; Edwards, D.R.; Mattice, J.D.; Wickliff, D.B. Effect of drying and rainfall intensity on cyromazine loss from surface-applied caged-layer manure. *J. Environ. Qual.* **1994**, *23*, 101–104. [CrossRef]
9. Rauert, C.; Kaserzon, S.L.; Veal, C.; Yeh, R.Y.; Mueller, J.F.; Thomas, K.V. The first environmental assessment of hexa(methoxymethyl)melamine and co-occurring cyclic amines in Australian waterways. *Sci. Total Environ.* **2020**, *743*, 140834. [CrossRef]
10. Alhelou, R.; Seiwert, B.; Reemtsma, T. Hexamethoxymethylmelamine—A precursor of persistent and mobile contaminants in municipal wastewater and the water cycle. *Water Res.* **2019**, *165*, 114973. [CrossRef]
11. Tyan, Y.-C.; Yang, M.-H.; Jong, S.-B.; Wang, C.-K.; Shiea, J. Melamine contamination. *Anal. Bioanal. Chem.* **2009**, *395*, 729–735. [CrossRef]
12. Tittlemier, S.A. Methods for the analysis of melamine and related compounds in foods: A review. *Food Addit. Contam. Part A* **2010**, *27*, 129–145. [CrossRef]
13. Puschner, B.; Reimschuessel, R. Toxicosis caused by melamine and cyanuric acid in dogs and cats: Uncovering the mystery and subsequent global implications. *Clin. Lab. Med.* **2011**, *31*, 181–199. [CrossRef] [PubMed]
14. Fan, Y.; Ma, X.; Li, Z.; Chen, M. Fast derivatization followed by gas chromatography–mass spectrometry for simultaneous detection of melamine, ammeline, ammelide, and cyanuric acid in fish and shrimp. *Food Anal. Methods* **2016**, *9*, 16–22. [CrossRef]
15. Wong, Y.-L.; Mok, C.-S. A single analytical procedure for the simultaneous and confirmatory determination of melamine and related compounds in various food matrices by isotope dilution gas chromatography-mass spectrometry (ID-GC-MS). *Anal. Methods* **2013**, *5*, 2305–2314. [CrossRef]
16. Miao, H.; Fan, S.; Zhou, P.P.; Zhang, L.; Zhao, Y.F.; Wu, Y.N. Determination of melamine and its analogues in egg by gas chromatography-tandem mass spectrometry using an isotope dilution technique. *Food Addit. Contam.* **2010**, *27*, 1497–1506. [CrossRef]
17. Tzing, S.-H.; Ding, W.-H. Determination of melamine and cyanuric acid in powdered milk using injection-port derivatization and gas chromatography–tandem mass spectrometry with furan chemical ionization. *J. Chromatogr. A* **2010**, *1217*, 6267–6273. [CrossRef]
18. Wang, H.; Lin, L.; Sun, Q.; Lin, Q.; Xiong, X.; Wu, K.; Yu, C.-P. Simultaneous determination of cyromazine, melamine and their biodegradation products by ion-pair high-performance liquid chromatography. *Int. J. Environ. Anal. Chem.* **2014**, *94*, 1173–1182. [CrossRef]
19. Sun, H.; Qin, X.; Ge, X.; Wang, L. Effective separation and sensitive determination of cyanuric acid, melamine and cyromazine in environmental water by reversed phase high-performance liquid chromatography. *Enivron. Technol.* **2011**, *32*, 317–323. [CrossRef]
20. Ehling, S.; Tefera, S.; Ho, I.P. High-performance liquid chromatographic method for the simultaneous detection of the adulteration of cereal flours with melamine and related triazine by-products ammeline, ammelide, and cyanuric acid. *Food Addit. Contam.* **2007**, *24*, 1319–1325. [CrossRef]
21. Vinas, P.; Campillo, N.; Férez-Melgarejo, G.; Hernández-Córdoba, M. Determination of melamine and derivatives in foods by liquid chromatography coupled to atmospheric pressure chemical ionisation mass spectrometry and diode array detection. *Anal. Lett.* **2012**, *45*, 2508–2518. [CrossRef]
22. Draher, J.; Ehling, S.; Cellar, N.; Reddy, T.; Henion, J.; Sousou, N. Determination of emerging nitrogenous economic adulterants in milk proteins by high-performance liquid chromatography/compact mass spectrometry. *Rapid Commun. Mass Spectrom.* **2016**, *30*, 1265–1272. [CrossRef] [PubMed]
23. Kolkman, A.; Vughs, D.; Sjerps, R.; Kooij, P.J.F.; van der Kooi, M.; Baken, K.; Louisse, J.; de Voogt, P. Assessment of highly polar chemicals in Dutch and Flemish drinking water and its sources: Presence and potential risks. *Environ. Sci. Technol. Water* **2021**, *1*, 928–937. [CrossRef]
24. Zhu, H.; Kannan, K. Melamine and cyanuric acid in foodstuffs from the United States and their implications for human exposure. *Environ. Int.* **2019**, *130*, 104950. [CrossRef] [PubMed]

25. García-Miguel, E.; Meza-Márquez, O.G.; Osorio-Revilla, G.; Téllez-Medina, D.I.; Jiménez-Martínez, C.; Cornejo-Mazón, M.; Hernández-Martínez, D.M.; Gallardo-Velazquez, T. Detection of cyanuric acid and melamine in infant formula powders bei mid-FTIR spectroscopy and multivariate analysis. *J. Food Qual.* **2018**, *2018*, 7926768. [CrossRef]
26. Braekevelt, E.; Lau, B.P.-Y.; Feng, S.; Ménard, C.; Tittlemier, S.A. Determination of melamine, ammeline, ammelide and cyanuric acid in infant formula purchased in Canada by liquid chromatography-tandem mass spectrometry. *Food Addit. Contam.* **2011**, *28*, 698–704. [CrossRef] [PubMed]
27. Lütjens, L.H.; Pawlowski, S.; Silvani, M.; Blumenstein, U.; Richter, I. Melamine in the environment: A critical review of available information. *Environ. Sci. Eur.* **2023**, *35*, 2. [CrossRef]
28. Warner, W.; Licha, T. Melamine—A PMT/vPvM substance as a generic indicator for anthropogenic activity and urbanisation? An explorative study on melamine in the water cycle and soil. *Chemosphere* **2025**, *370*, 143918. [CrossRef]
29. Zhu, H.; Kannan, K. Occurrence and distribution of melamine and its derivatives in surface water, drinking water, precipitation, wastewater, and swimming pool water. *Environ. Pollut.* **2019**, *258*, 113743. [CrossRef]
30. Liu, S.-S.; Cai, Q.-S.; Li, C.; Cheng, S.; Wang, Z.; Yang, Y.; Ying, G.-G.; Sweetman, A.J.; Chen, C.-E. In situ measurement of an emerging persistent, mobile and toxic (PMT) substance—Melamine and related triazines in waters by diffusive gradient in thin-films. *Water Res.* **2021**, *206*, 117752. [CrossRef]
31. Neuwald, I.J.; Hübner, D.; Wiegand, H.L.; Valkov, V.; Borchers, U.; Nödler, K.; Scheurer, M.; Hale, S.E.; Arp, H.P.H.; Zahn, D. Occurence, distribution, and environmental behavior of persistent, mobile, and toxic (PMT) and very persistent and very mobile (vPvM) substances in the sources of German drinking water. *Environ. Sci. Technol.* **2022**, *56*, 10857–10867. [CrossRef]
32. Schulze, S.; Zahn, D.; Montes, R.; Rodil, R.; Quintana, J.B.; Knepper, T.P.; Reemtsma, T.; Berger, U. Occurrence of emerging persistent and mobile organic contaminants in European water samples. *Water Res.* **2019**, *153*, 80–90. [CrossRef]
33. He, L.; Su, Y.; Shen, X.; Zheng, Y.; Guo, H.; Zeng, Z. Solid-phase extraction of melamine from aqueous samples using water-compatible molecularly imprinted polymers. *J. Sep. Sci.* **2009**, *32*, 3310–3318. [CrossRef]
34. Johannessen, C.; Helm, P.; Metcalfe, C.D. Detection of selected tire wear compounds in urban receiving waters. *Environ. Pollut.* **2021**, *287*, 117659. [CrossRef]
35. GTFCH (Gesellschaft für Toxikologische und Forensische Chemie). Requirements for the Validation of Analytical Methods. 2009. Available online: https://www.gtfch.org/cms/images/stories/files/Appendix%20B%20GTFCh%2020090601.pdf (accessed on 12 June 2025).
36. Gong, H.; Tang, S.; Zhang, T. Catalytic hydrolysis of waste residue from the melamine process and the kinetics of melamine hydrolysis in NaOH solution. *React. Kinet. Mech. Catal.* **2016**, *118*, 377–391. [CrossRef]
37. Johannessen, C.; Parnis, J.M. Environmental modelling of hexamethoxymethylmelamine, its transformation products, and precursor compounds: An emerging family of contaminants from tire wear. *Chemosphere* **2021**, *280*, 130914. [CrossRef] [PubMed]
38. Dsikowitzky, L.; Schwarzbauer, J. Hexa(methoxymethyl) melamine: An emerging contaminant in German rivers. *Water Environ. Res.* **2015**, *87*, 461–469. [CrossRef] [PubMed]
39. Johannessen, C.; Helm, P.; Metcalfe, C.D. Runoff of the tire-wear compound hexamethoxymethyl-melamine into urban watersheds. *Archives Environ. Contam. Toxicol.* **2022**, *82*, 162–170. [CrossRef]
40. Zhu, H.; Kannan, K. Determination of melamine and its derivatives in textiles and infant clothing purchased in the United States. *Sci. Total Environ.* **2020**, *710*, 136396. [CrossRef]
41. Judzentiene, A.; Zdaniauskiene, A.; Ignatjev, I.; Druteikiene, R. Evaluation of physico-chemical characteristics of cement superplasticizer based on polymelamine sulphonate. *Materials* **2024**, *17*, 1940. [CrossRef]
42. Wang, H.; Yang, X.; Xiong, W.; Liu, X.; Zhang, Z. Synthesis and the effects of new melamine superplasticizer on the properties of concrete. *Int. Sch. Res. Notices* **2013**, *1*, 708063. [CrossRef]
43. Seitz, W.; Winzenbacher, R. A survey on trace organic chemicals in a German water protection area and the proposal of relevant indicators for anthropogenic influences. *Environ. Monit. Assess.* **2017**, *189*, 244. [CrossRef]
44. An, H.; Li, X.; Yang, Q.; Wang, D.; Xie, T.; Zhao, J.; Xu, Q.; Chen, F.; Zhong, Y.; Yuan, Y.; et al. The behavior of melamine in biological wastewater treatment system. *J. Hazard. Mater.* **2017**, *322*, 445–453. [CrossRef] [PubMed]
45. Jutzi, K.; Cook, A.M.; Hütter, R. The degradative pathway of the s-triazine melamine. The steps to ring cleavage. *Biochem. J.* **1982**, *208*, 679–684. [CrossRef]
46. Takagi, K.; Fujii, K.; Yamazaki, K.; Harada, N.; Iwasaki, A. Biodegradation of melamine and its hydroxy derivatives by a bacterial consortium containing a novel *Nocardioides* species. *Appl. Microbiol. Biot.* **2012**, *94*, 1647–1656. [CrossRef] [PubMed]
47. Chen, Z.; Su, Y.; Chen, J.; Li, Z.; Wang, T. Study on the health risk of cyanuric acid in swimming pool water and its prevention and control measures. *Front. Public Health* **2024**, *11*, 1294842. [CrossRef] [PubMed]

molecules

MDPI

Article

An Innovative Strategy for Untargeted Mass Spectrometry Data Analysis: Rapid Chemical Profiling of the Medicinal Plant *Terminalia chebula* Using Ultra-High-Performance Liquid Chromatography Coupled with Q/TOF Mass Spectrometry–Key Ion Diagnostics–Neutral Loss Filtering

Jia Yu [1], Xinyan Zhao [2], Yuqi He [2], Yi Zhang [1,*] and Ce Tang [1,*]

[1] School of Ethnic Medicine, Chengdu University of Traditional Chinese Medicine, Chengdu 611137, China; yu77@cdutcm.edu.cn

[2] School of Pharmacy, Chengdu University of Traditional Chinese Medicine, Chengdu 611137, China; 19946734998@163.com (X.Z.); 18502828016@163.com (Y.H.)

* Correspondence: zhangyi@cdutcm.edu.cn (Y.Z.); tangce@cdutcm.edu.cn (C.T.)

Abstract: Structural characterization of natural products in complex herbal extracts remains a major challenge in phytochemical analysis. In this study, we present a novel post-acquisition data-processing strategy—key ion diagnostics–neutral loss filtering (KID-NLF)—combined with ultra-high-performance liquid chromatography–quadrupole time-of-flight mass spectrometry (UPLC-Q/TOF-MS) for systematic profiling of the medicinal plant *Terminalia chebula*. The strategy consists of four main steps. First, untargeted data are acquired in negative electrospray ionization (ESI^-) mode. Second, a genus-specific diagnostic ion database is constructed by leveraging characteristic fragment ions (e.g., gallic acid, chebuloyl, and HHDP groups) and conserved substructures. Third, MS/MS data are high-resolution filtered using key ion diagnostics and neutral loss patterns (302 Da for HHDP; 320 Da for chebuloyl). Finally, structures are elucidated via detailed spectral analysis. The methanol extract of *T. chebula* was separated on a C18 column using a gradient of acetonitrile and 0.1% aqueous formic acid within 33 min. This separation enabled detection of 164 compounds, of which 47 were reported for the first time. Based on fragmentation pathways and diagnostic ions (e.g., *m*/*z* 169 for gallic acid, *m*/*z* 301 for ellagic acid, and neutral losses of 152, 302, and 320 Da), the compounds were classified into three major groups: gallic acid derivatives, ellagitannins (containing HHDP, chebuloyl, or neochebuloyl moieties), and triterpenoid glycosides. KID-NLF overcomes key limitations of conventional workflows—namely, isomer discrimination and detection of low-abundance compounds—by exploiting genus-specific structural signatures. This strategy demonstrates high efficiency in resolving complex polyphenolic and triterpenoid profiles and enables rapid annotation of both known and novel metabolites. This study highlights KID-NLF as a robust framework for phytochemical analysis in species with high chemical complexity. It also paves the way for applications in quality control, drug discovery, and mechanistic studies of medicinal plants.

Keywords: *Terminalia chebula*; key ion diagnostics; neutral loss filtering; untargeted mass spectrometry; UPLC-Q-TOF/MS

https://doi.org/10.3390/molecules30112451

1. Introduction

The exploration of medicinal plants has gained significant momentum globally, driven by their long-established therapeutic history and untapped potential for drug discovery. A central aspect of this effort is the thorough analysis of the chemical composition of medicinal plants, as their efficacy, safety, and quality are inherently tied to their phytochemical profiles [1]. High-resolution mass spectrometry (HRMS), particularly when combined with ultra-high-performance liquid chromatography (UPLC), has become a cornerstone technology for analyzing complex plant extracts, facilitating the rapid identification of both known and novel compounds [2]. However, some limitations remain. For instance, the peak capacity is limited, making it challenging to fulfill the separation requirements of chemical components in the complex systems of medicinal plants [3]. Traditional identification strategies, which rely on chromatographic retention behavior and mass spectrometry cleavage rules, are constrained by a limited number of reference substances and inadequate structural coverage in dedicated databases [4].

Current strategies for LC-MS/MS data processing are typically classified into targeted and untargeted approaches [5]. Targeted methods rely on predefined compound libraries and fragmentation pathways, offering high specificity but limited capacity for discovering novel compounds. In contrast, untargeted methods capture MS/MS spectra for all detectable ions, providing a comprehensive chemical profile but resulting in large datasets that require advanced post-acquisition processing. In recent years, feature-based molecular networking (FBMN) [6] and ion identity molecular networking (IIMN) have enhanced metabolite grouping by integrating retention time and admixture data [7]. Additionally, tools such as SIRIUS 5, powered by deep learning, have increased annotation accuracy through fragment tree calculations [8]. However, these general strategies remain inadequate for highly specific structures. These include excimer ion matching errors, automatic matching of endogenous cleavage fragments to other components, and the low recognition accuracy of complex additive ions [9].

Terminalia chebula, known as the "king of Tibetan medicine", has been used for thousands of years in traditional Chinese and Ayurvedic medicine [10]. *T. chebula* is used in the treatment of asthma, bronchitis, hepatitis, dyspepsia, eye diseases, and hoarseness and to promote hair growth [11]. The flesh of the plant has been used to treat diarrhea, leprosy, and edema [12]. It improves appetite, reduces cholesterol and blood pressure, strengthens the immune system, prevents aging, and enhances resistance to infections [13]. In clinical practice, the therapeutic effects of specific preparations can be tailored to treat various diseases and optimized by combining them based on distinct cold and heat symptoms [14]. *T. chebula* contains a diverse array of chemical constituents, primarily including phenolic acids, tannins, triterpenoids, and flavonoids [15]. In *T. chebula*, 33% of the total phytoconstituents are hydrolysable tannins, with variation between 20 and 50%. These tannins contain phenolic carboxylic acids such as gallic acid, as well as gallotannins. Ellagitannins including punacalagin, casuarinin, corilagin, and terchebulin as well as others such as chebulanin, neochebulinic, chebulagic, and chebulinic acids are also present [16]. However, the chemical complexity of *T. chebula* poses a unique analytical challenge. Its ellagic tannins often contain multiple configurations, including HHDP, chebulyl, or neochebulyl groups. Numerous isomers with the same molecular weight exist, and their fragment pathways are complex. Most components still lack systematic annotation, which significantly hinders the establishment of quality control standards and the investigation of the material basis for its efficacy. Therefore, addressing the challenge of component analysis in *T. chebula* can reveal the chemical nature of its therapeutic potential and offer insights into the metabolomic analysis of similar complex plants.

To overcome the above limitations, combined with the additive ions that are easily formed by the *T. chebula* components and the characteristic fragments, key ion diagnostics–neutral loss filtering (KID-NLF) is presented: a post-acquisition data processing approach that leverages characteristic fragment ions and neutral loss patterns of target compounds. KID-NLF constructs a genus-specific diagnostic ion database based on known *T. chebula* metabolites and their fragmentation profiles. This enables the prioritization of structurally relevant ions in complex, untargeted MS/MS data, thereby improving both the speed and accuracy of compound identification. The approach consists of two key components: (1) key ion diagnosis (KID), which utilizes high-resolution MS/MS data to identify diagnostic fragment ions specific to gallic acid derivatives, ellagitannins, and triterpenoids; and (2) neutral loss filtering (NLF), which systematically detects characteristic mass losses (e.g., 302 Da for HHDP, 320 Da for chebuloyl) to isolate structurally related compounds. This approach addresses challenges in isomer discrimination and low-abundance compound detection by exploiting the unique structural signatures of *T. chebula* constituents. It offers a robust framework for profiling even the most complex plant extracts.

In this study, KID-NLF was applied to analyze methanol extracts of *T. chebula* using UPLC-Q/TOF-MS. The method enabled the rapid identification of 164 compounds, 47 of which are newly reported. KID-NLF overcomes limitations of existing workflows by combining targeted fragmentation analysis with genus-specific database curation. This provides a transformative strategy for phytochemical profiling, especially in species rich in polyphenols and triterpenoids. The strategy enhances understanding of *T. chebula*'s chemical diversity and provides a scalable framework for quality control and drug discovery in medicinal plant research.

2. Results and Discussion

2.1. Research Strategy

Despite observing only a few major peaks in the UHPLC/UV analysis of *T. chebula*, the enlarged chromatogram revealed numerous minor compounds (Figure 1). To conduct a comprehensive analysis of these compounds, a 70% methanol extract was examined using UPLC-Q-TOF/MS in a non-targeted method (without specifying the parent ion). Operating in (—)-ESI mode, the analysis detected numerous compounds, facilitating the acquisition of key quasi-molecular ions with enhanced sensitivity. During the scan, all parent ions underwent analysis in MS^E mode at a collision energy ranging from 20 to 40 eV. Using these settings, satisfactory secondary mass spectrometry data were obtained.

Subsequently, the data underwent processing using targeted KID-NLF efficiently identifying the parent ion and its respective daughter ion structure. The fundamental structure of *T. chebula* compounds, gallic acid, features similar skeletons and substitution patterns that facilitate the generation of KID and NLF in MS/MS spectra. KID-NLF were identified from the MS/MS spectra of reference substances, leading to the creation of a database cataloguing ions related to all reported *Terminalia* compound structures. High-resolution KID were next extracted via UNIFI software (Version 1.9) for MS diagnosis and MS/MS spectral data filtering, achieving rapid and efficient structural identification. Ultimately, further determination of the identified compounds' structures was accomplished through in-depth analysis of their high-resolution MS and MS/MS spectra.

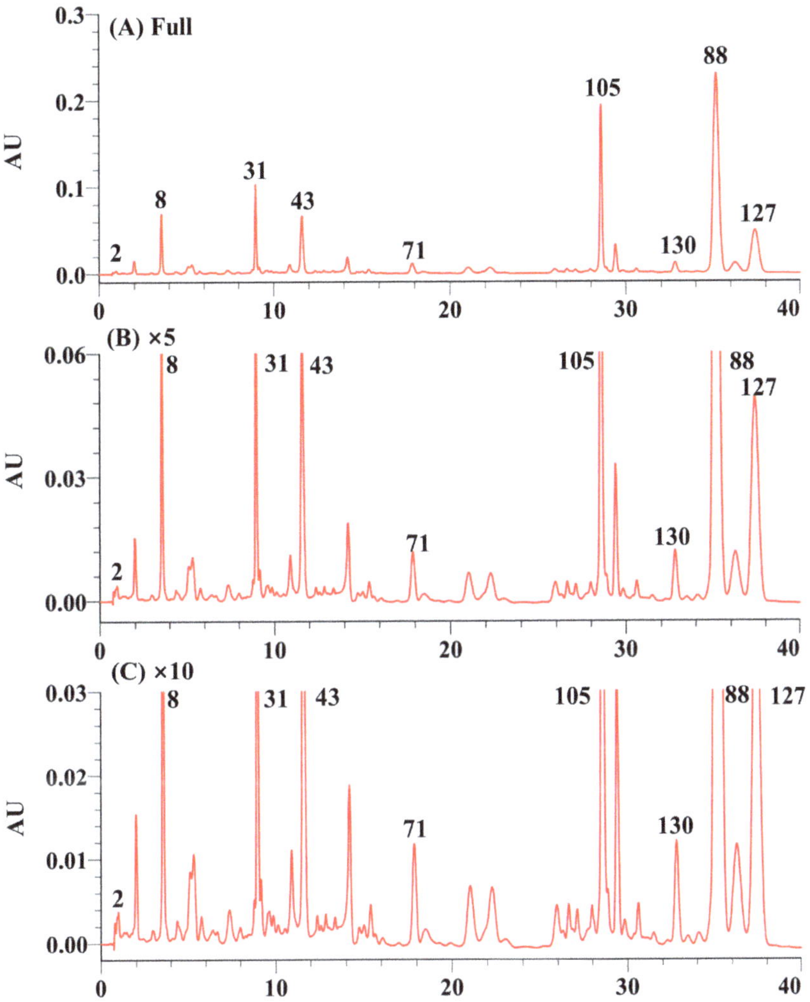

Figure 1. UHPLC/UV chromatograms of *Terminalia chebula* extract (254 nm).

In summary, the analytical strategy comprises four key steps: (1) acquisition of high-resolution MS and MS/MS data via UPLC-Q-TOF/MS; (2) creation of a diagnostic ion database informed by MS/MS fragments, similar skeletons, and substitution patterns of reported *T. chebula* compounds; (3) identification of precursor ions and significant fragments using high-resolution diagnostic ions and neutral loss filtering; and (4) further elucidation of target compound structures through detailed MS and MS/MS spectral analysis (Figure 2). According to the KID-NLF strategy, a total of 164 components were identified (Figure 3).

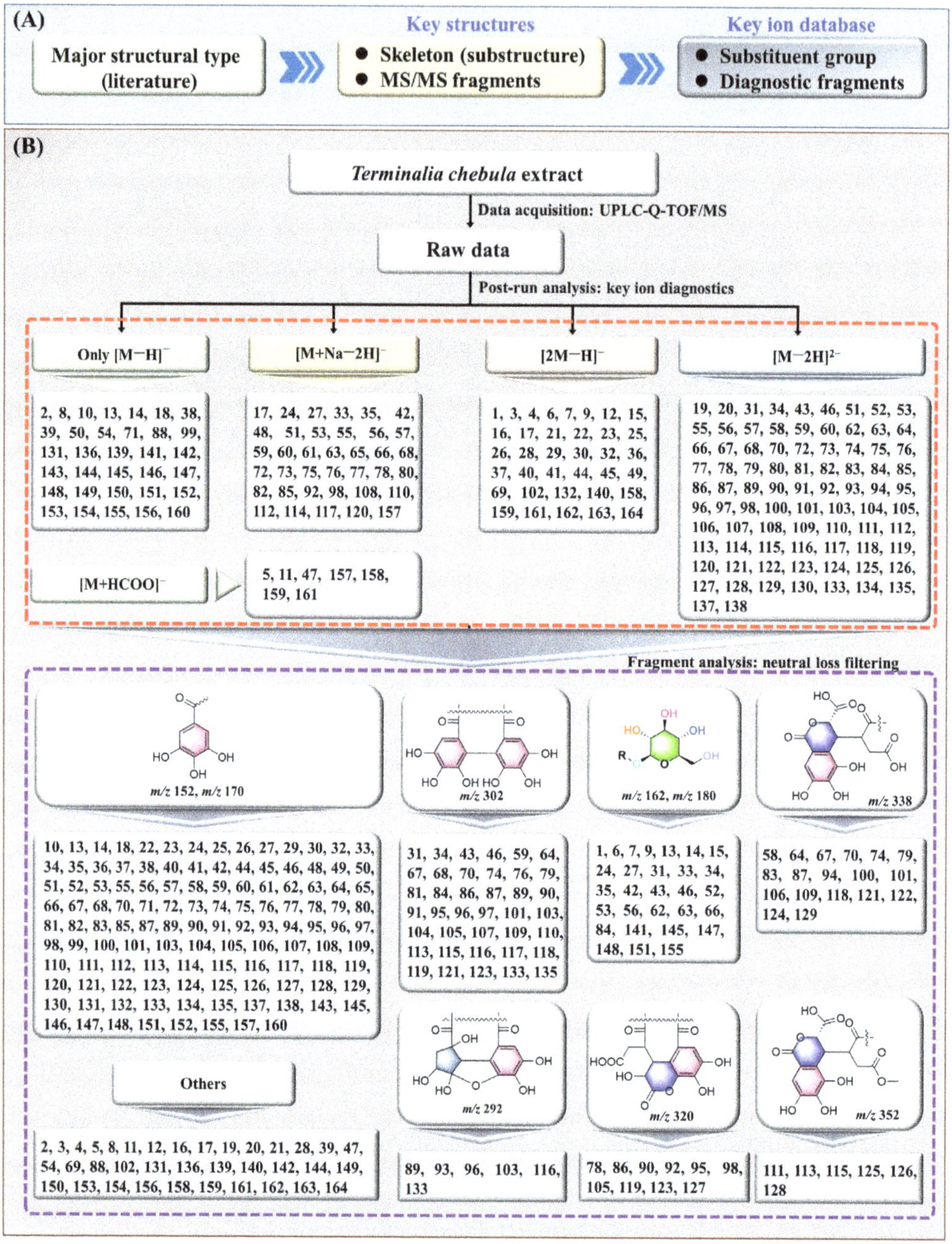

Figure 2. Construction of key ion database (**A**) and the identification of 164 chemical constituents in *T. chebula* using the KID-NLF strategy (**B**). The red dashed box marks components compliant with the KID strategy, the purple dashed box denotes those not only adhering to the KID strategy but also aligning with the NLF rules.

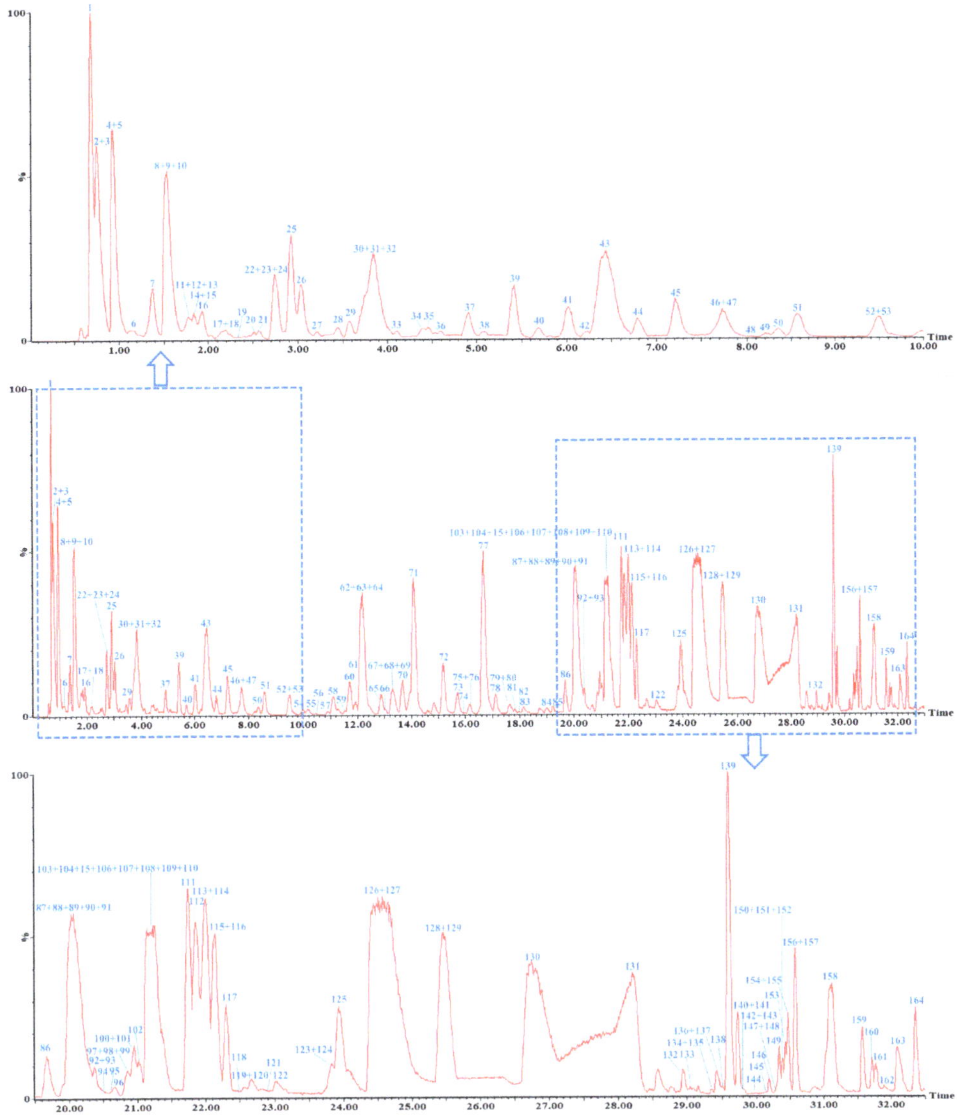

Figure 3. The representative base peak intensity (BPI) chromatogram of *T. chebula* in negative ion mode. The box and arrow denote the enlarged section.

2.2. Establishing a Diagnostic Ion Database

The process for establishing the database is illustrated in Figure 2A. Initially, the literature was reviewed to summarize and classify *T. chebula* compounds into three main structural types: gallic acid and its derivatives, ellagic acid and its derivatives, and terpenoids. Subsequently, characteristic diagnostic ions were identified for each type of compound.

For gallic acid and its derivatives, diagnosis involved identifying quasi-molecular ions, including $[M-2H]^{2-}$ and $[M-H]^-$ ions. These compounds frequently feature multiple galloyl groups or exhibit neutral loss of gallic acid, resulting in ions such as $[M-H-152]^-$ and $[M-H-170]^-$. Consequently, the accurate masses of $[M-H-152]^-$, $[M-H-170]^-$, and potential derivatives were utilized to develop a diagnostic ion database.

T. chebula contains ellagic acid and its derivatives, typically comprising HHDP (302 Da), chebuloyl (320 Da), neoche (338 Da), THDP (292 Da), DHHDP (318 Da), flavgallonyl (452 Da), and Gallagyl (602 Da) groups. These groups form complex combinations, possibly including HHDP alongside groups such as chebuloyl, neoche, THDP, DHHDP, Gallagyl, or flavgallonyl (Figure 4). In a similar manner, a diagnostic ion database for *T. chebula* ellagic acid compounds was created using the accurate masses of potential derivatives.

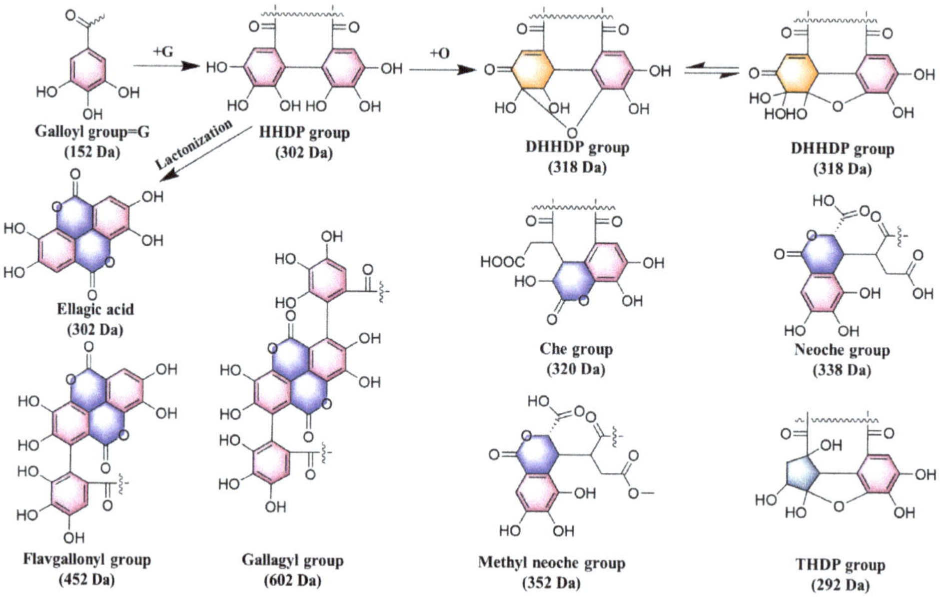

Figure 4. The main neutral loss fragment in *T. chebula*.

Terpenoids typically yield abundant deprotonated molecular ions $[M–H]^-$ in the primary mass spectrometer, with some compounds also producing $[2M–H]^-$ ions. Terpenoids in *T. chebula* commonly attach to sugar groups like glucose (Glc), galloyl, and glucoheptonic acid via various substitution methods. Aglycone fragments, which are relatively stable, typically lose sugar and galloyl segments (152 Da, 162 Da, 208 Da) under standard voltage. Analysis of terpenoid cleavage fragments reveals two primary types of aglycones, 503 Da and 487 Da, each with various configurations. Ultimately, the diagnostic ion database for *T. chebula* terpenoids was developed using the accurate masses of potential sugar derivatives.

2.3. Chemical Composition Analysis of T. chebula by Key Ion Diagnostics and Neutral Loss Filtering Using UPLC-Q-TOF/MS

Mass spectrometry analysis of gallic tannin or glucose gallate frequently reveals the neutral loss of multiple galloyl groups and gallic esters, as evidenced by the observation of fragments such as $[M–H–152]^-$ and $[M–H–170]^-$. These characteristic neutral loss fragments serve as valuable indicators for deducing the molecular structure. Additionally, the primary mass spectrum typically displays high-intensity $[M–2H]^{2-}$ peaks for macromolecular components containing multiple galloyl groups, while derivatives with fewer galloyl groups primarily generate $[M–H]^-$ or $[M+HCOO]^-$ adduct ions. However, misinterpretation may occur wherein the $[M–2H]^{2-}$ peak is incorrectly identified as $[M–H]^-$, leading to the misidentification of the true $[M–H]^-$ peak as $[2M–H]^-$. To mitigate the potential for such misjudgments, the secondary mass spectrum is generally employed for confirmation. A comparative analysis of the primary and secondary mass spectra reveals that fragments exhibiting higher intensity in the primary mass spectrum often disappear in the secondary mass spectrum, indicating their status as $[M–2H]^{2−}$ characteristic peaks. Concurrently, fragments that exhibit reduced intensity in the secondary mass spectrum can be more reliably assigned as $[M–H]^{−}$. This analytical approach significantly enhances the reliability of structural analysis for the target compound by effectively reducing the risk of misidentification.

By using the UPLC-Q-TOF/MS instrument, (Waters Corp., Manchester, UK) the *T. chebula* extract could be analyzed within 33 min, and the data were processed by KID-NLF. As a result, a total of 164 compounds were rapidly identified (Figures 2 and 3 and Table 1). They could be divided into three groups according to their structural types and MS/MS fragmentation pathways.

Table 1. Characterization of chemical constituents in *T. chebula* by UPLC-Q-TOF/MS. ([a] Isolated from *T. chebula* previously; [b] not isolated from *T. chebula* previously; [c] a newly discovered ingredient. [♠] Gallic acid derivatives; [♣] ellagitannins; [♥] triterpenoids; [♦] others. [*] Identified by reference standards; [#] MS/MS self-built library matching. In MS/MS fragment ions, bold characters with horizontal lines are KID, and bold characters only are NLF.).

No.	Identification	Formula	t_R (min)	Experimental (*m*/*z*)	Adducts	Error (ppm)	MS/MS Fragment Ions (*m*/*z*)
1 [b,#,♠]	3-*O*-galloyl-glucose [17]	$C_{13}H_{16}O_{10}$	0.71	331.0662	$[M–H]^{−}$	−0.9	**663.1391**, 271.0435, 211.0234, **169.0134**, 125.0235
2 [a,*,♦]	Shikimic acid [18]	$C_7H_{10}O_5$	0.75	173.0445	$[M–H]^{−}$	−2.9	155.0342, 137.0235, 93.0341
3 [a,#,♣]	Neochebulic acid [19]	$C_{14}H_{12}O_{11}$	0.76	355.0298	$[M–H]^{−}$	−0.8	**711.0679**, 337.0191, 293.0295, 249.0401, $205.0497^{−}$
4 [a,#,♣]	Chebulic acid [20]	$C_{14}H_{12}O_{11}$	0.93	355.0303	$[M–H]^{−}$	0.6	**711.0682**, 337.0194, 293.0297, 249.0402, $205.0501^{−}$
5 [b,#,♠]	3-galloylquinic acid [21]	$C_{14}H_{16}O_{10}$	1.05	343.0661	$[M–H]^{−}$	−1.2	**389.0355**, 191.0572, **169.0141**, 125.0240
6 [a,#,♠]	1-*O*-galloyl-glucose [22]	$C_{13}H_{16}O_{10}$	1.17	331.0665	$[M–H]^{−}$	0.0	**663.1403**, $271.0457^{−}$, 211.0239, **169.0140**, 125.0238
7 [a,#,♠]	6-*O*-galloyl-glucose [20]	$C_{13}H_{16}O_{10}$	1.38	331.0666	$[M–H]^{−}$	0.3	**663.1408**, 271.0455, 211.0244, **169.0135**, 125.0238
8 [a,*,♠]	Gallic acid [22]	$C_7H_6O_5$	1.53	169.0139	$[M–H]^{−}$	1.2	125.0238
9 [a,#,♠]	2-*O*-galloyl-glucose [23]	$C_{13}H_{16}O_{10}$	1.55	331.0665	$[M–H]^{−}$	0.0	**663.1407**, 271.0455, 211.0244, **169.0134**, 125.0236
10 [a,#,♣]	Gemin D [20]	$C_{27}H_{22}O_{18}$	1.59	633.0726	$[M–H]^{−}$	−0.3	**463.0544**, **300.9984**, 275.0194, **169.0135**, 125.0234
11 [b,#,♠]	5-galloylquinic acid [21]	$C_{14}H_{16}O_{10}$	1.72	343.0662	$[M–H]^{−}$	−0.9	**389.0356**, 191.0542, **169.0138**, 125.0236
12 [a,#,♣]	Isochebulic acid [19]	$C_{14}H_{12}O_{11}$	1.78	355.0299	$[M–H]^{−}$	−0.6	**711.0686**, 337.0196, 293.0303, 249.0401, 205.0499
13 [a,#,♣]	Punicalin α [24]	$C_{34}H_{22}O_{22}$	1.78	781.0529	$[M–H]^{−}$	0.6	**600.9893**, **448.9792**, **300.9988**
14 [a,#,♣]	Punicalin β [24]	$C_{34}H_{22}O_{22}$	1.85	781.0524	$[M–H]^{−}$	−0.4	**600.9888**, **448.9778**, **300.9974**
15 [a,#,♠]	4-*O*-galloyl-glucose [23]	$C_{13}H_{16}O_{10}$	1.88	331.06663	$[M–H]^{−}$	−0.6	**663.1405**, 271.0455, 211.0242, **169.0132**, 125.0232
16 [b,#,♦]	Caffeic acid 3,4-*O*-Di glucuronide [25]	$C_{21}H_{24}O_{16}$	1.93	531.0993	$[M–H]^{−}$	1.3	**1063.2059**, **355.0307**, 337.0200, 179.0710, 161.0603, 135.0446
17 [b,#,♠]	4-galloylquinic acid [21]	$C_{14}H_{16}O_{10}$	2.19	343.0665	$[M–H]^{−}$	0	**709.0888**, **687.1402**, **389.0350**, 191.0535, **169.0132**, 125.0238
18 [b,#,♣]	Isostrictinin [26]	$C_{27}H_{22}O_{18}$	2.19	633.0728	$[M–H]^{−}$	0.2	**463.0517**, **300.9981**, 275.0190, **169.0136**, 125.0235
19 [a,#,♣]	Punicacortein C [20]	$C_{48}H_{28}O_{30}$	2.33	1083.0585	$[M–H]^{−}$	−0.2	**541.0242**, 1065.0491, 1021.0580, **600.9891**, **499.0722**, **300.9983**, **169.0136**, 125.0235

Table 1. *Cont.*

No.	Identification	Formula	t_R (min)	Experimental (*m*/*z*)	Adducts	Error (ppm)	MS/MS Fragment Ions (*m*/*z*)
20 [a,#,♣]	Punicacortein D [20]	$C_{48}H_{28}O_{30}$	2.52	1083.0599	$[M–H]^-$	1.1	**541.0250**, 1065.0491, 1021.0573, **600.9891**, **499.0722**, **300.9983**, **169.0136**, 125.0235
21 [a,#,♣]	7′-*O*-methyl chebulate [20]	$C_{15}H_{14}O_{11}$	2.57	369.0460	$[M–H]^-$	0.5	**739.1009**, 351.0338, 325.0563, 307.0460
22 [a,#,♠]	4-*O*-galloyl-shikimic acid [20]	$C_{14}H_{14}O_9$	2.74	325.0557	$[M–H]^-$	−0.9	**651.1190**, 307.0451$^-$, 173.0440, **169.0136**, 155.0342, 125.0235
23 [a,#,♠]	1,4-di-*O*-galloyl-β-D-glucose [23]	$C_{20}H_{20}O_{14}$	2.78	483.0778	$[M–H]^-$	0.6	**967.1592**, **331.0670**, **313.0560**, 211.0242, 193.0137, **169.0135**, 125.0234
24 [a,#,♣]	Chebumeinin A [27]	$C_{27}H_{26}O_{20}$	2.78	669.0943	$[M–H]^-$	0.6	**691.0753**, **517.0834**, **499.0733**, **337.0200**, 293.0300, 249.0397, 205.0500
25 [a,#,♠]	5-*O*-galloyl-shikimic acid [20]	$C_{14}H_{14}O_9$	2.92	325.0559	$[M–H]^-$	−0.3	**651.1192**, 307.0451, 173.0440, **169.0136**, 155.0342, 125.0235
26 [a,#,♠]	3-*O*-galloyl-shikimic acid [20]	$C_{14}H_{14}O_9$	3.04	325.0559	$[M–H]^-$	−0.3	**651.1185**, 307.0451$^-$, 173.0440, **169.0136**, 155.0342, 125.0235
27 [a,#,♣]	Chebumeinin B [27]	$C_{27}H_{26}O_{20}$	3.21	669.0939	$[M–H]^-$	0.0	**691.0773**, **517.0753**, **499.0724**, **337.0219**, 293.0300, 249.0402, 205.0500
28 [a,#,♣]	6′-*O*-methyl chebulate [20]	$C_{15}H_{14}O_{11}$	3.45	369.0453	$[M–H]^-$	−1.4	**739.0991**, 351.0316, 325.0558, 307.0438
29 [a,#,♣]	Valoneic acid dilactone [28]	$C_{21}H_{10}O_{13}$	3.57	469.0050	$[M–H]^-$	−0.6	**939.0164**, 425.0144, 407.0038, **300.9969**, **299.9902**, **169.0137**, 125.0237
30 [a,#,♠]	2,4-di-*O*-galloyl-β-D-glucose [29]	$C_{20}H_{20}O_{14}$	3.77	483.0780	$[M–H]^-$	1.0	**967.1611**, **331.0666**, **313.0543**, 211.0245, 193.0131, **169.0134**, 125.0236
31 [a,∗,♣]	Punicalagin α [20]	$C_{48}H_{28}O_{30}$	3.84	1083.0583	$[M–H]^-$	−0.4	**541.0252**, **781.0526**, **600.9891**, **300.9983**, **169.0136**, 125.0235
32 [b,#,♠]	3,4-di-*O*-galloyl-β-D-glucose [30]	$C_{20}H_{20}O_{14}$	3.92	483.0771	$[M–H]^-$	−0.8	**967.1619**, **331.0651**, **313.0544**, 271.0446, 211.0243, 193.0128, **169.0134**, 125.0235
33 [a,#,♣]	Phyllanemblinin D [24]	$C_{27}H_{26}O_{20}$	4.10	669.0940	$[M–H]^-$	0.1	**691.0770**, **517.0822**, **499.0734**, **337.0210**, 293.0293, 249.0396, 205.0501
34 [b,#,♣]	Rhoipteleanin G [31]	$C_{48}H_{30}O_{30}$	4.39	1085.0745	$[M–H]^-$	0.1	**542.0319**, **783.0687**, **631.0569**, **450.9943**, **300.9985**
35 [a,#,♣]	Phyllanemblinin F [20]	$C_{27}H_{26}O_{20}$	4.45	669.0939	$[M–H]^-$	0.0	**691.0776**, **499.0734**, **337.0196**, 293.0295, 249.0397, 205.0500
36 [a,#,♠]	2,6-di-*O*-galloyl-β-D-glucose [32]	$C_{20}H_{20}O_{14}$	4.59	483.0778	$[M–H]^-$	0.6	**967.1616**, **331.0658**, **313.0567**, 271.0459, 211.0241, 193.0138, **169.0134**, 125.0239
37 [a,#,♠]	4,6-di-*O*-galloyl-β-D-glucose [23]	$C_{20}H_{20}O_{14}$	4.92	483.0776	$[M–H]^-$	0.2	**967.1622**, **331.0672**, **313.0565**, 271.0455, 211.0243, 193.0134, **169.0134**, 125.0236
38 [a,#,♣]	Strictinin [26]	$C_{27}H_{22}O_{18}$	5.08	633.0728	$[M–H]^-$	0.0	**463.0482**, **300.9982**, 275.0190, **169.0135**, 125.0237
39 [a,#,♠]	Methyl gallate [20]	$C_8H_8O_5$	5.41	183.0290	$[M–H]^-$	−1.6	**169.0134**, 168.0053, 125.0230, 124.0160
40 [b,#,♠]	1,2-di-*O*-galloyl-β-D-glucose [33]	$C_{20}H_{20}O_{14}$	5.69	483.0776	$[M–H]^-$	0.2	**967.1614**, **331.0665**, **313.0554**, 271.0451, 211.0243, 193.0136, **169.0134**, 125.0237

Table 1. *Cont.*

No.	Identification	Formula	t_R (min)	Experimental (*m*/*z*)	Adducts	Error (ppm)	MS/MS Fragment Ions (*m*/*z*)
41 [a,#,♠]	2,3-di-*O*-galloyl-β-D-glucose [23]	$C_{20}H_{20}O_{14}$	6.01	483.0776	[M–H]⁻	0.2	**967.1624**, **331.0657**, **313.0558**, 271.0454, 211.0242, 193.0136, **169.0133**, 125.0237
42 [a,#,♣]	Phyllanemblinin E [20]	$C_{27}H_{26}O_{20}$	6.23	669.0933	[M–H]⁻	−0.9	**691.0761**, **499.0725**, **337.0203**, 293.0300, 249.0397, 205.0497
43 [a,∗,♣]	Punicalagin β [20]	$C_{48}H_{28}O_{30}$	6.43	1083.0586	[M–H]⁻	−0.1	**541.0252**, **781.0526**, **600.9891**, **300.9983**, **169.0136**
44 [a,#,♠]	1,6-di-*O*-galloyl-β-D-glucose [20]	$C_{20}H_{20}O_{14}$	6.79	483.0780	[M–H]⁻	1.0	**967.1631**, **331.0668**, **313.0558**, 271.0455, 211.0242, 193.0136, **169.0134**, 125.0237
45 [a,#,♠]	3,6-di-*O*-galloyl-β-D-glucose [20]	$C_{20}H_{20}O_{14}$	7.21	483.0782	[M–H]⁻	1.4	**967.1636**, **331.0666**, **313.0560**, 271.0458, 211.0243, **169.0135**, 125.0238
46 [a,#,♣]	Terflavin a [20]	$C_{48}H_{30}O_{30}$	7.73	1085.0756	[M–H]⁻	1.1	**542.0331**, **783.0696**, **631.0578**, **450.9946**, **300.9987**
47 [a,#,♦]	Brevifolin carboxylic acid [20]	$C_{13}H_8O_8$	7.73	291.0145	[M–H]⁻	1.4	337.0199, 247.0232, 203.0342
48 [a,#,♠]	3,4-di-*O*-galloylshikimic acid [29]	$C_{21}H_{18}O_{13}$	8.03	477.0672	[M–H]⁻	0.6	**499.0485**, **325.0577**, **307.0452**, **169.0135**, 137.0237, 125.0238
49 [a,#,♠]	1,3-di-*O*-galloyl-β-D-glucose [32]	$C_{20}H_{20}O_{14}$	8.21	483.0777	[M–H]⁻	0.4	**967.1615**, **331.0655**, **313.0557**, 271.0454, 211.0240, **169.0136**, 125.0237
50 [b,#,♣]	Hippomanin A [34]	$C_{27}H_{22}O_{18}$	8.35	633.0732	[M–H]⁻	0.6	**463.0542**, **300.9984**, 275.0194, **169.0138**, 125.0237
51 [b,#,♠]	1,2,4-tri-*O*-galloyl-β-D-glucose [35]	$C_{27}H_{24}O_{18}$	8.57	635.0885	[M–H]⁻	0.2	**317.0391**, **657.0715**, **483.0771**, **465.0676**, **313.0557**, **295.0457**, **169.0135**, 125.0235
52 [a,#,♣]	Amlaic acid [32]	$C_{27}H_{24}O_{19}$	9.48	651.0836	[M–H]⁻	0.3	**325.0374**, 633.0739, **481.0624**, **337.0210**, **319.0083**, 275.0190, **169.0135**
53 [a,#,♣]	Methyl neochebulanin [20](4-*O*-methyl neochebulate-1-*O*-galloyl-glucose)	$C_{28}H_{28}O_{20}$	9.48	683.1100	[M–H]⁻	0.6	**705.0916**, **341.0500**, **513.0892**, **351.0351**, 307.0457, 263.0559, 219.0293, 204.0395,
54 [b,#,♦]	Phelligridin J [36]	$C_{13}H_6O_8$	9.98	288.9984	[M–H]⁻	1.4	245.0086
55 [b,#,♠]	1,3,4-tri-*O*-galloyl-β-D-glucose [37]	$C_{27}H_{24}O_{18}$	10.19	635.0891	[M–H]⁻	1.1	**657.0726**, **317.0398**, **483.0783**, **465.0667**, **313.0562**, **295.0447**, **169.0134**, 125.0237
56 [c,♣]	2-*O*-methyl neochebulate-1-*O*-galloyl-glucose	$C_{28}H_{28}O_{20}$	10.48	683.1098	[M–H]⁻	0.3	**705.0890**, **341.0493**, **513.0911**, **351.0357**, 307.0466, 263.0541, 219.0296, 204.0395
57 [a,#,♠]	1,2,6-tri-*O*-galloyl-β-D-glucose [23]	$C_{27}H_{24}O_{18}$	10.90	635.0886	[M–H]⁻	0.3	**657.0710**, **317.0399**, 483.0784, 465.0669, 313.0558, 295.0454, 169.0134, 125.0239
58 [b,#,♣]	Carpinusnin [38]	$C_{41}H_{34}O_{29}$	11.10	989.1116	[M–H]⁻	0.8	**494.0514**, **651.0831**, **481.0624**, **337.0194**
59 [a,#,♣]	Tercatain [20]	$C_{34}H_{26}O_{22}$	11.56	785.0833	[M–H]⁻	−0.5	**807.0775**, **392.0367**, **615.0608**, **483.0765**, **463.0505**, **445.0401**, **300.9980**
60 [a,#,♠]	3,4,6-tri-*O*-galloyl-β-D-glucose [20]	$C_{27}H_{24}O_{18}$	11.69	635.0884	[M–H]⁻	0	**657.0705**, **317.0396**, **483.0775**, **465.0668**, **313.0557**, **295.0449**, **169.0133**, 125.0237

Table 1. *Cont.*

No.	Identification	Formula	t_R (min)	Experimental (*m*/*z*)	Adducts	Error (ppm)	MS/MS Fragment Ions (*m*/*z*)
61 [b,#,♠]	3,5-di-*O*-galloylshikimic acid [39]	$C_{21}H_{18}O_{13}$	11.92	477.0671	[M–H]⁻	0.4	**499.0492**, **325.0577**, **307.0466**, **169.0133**, 137.0235, 125.0236
62 [a,#,♣]	Chebulanin [20]	$C_{27}H_{24}O_{19}$	12.17	651.0833	[M–H]⁻	−0.2	**325.0369**, 633.0728, **481.0620**, **337.0202**, **319.0089**, 275.0195, **169.0136**, 125.0236
63 [c,♣]	6-*O*-methyl neochebulate-1-*O*-galloyl-glucose	$C_{28}H_{28}O_{20}$	12.17	683.1091	[M–H]⁻	−0.7	**705.0886**, **341.0510**, **513.0881**, **351.0343**, 307.0459, 263.0560, 219.0291, 204.0392
64 [c,♣]	4-galloyl-6-neochebuloyl-2,3-HHDP-glucose	$C_{41}H_{32}O_{28}$	12.17	971.1006	[M–H]⁻	0.4	**485.0460**, 953.0898, 935.0793, **801.0782**, **669.0914**, **633.0727**, **499.0726**, **463.0513**, 337.0202, **300.9981**
65 [b,#,♠]	4,5-di-*O*-galloylshikimic acid [39]	$C_{21}H_{18}O_{13}$	12.89	477.0672	[M–H]⁻	0.6	**499.0485**, **325.0577**, **307.0452**, **169.0135**, 137.0237, 125.0238
66 [c,♣]	3-*O*-methyl neochebulate-1-*O*-galloyl-glucose	$C_{28}H_{28}O_{20}$	13.02	683.1088	[M–H]⁻	−1.2	**705.0922**, **341.0499**, **513.0967**, **351.0341**, 307.0438, 263.0558, 219.0274, 204.0385
67 [c,♣]	1-galloyl-2-neochebuloyl-4,6-HHDP-glucose	$C_{41}H_{32}O_{28}$	13.29	971.1009	[M–H]⁻	0.7	**485.0450**, 953.0893, 935.0793, **801.0742**, **669.0939**, **633.0729**, **499.0729**, **463.0508**, **337.0195**, **300.9981**
68 [a,#,♣]	Tellimagrandin I [20]	$C_{34}H_{26}O_{22}$	13.32	785.0848	[M–H]⁻	1.4	**807.0668**, **392.0373**, **633.0729**, **483.0779**, **463.0504**, **445.0413**, **300.9980**
69 [a,#,♦]	Urolithin M_5 [40]	$C_{13}H_8O_7$	13.32	275.0195	[M–H]⁻	1.1	**551.0461**, 257.0085, 229.0139, 201.0183
70 [c,♣]	1-galloyl-3-neochebuloyl-4,6-HHDP-glucose	$C_{41}H_{32}O_{28}$	13.67	971.1014	[M–H]⁻	1.2	**485.0457**, 953.0887, 935.0786, **801.0779**, **669.0923**, **633.0719**, **499.0723**, **463.0506**, **337.0191**, **300.9980**
71 [a,*,♣]	Corilagin [20]	$C_{27}H_{22}O_{18}$	14.07	633.0724	[M–H]⁻	−0.6	**463.0514**, **300.9985**, **275.0189**, **169.0135**, 125.0235
72 [a,#,♠]	1,4,6-tri-*O*-galloyl-β-D-glucose [23]	$C_{27}H_{24}O_{18}$	15.17	635.0880	[M–H]⁻	−0.6	**317.0391**, **657.0696**, **483.0782**, **465.0675**, **313.0558**, **295.0453**, **169.0134**, 125.0236
73 [b,#,♠]	2,3,4-tri-*O*-galloyl-β-D-glucose [41]	$C_{27}H_{24}O_{18}$	15.71	635.0881	[M–H]⁻	−0.5	**317.0393**, **657.0696**, **483.0773**, **465.0668**, **313.0555**, **295.0451**, **169.0133**, 125.0238
74 [c,♣]	1-galloyl-2-neochebuloyl-3,6-HHDP-glucose	$C_{41}H_{32}O_{28}$	15.96	971.1000	[M–H]⁻	−0.2	**485.0450**, 953.0904, 935.0748, **801.0808**, **669.1053**, **633.0721**, **499.0756**, **463.0502**, **337.0193**, **300.9983**
75 [b,#,♠]	2,4,6-tri-*O*-galloyl-β-D-glucose [42]	$C_{27}H_{24}O_{18}$	16.16	635.0881	[M–H]⁻	−0.5	**317.0395**, **657.0709**, **483.0776**, **465.0690**, **313.0558**, **295.0443**, **169.0134**, 125.0240
76 [b,#,♣]	1,3-di-*O*-galloyl-4,6-HHDP-glucose [43]	$C_{34}H_{26}O_{22}$	16.25	785.0848	[M–H]⁻	1.4	**807.0668**, **392.0373**, **633.0729**, **483.0779**, **463.0504**, **445.0413**, **300.9980**
77 [a,#,♠]	1,3,6-tri-*O*-galloyl-β-D-glucose [22]	$C_{27}H_{24}O_{18}$	16.68	635.0872	[M–H]⁻	−1.9	**317.0391**, **657.0693**, 483.0772, 465.0667, **313.0555**, **295.0449**, **169.0134**, 125.0234
78 [a,#,♣]	1,3-di-*O*-galloyl-2,4-chebuloyl-D-glucose [44]	$C_{34}H_{28}O_{23}$	17.10	803.0942	[M–H]⁻	−0.1	**825.0756**, **401.0424**, 785.0836, **633.0723**, 589.0815,533.0569, **483.0198**, **313.0563**

Table 1. *Cont.*

No.	Identification	Formula	t_R (min)	Experimental (*m*/*z*)	Adducts	Error (ppm)	MS/MS Fragment Ions (*m*/*z*)
79 [a,#,♣]	Neochebulagic acid [20](1-galloyl-4-neochebuloyl-3,6-HHDP-glucose)	$C_{41}H_{32}O_{28}$	17.57	971.0999	$[M–H]^-$	−0.3	**485.0457**, 953.0826, 935.0793, **801.0682**, **669.1024**, **633.0735**, **499.0777**, **463.0483**, **337.0194**, **300.9984**
80 [a,#,♠]	1,2,3,6-tetra-*O*-galloyl-β-D-glucose [20]	$C_{34}H_{28}O_{22}$	17.62	787.0999	$[M–H]^-$	0.6	**809.0811**, **393.0450**, **635.0889**, **617.0783**, **483.0782**, **465.0676**, **447.0562**, **313.0555**, **295.0450**
81 [a,#,♣]	Tellimagrandin II [18]	$C_{41}H_{30}O_{26}$	17.82	937.0959	$[M–H]^-$	1.3	**468.0425**, **767.0801**, **635.0894**, **465.0661**, **313.0561**, **300.9982**
82 [b,#,♠]	2,3,6-tri-*O*-galloyl-β-D-glucose [35]	$C_{27}H_{24}O_{18}$	17.97	635.0879	$[M–H]^-$	−0.8	**317.0401**, **657.0699**, **483.0765**, **465.0678**, **313.0568**, **295.0454**, **169.0136**, 125.0237
83 [c,♣]	1,2,3-tri-*O*-galloyl-4-neochebuloyl-D-glucose	$C_{41}H_{34}O_{28}$	18.15	973.1159	$[M–H]^-$	0.1	**486.0532**, **803.0942**, **635.0881**, **633.0729**, **483.0779**, **465.0667**, **463.0511**, **337.0193**, **300.9985**
84 [a,#,♣]	Terchebulin [45]	$C_{48}H_{28}O_{30}$	19.00	1083.0604	$[M–H]^-$	1.6	**541.0252**, **781.0580**, **600.9910**, **448.9790**, **300.9985**, **169.0136**, 125.0236
85 [a,#,♠]	1,2,3-tri-*O*-galloyl-β-D-glucose [23]	$C_{27}H_{24}O_{18}$	19.24	635.0880	$[M–H]^-$	−0.6	**657.0691**, **317.0393**, **483.0773**, **465.0668**, **313.0558**, **295.0443**, **169.0133**, 125.0238
86 [c,♣]	1-*O*-galloyl-3,4-chebuloyl-2,6-HHDP-D-glucose	$C_{41}H_{30}O_{27}$	19.65	953.0919	$[M–H]^-$	2.4	**476.0416**, **783.0739**, **651.0840**, **633.0735**, **481.0611**, **463.0504**, **337.0199**, **331.0665**, **319.0095**, **300.9988**
87 [c,♣]	1-galloyl-4-neochebuloyl-2,3-HHDP-glucose	$C_{41}H_{32}O_{28}$	19.92	971.1003	$[M–H]^-$	0.1	**485.0459**, 953.0898, 935.0773, **801.0789**, **669.0975**, **633.0717**, **499.0704**, **463.0513**, **337.0197**, **300.9984**
88 [a,*,♣]	Ellagic acid [20]	$C_{14}H_6O_8$	20.04	300.9987	$[M–H]^-$	1.0	283.9958, 257.0087, 229.0122, 201.0182, 185.0236, 173.0223, 145.0288, 117.0326
89 [c,♣]	1-*O*-galloyl-3,4-THDP-2,6-HHDP-D-glucose	$C_{40}H_{30}O_{26}$	20.13	925.0954	$[M–H]^-$	0.8	**462.0426**, **773.0887**, **755.0728**, **633.0735**, **463.0518**, **300.9986**
90 [c,♣]	1-*O*-galloyl-3,6-chebuloyl-2,4-HHDP-D-glucose	$C_{41}H_{30}O_{27}$	20.27	953.0895	$[M–H]^-$	−0.1	**476.0406**, **783.0743**, **651.0848**, **633.0744**, **481.0629**, **463.0519**, **337.0198**, **331.0643**, **319.0092**, **300.9991**
91 [b,#,♣]	Punicafolin [38]	$C_{41}H_{30}O_{26}$	20.29	937.0961	$[M–H]^-$	1.5	**468.0435**, **767.0743**, **635.0884**, **465.0688**, **313.0545**, **300.9987**
92 [c,♣]	1,6-di-*O*-galloyl-2,4-chebuloyl-D-glucose	$C_{34}H_{28}O_{23}$	20.37	803.0957	$[M–H]^-$	1.7	**401.0431**, **825.0756**, 785,0836, **633.0723**, 589.0815, 533.0569, **483.0198**, **313.0563**
93 [b,#,♣]	Phyllantusiin C [30]	$C_{40}H_{30}O_{26}$	20.37	925.0955	$[M–H]^-$	−0.3	**462.0449**, **773.0849**, **755.0762**, **633.0731**, **463.0514**, **300.9985**
94 [a,♣]	1,3,6-tri-*O*-galloyl-4-neochebuloyl-D-glucose [29]	$C_{41}H_{34}O_{28}$	20.50	973.1162	$[M–H]^-$	0.4	**486.0536**, **803.0956**, **635.0879**, **633.0724**, **483.0755**, **465.0685**, **463.0514**, **337.0192**, **300.9984**
95 [c,♣]	1-*O*-galloyl-4,6-chebuloyl-2,3-HHDP-D-glucose	$C_{41}H_{30}O_{27}$	20.60	953.0893	$[M–H]^-$	−0.3	**476.0412**, **783.0712**, **651.0827**, **633.0726**, **481.0653**, **463.0526**, **337.0191**, **331.0658**, **319.0093**, **300.9987**

Table 1. *Cont.*

No.	Identification	Formula	t_R (min)	Experimental (*m*/*z*)	Adducts	Error (ppm)	MS/MS Fragment Ions (*m*/*z*)
96 [c,♣]	1-*O*-galloyl-3,6-THDP-2,4-HHDP-D-glucose	$C_{40}H_{30}O_{26}$	20.77	925.0959	$[M–H]^-$	1.3	462.0430, 773.0862, 755.0744, 633.0735, 463.0521, 300.9988
97 [b,#,♣]	Davidiin [46]	$C_{41}H_{30}O_{26}$	20.86	937.0964	$[M–H]^-$	1.8	468.0435, 767.0731, 635.0885, 465.0673, 313.0563, 300.9987
98 [c,♣]	3,6-di-*O*-galloyl-2,4-chebuloyl-D-glucose	$C_{34}H_{28}O_{23}$	20.90	803.0961	$[M–H]^-$	2.2	825.0776, 401.0432, 785.0864, 633.0743, 589.0851, 533.0582, 483.0163, 313.0558
99 [c,♠]	3,4,5-tri-*O*-galloyl shikimic acid	$C_{28}H_{22}O_{18}$	20.90	629.0789	$[M–H]^-$	1.6	477.0655, 459.0561, 325.0501, 307.0443, 289.0357, 169.0138
100 [c,♣]	1,2,6-tri-*O*-galloyl-4-neochebuloyl-D-glucose	$C_{41}H_{34}O_{28}$	20.94	973.1168	$[M–H]^-$	1.0	486.0546, 803.0958, 635.0894, 633.0729, 483.0772, 465.0677, 463.0529, 337.0198, 300.9986
101 [c,♣]	2-galloyl-3-neochebuloyl-4,6-HHDP-glucose	$C_{41}H_{32}O_{28}$	20.94	971.1013	$[M–H]^-$	1.1	485.0463, 953.0900, 935.0801, 801.0813, 669.0943, 633.0731, 499.0749, 463.0515, 337.0198, 300.9986
102 [a,#,♣]	Eschweilenol C [20]	$C_{20}H_{16}O_{12}$	21.01	447.0573	$[M–H]^-$	2.0	895.1224, 300.9984
103 [c,♣]	1-*O*-galloyl-4,6-THDP-2,3-HHDP-D-glucose	$C_{40}H_{30}O_{26}$	21.07	925.0956	$[M–H]^-$	1.0	462.0424, 773.0867, 755.0684, 633.0731, 463.0518, 300.9983
104 [c,♣]	1,3,4-tri-*O*-galloyl-2,6-HHDP-glucose	$C_{41}H_{30}O_{26}$	21.10	937.0964	$[M–H]^-$	1.8	468.0437, 767.0753, 635.0886, 465.0679, 313.0557, 300.9991
105 [a,*,♣]	Chebulagic acid [22]	$C_{41}H_{30}O_{27}$	21.15	953.0896	$[M–H]^-$	1.7	476.0415, 783.0693, 651.0840, 633.0735, 481.0627, 463.0518, 337.0198, 331.0667, 319.0092, 300.9990
106 [c,♣]	2,3,6-tri-*O*-galloyl-4-neochebuloyl-D-glucose	$C_{41}H_{34}O_{28}$	21.29	973.1172	$[M–H]^-$	1.4	486.0541, 803.0955, 635.0891, 633.07235, 483.0763, 465.0677, 463.0523, 337.0198, 300.9992
107 [b,#,♣]	Pterocarinin C [47]	$C_{41}H_{30}O_{26}$	21.34	937.0959	$[M–H]^-$	1.3	468.0432, 767.0742, 635.0887, 465.0670, 313.0559, 300.9989
108 [a,#,♠]	1,2,3,4-tetra-*O*-galloyl-β-D-glucose [23]	$C_{34}H_{28}O_{22}$	21.35	787.0999	$[M–H]^-$	0.6	809.0834, 393.0457, 635.0889, 617.0782, 483.0770, 465.0676, 447.0555, 313.0556, 295.0450
109 [c,♣]	1-galloyl-6-neochebuloyl-2,3-HHDP-glucose	$C_{41}H_{32}O_{28}$	21.39	971.1014	$[M–H]^-$	1.2	485.0464, 953.0909, 935.0782, 801.0809, 669.1044, 633.0738, 499.0724, 463.0503, 337.0197, 300.9990
110 [b,#,♣]	1,6-di-*O*-galloyl-2,3-HHDP-glucose [48]	$C_{34}H_{26}O_{22}$	21.47	785.0848	$[M–H]^-$	1.4	807.0668, 392.0373, 633.0729, 483.0779, 463.0504, 445.0413, 300.9980
111 [c,♣]	1,2,3-tri-*O*-galloyl-4-methyl neochebuloyl-glucose	$C_{42}H_{36}O_{28}$	21.74	987.1318	$[M–H]^-$	0.3	493.0620, 817.1097, 635.0881, 465.0668, 351.0352, 295.0450
112 [a,#,♠]	1,2,4,6-tetra-*O*-galloyl-β-D-glucose [23]	$C_{34}H_{28}O_{22}$	21.85	787.0994	$[M–H]^-$	0.0	809.0809, 393.0447, 635.0881, 617.0773, 483.0768, 465.0674, 447.0560, 313.0555, 295.0451

Table 1. *Cont.*

No.	Identification	Formula	t_R (min)	Experimental (*m*/*z*)	Adducts	Error (ppm)	MS/MS Fragment Ions (*m*/*z*)
113 [a,#,♣]	1-*O*-galloyl-3,6-HHDP-4-6′ methyl neochebuloyl-glucose [49]	$C_{42}H_{34}O_{28}$	21.94	985.1156	[M–H]⁻	−0.2	492.0533, 815.0938, 683.0846, 633.0719, 513.0870, 463.0513, 351.0342, 300.9983
114 [a,#,♠]	1,3,4,6-tetra-*O*-galloyl-β-D-glucose [20]	$C_{34}H_{28}O_{22}$	22.00	787.0995	[M–H]⁻	0.1	809.0819, 393.0447, 635.0874, 617.0775, 483.0771, 465.0668, 447.0558, 313.0556, 295.0449
115 [c,♣]	2-*O*-galloyl-3,6-HHDP-4-6′ methyl neochebuloyl-glucose	$C_{42}H_{34}O_{28}$	22.13	985.1154	[M–H]⁻	−0.4	492.0534, 815.0939, 683.0846, 633.0718, 513.0867, 463.0512, 351.0346, 300.9983
116 [c,♣]	1-*O*-galloyl-2,3-THDP-4,6-HHDP-D-glucose	$C_{40}H_{30}O_{26}$	22.20	925.0941	[M–H]⁻	−0.6	462.0423, 773.0825, 755.0762, 633.0716, 463.0508, 300.9982
117 [a,#,♣]	1,2-di-*O*-galloyl-4,6-HHDP-glucose [50]	$C_{34}H_{26}O_{22}$	22.29	785.0833	[M–H]⁻	−0.5	807.0775, 392.0367, 615.0608, 483.0765, 463.0505, 445.0401, 300.9980
118 [c,♣]	2-galloyl-4-neochebuloyl-3,6-HHDP-glucose	$C_{41}H_{32}O_{28}$	22.49	971.1006	[M–H]⁻	0.4	485.0472, 953.0893, 935.0806, 801.0945, 669.0826, 633.0731, 499.0716, 463.0505, 337.0193, 300.9982
119 [c,♣]	1-*O*-galloyl-2,6-chebuloyl-3,4-HHDP-D-glucose	$C_{41}H_{30}O_{27}$	22.65	953.0883	[M–H]⁻	−1.4	476.0383, 783.0655, 651.0851, 633.0722, 481.0618, 463.0501, 337.0194, 331.0634, 319.0079, 300.9979
120 [b,#,♠]	2,3,4,6-tetra-*O*-galloyl-β-D-glucose [51]	$C_{34}H_{28}O_{22}$	22.66	787.0986	[M–H]⁻	−1.0	809.0819, 393.0447, 635.0871, 617.0767, 483.0764, 465.0662, 447.0559, 313.0554, 295.0446
121 [c,♣]	6-galloyl-4-neochebuloyl-2,3-HHDP-glucose	$C_{41}H_{32}O_{28}$	22.84	971.0999	[M–H]⁻	−0.3	485.0458, 953.0894, 935.0800, 801.0879, 669.0791, 633.0726, 499.0721, 463.0498, 337.0190, 300.9981
122 [c,♣]	1,2,3-tri-*O*-galloyl-4,6-neochebuloyl-glucose	$C_{41}H_{32}O_{27}$	23.03	955.1048	[M–H]⁻	−0.5	477.0480, 785.0828, 617.0756, 465.0661, 447.0555, 337.0194
123 [c,♣]	1-*O*-galloyl-2,3-chebuloyl-4,6-HHDP-D-glucose	$C_{41}H_{30}O_{27}$	23.76	953.0882	[M–H]⁻	−1.5	476.0395, 783.0699, 651.0839, 633.0712, 481.0641, 463.0492, 337.0190, 331.0685, 319.0087, 300.9981
124 [c,♣]	1,2,4-tri-*O*-galloyl-3,6-neochebuloyl-glucose	$C_{41}H_{32}O_{27}$	23.82	955.1045	[M-H]⁻	−0.8	477.0481, 785.0828, 617.0765, 465.0669, 447.0562, 337.0189
125 [a,#,♣]	1,3,6-tri-*O*-galloyl-4-methyl neochebuloyl-glucose [20]	$C_{42}H_{36}O_{28}$	23.93	987.1302	[M–H]⁻	1.1	493.0611, 817.1092, 635.0872, 465.0662, 351.0343, 295.0445
126 [c,♣]	1,2,6-tri-*O*-galloyl-4-methyl neochebuloyl-glucose	$C_{42}H_{36}O_{28}$	24.36	987.1313	[M–H]⁻	−0.2	493.5631, 817.1094, 635.0876, 465.0673, 351.0338, 295.0440

Table 1. *Cont.*

No.	Identification	Formula	t_R (min)	Experimental (*m*/*z*)	Adducts	Error (ppm)	MS/MS Fragment Ions (*m*/*z*)
127 [a,*,♣]	Chebulinic acid [20](1,3,6-tri-*O*-galloyl-2,4-chebuloyl-glucose)	$C_{41}H_{32}O_{27}$	24.57	955.1054	[M–H]⁻	0.1	**477.0486**, **785.0839**, **635.0873**, 617.0771, **465.0668**, **447.0558**, 337.0190
128 [c,♣]	2,3,6-tri-*O*-galloyl-4-methyl neochebuloyl-glucose	$C_{42}H_{36}O_{28}$	25.47	987.1324	[M–H]⁻	0.9	**493.0623**, **817.1107**, **635.0883**, **465.0674**, 351.0348, **295.0458**
129 [c,♣]	1,4,6-tri-*O*-galloyl-2,3-neochebuloyl-glucose	$C_{41}H_{32}O_{27}$	25.47	955.1041	[M–H]⁻	−1.3	**477.0491**, **785.0850**, **617.0781**, **465.0674**, **447.0564**, **337.0197**
130 [a,*,♠]	1,2,3,4,6-penta-*O*-galloyl-β-D-glucose [22]	$C_{41}H_{32}O_{26}$	26.73	939.1113	[M–H]⁻	1.0	**469.0515**, **787.1005**, **769.0900**, **617.0784**, **599.0676**, **465.0677**, **447.0568**, **429.0460**, **313.0565**, **295.0456**, **277.0352**, **259.0245**, **169.0136**, 125.0237
131 [a,#,♣]	Terchebin [52]	$C_{41}H_{30}O_{27}$	28.19	953.0901	[M–H]⁻	0.5	935.0799, 917.0695, **617.0781**, **465.0671**, **316.9932**, **295.0452**
132 [a,#,♣]	4-*O*-(4″-*O*-galloyl-α-rhamnopyranosyl) ellagic acid [49]	$C_{27}H_{20}O_{16}$	28.94	599.0673	[M–H]⁻	0.0	**1199.1431**, **621.0486**, **447.0560**, **429.0441**, **300.9985**
133 [c,♣]	1-*O*-galloyl-2,6-THDP-3,4-HHDP-D-glucose	$C_{40}H_{30}O_{26}$	29.04	925.0959	[M–H]⁻	1.3	**462.0432**, **773.0815**, **755.0752**, **633.0731**, **463.0506**, **300.9982**
134 [a,#,♣]	4-*O*-(2″,3″-di-*O*-galloyl-α-L-rhamnosyl) ellagic acid [20]	$C_{34}H_{24}O_{20}$	29.31	751.0782	[M–H]⁻	−0.1	**375.0342**, **599.0668**, **581.0553**, **449.0718**, **411.0366**, **300.9985**
135 [b,#,♣]	Nupharin A [53]	$C_{41}H_{30}O_{26}$	29.35	937.0955	[M–H]⁻	0.9	**468.0433**, **767.0732**, **635.0894**, **465.0674**, **313.0573**, **300.9988**
136 [c,♠]	1-*O*-cinnamoyl-6-*O*-galloyl-glucose	$C_{22}H_{22}O_{11}$	29.39	461.1084	[M–H]⁻	0	313.0558, **169.0135**, 147.0436
137 [a,#,♣]	4-*O*-(2″,4″-di-*O*-galloyl-α-L-rhamnosyl) ellagic acid [20]	$C_{34}H_{24}O_{20}$	29.42	751.0782	[M–H]⁻	−0.1	**375.0342**, **599.0668**, **581.0553**, **449.0718**, **411.0366**, **300.9985**
138 [a,#,♣]	4-*O*-(3″,4″-di-*O*-galloyl-α-rhamnopyranosyl) ellagic acid [49]	$C_{34}H_{24}O_{20}$	29.50	751.0776	[M–H]⁻	−0.9	**375.0338**, **599.0660**, **581.0549**, **449.0713**, **411.0292**, **300.9977**
139 [c,♥]	Arjungenin-24-*O*-glucoheptonic acid	$C_{37}H_{60}O_{13}$	29.59	711.3952	[M–H]⁻	−0.6	**503.3379**
140 [b,#,♠]	1-*O*-galloyl-2-*O*-cinnamoyl-glucose [54]	$C_{22}H_{22}O_{11}$	29.73	461.1092	[M–H]⁻	1.7	**923.2247**, 313.0561, **169.0133**, 147.0443
141 [a,#,♥]	Quercotriterpenoside I [55]	$C_{43}H_{62}O_{15}$	29.73	817.4018	[M–H]⁻	1.0	**655.3484**, **503.3366**
142 [b,#,♠]	1-*O*-galloyl-6-*O*-cinnamoyl-glucose [54]	$C_{22}H_{22}O_{11}$	29.81	461.1077	[M–H]⁻	−1.5	313.0562, **169.0132**, 147.0443
143 [a,#,♠]	1,2-*O*-galloyl-6-*O*-cinnamoyl-glucose [20]	$C_{29}H_{26}O_{15}$	29.82	613.1190	[M–H]⁻	−0.5	465.0682, **461.1077**, 313.0560, **169.0131**, 147.0443

Table 1. *Cont.*

No.	Identification	Formula	t_R (min)	Experimental (*m*/*z*)	Adducts	Error (ppm)	MS/MS Fragment Ions (*m*/*z*)
144 [b,#,♠]	1-*O*-cinnamoyl-2-*O*-galloyl-glucose [56]	$C_{22}H_{22}O_{11}$	29.95	461.10978	[M–H]⁻	−1.3	313.0549, **169.0132**, 147.0434
145 [c,♥]	Madecassic acid-24-galloyl-28-glucose	$C_{43}H_{62}O_{15}$	30.13	817.4008	[M–H]⁻	−0.2	**655.3470**, **503.3369**
146 [b,#,♠]	1-*O*-cinnamoyl-2,6-*O*-galloyl-glucose [57]	$C_{29}H_{26}O_{15}$	30.19	613.1190	[M–H]⁻	−0.5	465.0667, **461.1085**, 313.0558, **169.0131**, 147.0441
147 [a,#,♥]	Terminolic acid-24-galloyl-28-glucose [56]	$C_{43}H_{62}O_{15}$	30.27	817.3998	[M–H]⁻	−1.5	**655.3473**, **503.3361**
148 [c,♥]	Rotundic acid-24-galloyl-28-glucose	$C_{43}H_{62}O_{14}$	30.30	801.4066	[M–H]⁻	0.6	**639.3536**, **487.3401**
149 [c,♥]	Madecassic acid-24-*O*-glucoheptonic acid	$C_{37}H_{60}O_{13}$	30.34	711.3954	[M–H]⁻	−0.3	**503.3372**
150 [c,♥]	Rotundic acid-24-*O*-glucoheptonic acid	$C_{37}H_{60}O_{12}$	30.37	695.3983	[M–H]⁻	−3.5	**487.3413**
151 [c,♥]	Asiatic acid-24-galloyl-28-glucose	$C_{43}H_{62}O_{14}$	30.39	801.4073	[M–H]⁻	1.5	**639.3538**, **487.3424**
152 [a,#,♠]	1,6-*O*-galloyl-2-*O*-cinnamoyl-glucose [20]	$C_{29}H_{26}O_{15}$	30.39	613.1200	[M–H]⁻	1.1	465.0645, **461.1104**, 313.0572, **169.0134**, 147.0451
153 [c,♥]	Terminolic acid-24-*O*-glucoheptonic acid	$C_{37}H_{60}O_{13}$	30.43	711.3954	[M–H]⁻	−0.3	**503.3377**
154 [c,♥]	Asiatic acid-24-*O*-glucoheptonic acid	$C_{37}H_{60}O_{12}$	30.47	695.4011	[M–H]⁻	0.6	**487.3428**
155 [c,♥]	Arjunolic acid-24-galloyl-28-glucose	$C_{43}H_{62}O_{14}$	30.51	801.4060	[M–H]⁻	1–0.1	**639.3539**, **487.3427**
156 [c,♥]	Arjunolic acid-24-*O*-glucoheptonic acid	$C_{37}H_{60}O_{12}$	30.57	695.4019	[M–H]⁻	0.6	**487.3432**
157 [a,#,♠]	1,2,3-tri-*O*-galloyl-6-*O*-cinnamoyl-β-D-glucose [20]	$C_{36}H_{30}O_{19}$	30.57	765.1310	[M–H]⁻	0.9	**811.4279**, **787.1131**, 635.0932, 617.0841, **613.1192**, **595.1093**, **443.0975**
158 [a,#,♥]	Arjungenin [40]	$C_{30}H_{48}O_6$	31.12	503.3384	[M–H]⁻	2.2	**1007.6838**, **549.3438**
159 [a,#,♥]	Madecassic acid [28]	$C_{30}H_{48}O_6$	31.56	503.3387	[M–H]⁻	2.8	**1007.6841**, **549.3439**
160 [a,#,♥]	23-galloyl-arjunolic acid [58]	$C_{37}H_{52}O_9$	31.71	639.3547	[M–H]⁻	2.2	**487.3378**
161 [a,#,♥]	Terminolic acid [55]	$C_{30}H_{48}O_6$	31.77	503.3387	[M–H]⁻	2.8	**1007.6847**, **549.3439**
162 [b,#,♥]	Rotundic acid [59]	$C_{30}H_{48}O_5$	31.90	487.3437	[M–H]⁻	2.9	**975.6937**
163 [a,#,♥]	Asiatic acid [32]	$C_{30}H_{48}O_5$	32.09	487.3438	[M–H]⁻	3.1	**975.6949**
164 [a,#,♥]	Arjunolic acid [18]	$C_{30}H_{48}O_5$	32.35	487.3434	[M–H]⁻	2.3	**975.6979**

2.3.1. Gallic Acid Derivatives

In *T. chebula*, the primary compounds of gallic acid and its derivatives consist of simple gallic acid components and their acyl esters. The main constituents identified are gallic acid (**8**) and methyl gallate (**39**). These compounds are characterized by multiple phenolic hydroxyl and carboxyl groups, which render them susceptible to the loss of molecules such as CO_2, CO, and H_2O during cleavage. A relatively high content of gallic acid is present in myrobalan, with two significant ion peaks observed at m/z 169 and m/z 125 in the primary mass spectrum. Based on the peak elution time and ionic strength of the compounds, m/z 169 has been confirmed as the deprotonated molecular ion $[M–H]^-$, while m/z 125 is identified as the main fragment ion resulting from in-source fragmentation, with the molecular formula $C_7H_6O_5$. Further analysis indicates that the primary secondary fragment is the ion at m/z 125.0238 $[M–H–CO_2]^-$, formed by the removal of a carboxyl group (CO_2, $\Delta m = 44$). By comparing with the mass spectrometry data of the reference substance, peak **8** has been confirmed as gallic acid. The primary mass spectrum of peak **39** exhibits a prominent abundance at m/z 183, which is identified as a deprotonated molecular ion $[M–H]^-$, corresponding to the molecular formula $C_8H_8O_5$. In the secondary mass spectrum, in addition to the m/z 169 and m/z 125 fragments of gallic acid, characteristic fragments m/z 168 $[M–H–CH_3]^-$ and m/z 124.0160 $[M–H–CH_3-CO_2]^-$ have also been detected. Consequently, peak **39** has been confirmed as methyl gallate.

Simple galloyl ester compounds are mainly divided into three categories: gallic acid bound to glucose (Glc), gallic acid bound to quinic acid, and gallic acid bound to shikimic acid. In this study, 38 gallotannin components were identified. These components were classified according to the number of galloyl groups attached to the Glc molecules. The categories include mono-galloyl glucose (**1**, **6**, **7**, **9**, **15**), di-galloyl glucose (**23**, **30**, **32**, **36**, **37**, **40**, **41**, **44**, **45**, **49**), tri-galloyl glucose (**51**, **55**, **57**, **60**, **72**, **73**, **75**, **77**, **82**, **85**), tetra-galloyl glucose (**80**, **108**, **112**, **114**, **120**), and penta-galloyl glucose (**130**). Characteristic neutral losses were observed as galloyl and gallic acid moieties were gradually removed (Figure 5).

The mass spectrum of penta-galloyl glucose (m/z 939) demonstrated a sequential loss of galloyl groups, resulting in the formation of tetra-galloyl glucose (m/z 787), tri-galloyl glucose (m/z 635), di-galloyl glucose (m/z 483), and mono-galloyl glucose (m/z 331). Additionally, the primary intermediate ions observed for di-galloyl glucose and mono-galloyl glucose were m/z 271 and m/z 211, respectively. These fragment ions were attributed to the continuous loss of -CHOH groups from the glucose moiety, indicating that mono-galloyl glucose undergoes fragmentation to yield $[M–H–60]^-$ and $[M–H–60–60]^-$.

Based on the number and structure of galloyl and cinnamoyl groups, a series of isomers can be distinguished, and their peaks can be clearly identified. These compounds are categorized as cinnamoyl-mono-galloyl glucose (**136**, **140**, **142**, **144**), cinnamoyl-di-galloyl glucose (**143**, **146**, **152**), and cinnamoyl-tri-galloyl glucose (**157**). In the mass spectrum, cinnamoyl-tri-galloyl glucose (m/z 765) exhibits a gradual loss of the gallic acid component, leading to the formation of cinnamoyl-di-galloyl glucose (m/z 595), cinnamoyl-mono-galloyl glucose (m/z 425), and cinnamoyl glucose (m/z 255) (Figure 6). Additionally, characteristic diagnostic ions of m/z 169, attributed to gallic acid, are observed in the mass spectrometry analysis. Ion fragments of m/z 125 are generated through the neutral loss of carboxyl groups (CO_2, $\Delta m = 44$), while m/z 103 is formed by the loss of carboxyl groups (CO_2, $\Delta m = 44$) from cinnamic acid, serving as a distinctive fragment for cinnamic acid. For the position isomers, the calculated lipophilicity parameter (ClogP) was used to estimate the retention time of isomers in the reversed-phase column as the basis for differentiation. Generally, compounds with a larger ClogP value would retain longer. The structures of these isomers were ultimately assigned by combining peak times with calculated ClogP values (Table S1).

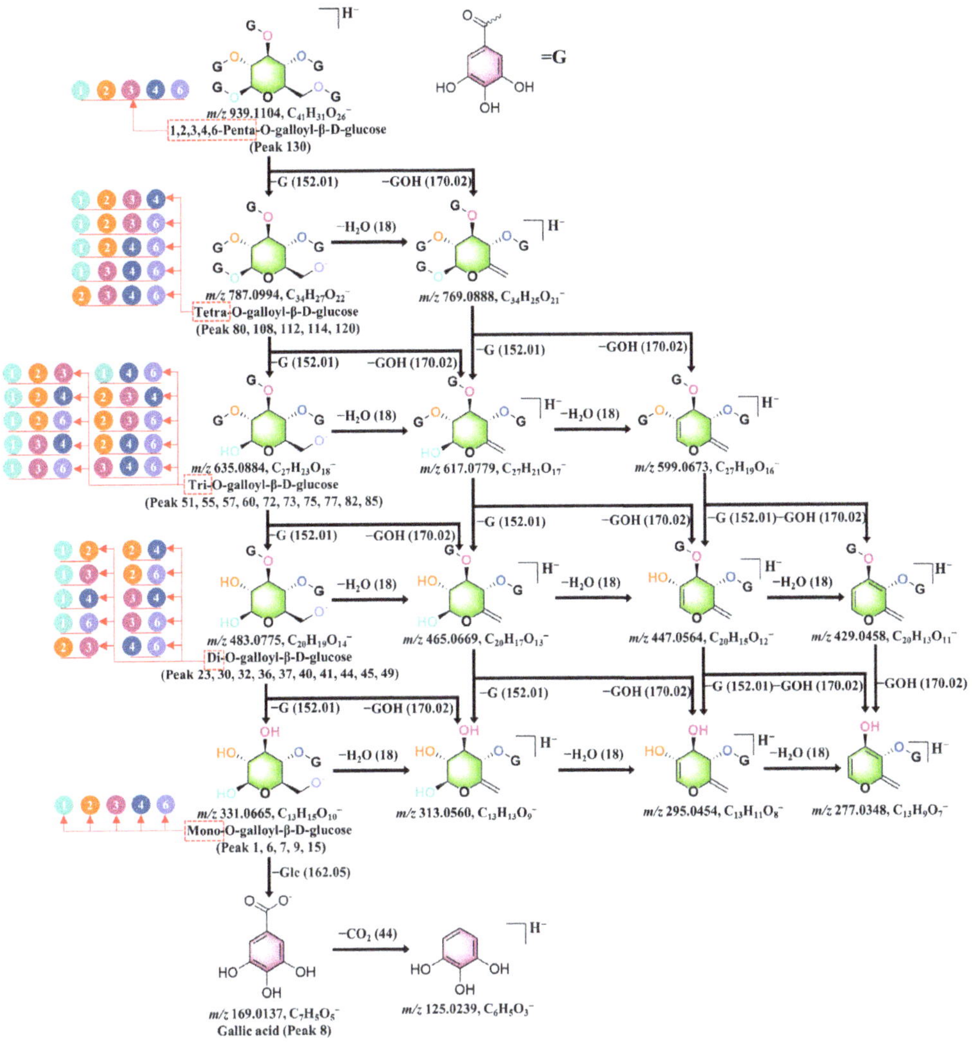

Figure 5. Schematic putative fragmentation pattern of gallotannins. The positions of galloyl groups on the sugar structure are indicated by numbers in circular shapes of different colors. The number of galloyl groups in glycoside is shown by red dashed boxes.

Another type of simple galloyl ester is formed by the combination of gallic acid and shikimic acid. Based on the number of galloyl groups, these compounds can be categorized as mono-galloyl shikimic acid, di-galloyl shikimic acid, and tri-galloyl shikimic acid. In this study, a total of seven related compounds were identified and classified according to the number of galloyl groups connected to shikimic acid, including mono-galloyl shikimic acid (**22**, **25**, **26**), di-galloyl shikimic acid (**48**, **61**, **65**), and tri-galloyl shikimic acid (**99**). Prominent characteristic fragment ions were observed in the mass spectrum through the sequential elimination of galloyl and shikimic acid moieties. Tri-galloyl shikimic acid (*m*/*z* 629) demonstrated continuous mass loss of the galloyl moiety, leading to the formation of di-galloyl shikimic acid (*m*/*z* 477) and mono-galloyl shikimic acid (*m*/*z* 325). A fragment ion of *m*/*z* 169 was detected for all components and identified as $[M-H]^-$ of gallic acid, resulting in a fragment of *m*/*z* 125 through the neutral loss of a carboxyl group (CO_2,

Δm = 44). In addition, fragment ions of *m*/*z* 155 and *m*/*z* 137 were attributed to the loss of water molecules (H_2O, Δm = 18) from shikimic acid. Similarly, further removal of the carboxyl group (CO_2, Δm = 44) after water loss resulted in fragment ions of *m*/*z* 111 and *m*/*z* 93, which are considered characteristic fragments of shikimic acid. Finally, by analyzing the arrangement of galloyl groups at different positions on shikimic acid, a series of isomers was identified (Figure 7). The assignment of each peak was determined by examining peak times and ClogP values (Table S1).

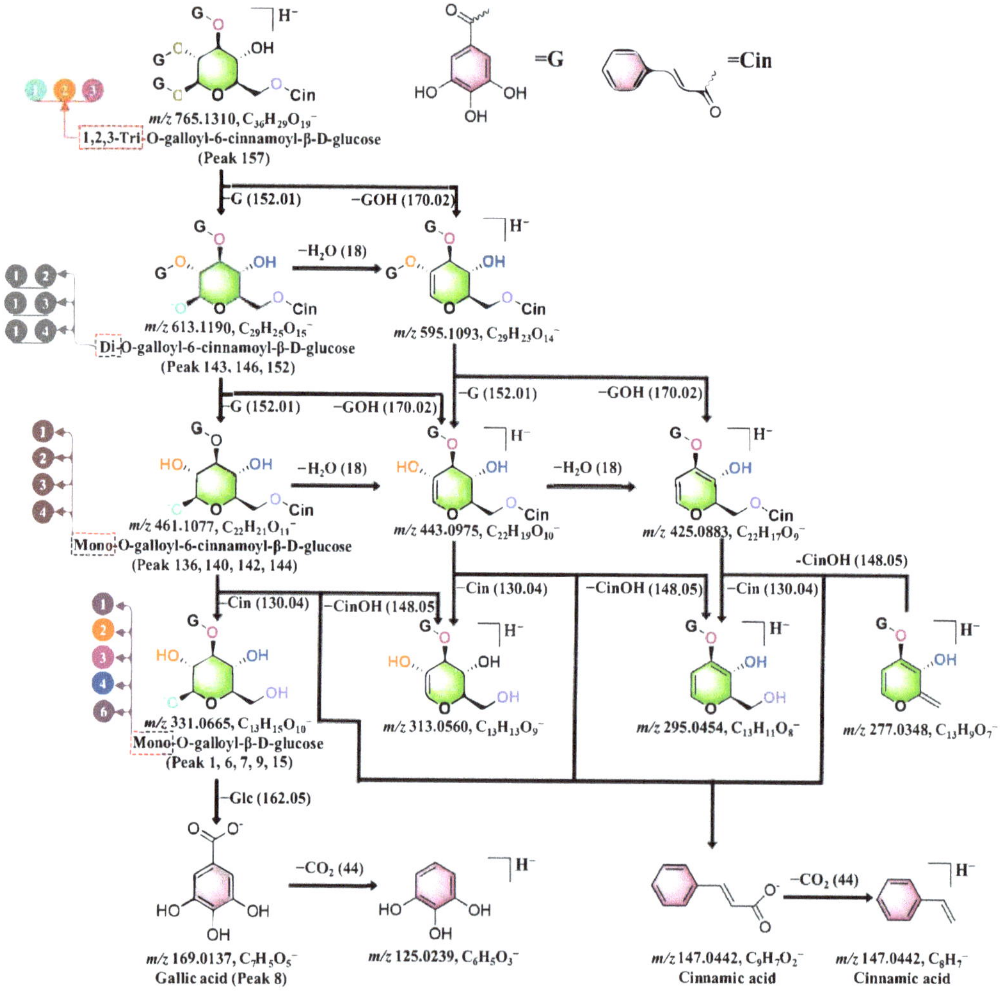

Figure 6. Schematic putative fragmentation pattern of galloyl derivatives of cinnamic acid. The positions of galloyl groups on the sugar structure are indicated by numbers in circular shapes of different colors. The number of galloyl groups in glycoside is shown by red dashed boxes.

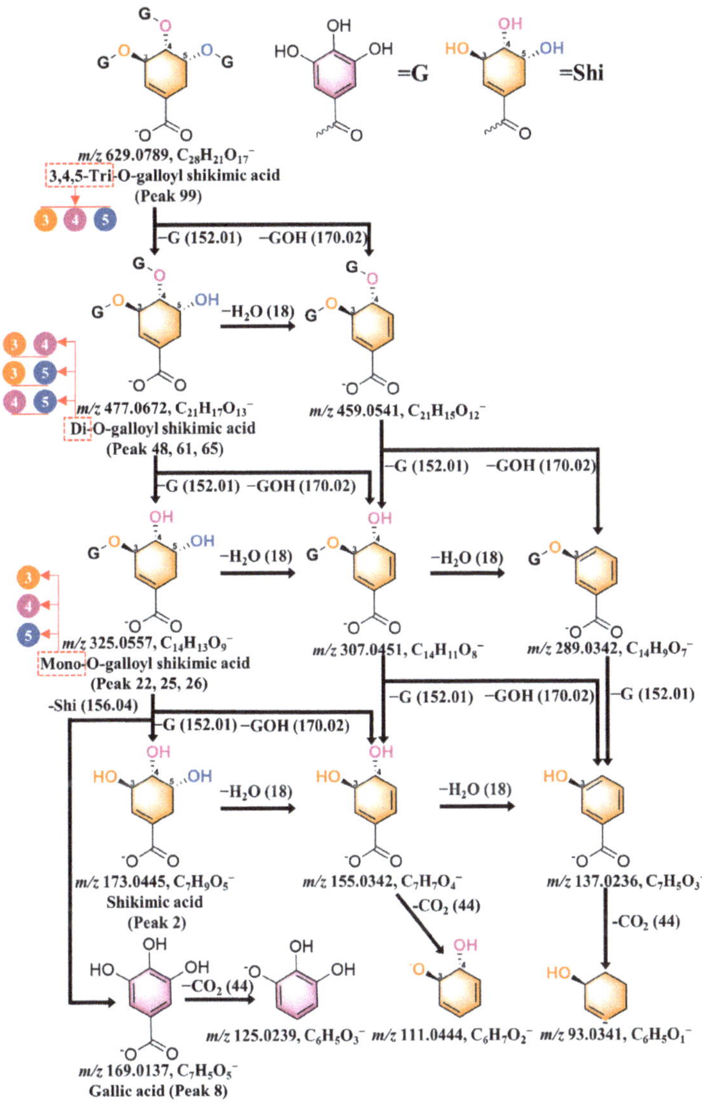

Figure 7. Schematic putative fragmentation pattern of galloyl derivatives of shikimic acid. The positions of galloyl groups on the shikimic acid structure are indicated by numbers in circular shapes of different colors. The number of galloyl groups in shikimic acid is shown by red dashed boxes.

The final type of simple galloyl ester is formed by the combination of gallic acid and quinic acid. This compound typically produces a characteristic fragment at m/z 191. While components with two or three galloyl groups combined with quinic acid may exist, none were identified in this study. A total of three isomers of mono-galloyl quinic acid were identified. In the primary mass spectrum, peaks **5**, **11**, and **17** all displayed ion peaks at m/z 389 and m/z 343. Notably, the intensity of peak **17** was significantly higher than that of peaks **5** and **11**, approximately five times greater. Additionally, peak **17** presented fragments at m/z 709 and m/z 687. In the secondary mass spectrum, the ion peak intensities of m/z 709, m/z 687, and m/z 389 were all reduced. Based on these observations, it was concluded that m/z 709 corresponds to the $[2M+Na-2H]^-$ ion peak, m/z 687 corresponds to the [2M–H]$^-$ ion peak, and *m/z* 389 represents the adduct ion [M+HCOO]$^-$, with the corresponding molecular formula being $C_{16}H_{14}O_{10}$. Fragment ion peaks of *m/z* 191 [quinic acid–H]$^-$ and *m/z* 169 [gallic acid–H]$^-$ can be generated through the neutral loss of quinic acid or gallic acid. By incorporating relevant literature, the order of the peaks was further clarified, leading to the identification of peak **5** as 3-galloyl quinic acid, peak **11** as 5-galloyl quinic acid, and peak **17** as 4-galloyl quinic acid.

2.3.2. Ellagitannins

Ellagitannins are a significant class of polyphenolic compounds containing one or more hexahydroxydiphenoyl (HHDP) groups or their oxidized forms, such as dehydrohexahydroxydiphenoyl (DHHDP) and chebuloyl (Che). Upon hydrolysis, they yield stable ellagic acid. In negative ion mode, ellagic acid tannins exhibit characteristic fragment ion losses, including HHDP (302 Da), DHHDP (318 Da), and chebuloyl (320 Da) (Figure 4).

Ellagic Acid and Its Simple Derivatives

The simple derivatives of ellagic acid primarily consist of ellagic acid conjugated with rhamnose to form glycosides at the 1-position, with additional hydroxyl groups on the rhamnose linked to multiple galloyl groups. MS analysis reveals that peaks **134**, **137**, and **138** display identical ions at *m/z* 375 and *m/z* 751 in the primary MS. The disappearance of *m/z* 375 ions in the secondary MS suggests that *m/z* 375 corresponds to the [M–2H]$^{2-}$ ion peak. The *m/z* 751 ion is identified as the [M–H]$^-$ ion peak, corresponding to the molecular formula $C_{34}H_{24}O_{20}$. In the secondary mass spectrum, fragment ions were observed at *m/z* 599 [M–H–152]$^-$, *m/z* 581 [M–H–170]$^-$, *m/z* 449 [M–H–302]$^-$, *m/z* 411 [M–H–170–170]$^-$, and *m/z* 301 [M–H–152–152–146]$^-$/[ellagic acid–H]$^-$, generated by the sequential loss of galloyl, gallic, and ellagic acid groups. The polarity was assessed using ClogP values (Table S1), leading to the identification of peak **134** as 4-*O*-(2″,3″-di-*O*-galloyl-α-L-rhamnosyl) ellagic acid, peak **137** as 4-*O*-(2″,4″-di-*O*-galloyl-α-L-rhamnosyl) ellagic acid and peak **138** as 4-*O*-(3″,4″-di-*O*-galloyl-α-rhamnopyranosyl) ellagic acid. Similar cleavage patterns were observed for peaks **102** and **132**, where peak **102** was identified as Eschweilenol C and peak **132** as 4-*O*-(4″-*O*-galloyl-α-rhamnopyranosyl) ellagic acid. Additionally, ellagic acid can directly conjugate with gallic acid. In the primary mass spectrum of peak **29**, ion peaks were observed at *m/z* 469 and *m/z* 939. The intensity of the *m/z* 939 ion peak diminished in the secondary mass spectrum, indicating that *m/z* 939 corresponds to the [2M–H]$^-$ ion peak, while *m/z* 469 corresponds to the [M–H]$^-$ ion peak, with the molecular formula $C_{21}H_{10}O_{13}$. Fragment ion peaks at *m/z* 425 [M–H–CO_2]$^-$, *m/z* 407 [M–H–CO_2–H_2O]$^-$, and *m/z* 299 [M–H–gallic acid]$^-$ were generated by the loss of CO_2 and H_2O. Peak **29** was identified as Valoneic acid dilactone. *Terminalia chebula* contains a significant amount of free ellagic acid. Numerous studies have focused on the quantification of ellagic acid using HPLC/UPLC. Peak **88** was identified as ellagic acid through comparison with reference standards.

Simple Tannins Containing a Single HHDP Group

Simple tannins containing a single HHDP group are primarily composed of compounds with one HHDP group and may include one or more attached galloyl groups. Based on the number of galloyl groups, these compounds can be classified into three types: mono-galloyl-HHDP glucose (**10**, **18**, **38**, **50**, **71**), di-galloyl-HHDP glucose (**59**, **68**, **76**, **110**, **117**), and tri-galloyl-HHDP glucose (**81**, **91**, **97**, **104**, **107**, **135**). In mass spectrometry, these compounds exhibit a characteristic cleavage pattern, initially losing the HHDP group, followed by sequential losses of galloyl groups. The mass spectrum shows a characteristic neutral loss of 302 Da, indicative of HHDP group cleavage. For instance, tri-galloyl-HHDP glucose (*m/z* 937) first loses the HHDP group to generate tri-galloyl glucose

(m/z 635), which then fragments following the cleavage pathway typical for tri-galloyl glucose. Similarly, mono-galloyl-HHDP glucose and di-galloyl-HHDP glucose exhibit comparable fragmentation patterns. Additionally, this class of compounds also produces a characteristic fragment at m/z 275, primarily generated by the elimination of the HHDP group, followed by the loss of a galloyl group and subsequently glucose.

Chebulic Acid and Its Simple Derivatives

The primary chebulic acids found in *Terminalia chebula* are neochebulic acid (**3**), chebulic acid (**4**), and isochebulic acid (**12**). These compounds are rich in phenolic hydroxyl and carboxyl groups, making them prone to losing CO_2, H_2O, and other groups during mass spectrometry fragmentation. In the primary mass spectrum, ion peaks at m/z 711, m/z 355, and m/z 337 correspond to peaks **3**, **4**, and **12**, respectively. These ion peaks are also observed in the secondary mass spectrum. The m/z 711 peak is presumed to be the $[2M–H]^-$ ion, m/z 355 as the $[M–H]^-$ ion, and m/z 337 as a fragment with the molecular formula $C_{14}H_{12}O_{11}$. During pyrolysis, the sequential loss of H_2O and CO_2 generates fragment ion peaks at m/z 293 $[M–H–H_2O–CO_2]^-$, m/z 249 $[M–H–H_2O–2CO_2]^-$, and m/z 205 $[M–H–H_2O–3CO_2]^-$. The polarity was assessed using ClogP values (Table S1), and peak order was established by comparison with reference standards. Consequently, peak **3** was identified as isochebulidic acid, peak **4** as chebulidic acid, and peak **12** as neochebulidic acid. In the primary mass spectrum, peaks **21** and **28** exhibit ion peaks at m/z 369 and m/z 739, respectively. In the secondary mass spectrum, the intensity of the m/z 739 ion peak diminishes, suggesting that m/z 739 corresponds to a $[2M–H]^-$ ion, while m/z 369 corresponds to a $[M–H]^-$ ion, with the molecular formula $C_{15}H_{14}O_{11}$. During pyrolysis, these compounds produce fragment ion peaks at m/z 351 $[M–H–H_2O]^-$, m/z 325 $[M–H–CO_2]^-$, and m/z 307 $[M–H–H_2O–CO_2]^-$ through the sequential loss of H_2O and CO_2. Peak **21** was identified as 7′-*O*-methylchebulate and peak **28** as 6′-*O*-methylchebulate.

Simple Tannins Containing a Single Chebuloyl or Neoche Group

In this study, the primary types of simple tannin-like components within the chebuloyl, neoche, and methylneoche groups were identified through mass spectrometry. These components primarily consist of a chebuloyl (320 Da), neoche (338 Da), or methylneoche (352 Da) group and their related derivatives, potentially containing multiple galloyl groups. The molecular weight difference between the chebuloyl and neoche groups is equivalent to one H_2O molecule (18 Da). The key distinction between these groups lies in the attachment: the chebuloyl group is linked to two glycohydroxyl groups, while the neoche group is linked to only one. Based on the number of galloyl groups, these compounds can be further categorized as mono galloyl-neoche glucose (**24, 27, 33, 35, 42, 52, 53, 56, 62, 63, 66**), bisgalloyl-chebuloyl glucose (**78, 92, 98**), and trigaloyl-chebuloyl/neoche glucose (**83, 94, 100, 106, 111, 122, 124, 125, 126, 127, 128, 129**). These compounds exhibited neutral losses of 320 Da, 338 Da, or 352 Da in the mass spectrum. This indicates that these compounds are prone to losing chebuloyl, neoche, or methylneoche groups during mass spectrometric analysis. Specifically, the fragmentation of trigaloyl-chebuloyl/neoche glucose results in the loss of the chebuloyl/neoche group, forming trigaloyl glucose (m/z 635), which then undergoes further cleavage following the pattern typical of trigaloyl glucose. Likewise, monogalloyl-chebuloyl/neoche glucose displayed a comparable fragmentation pattern.

Tannins Containing HHDP, Chebuloyl, Neoche, and Other Groups

Terminalia chebula contains various tannins, which typically include groups such as HHDP (302 Da), chebuloyl (320 Da), neoche (338 Da), THDP (292 Da), DHHDP (318 Da), flavogallonyl (452 Da), and Gallagyl (602 Da). The combinations of these groups are complex and varied, potentially including HHDP along with other groups such as chebuloyl, neoche, THDP, DHHDP, Gallagyl, or flavogallonyl. Different components can be distinguished and identified based on the characteristic neutral loss of these groups and their related fragments.

In mass spectrometry, peaks **19**, **20**, **31**, **43**, and **84** display identical ions at *m*/*z* 541 and 1083 in the primary mass spectrum. The *m*/*z* 541 ion disappears in the secondary mass spectrum and is identified as $[M–2H]^{2-}$, while *m*/*z* 1083 is identified as the $[M–H]^-$ ion with the molecular formula $C_{48}H_{28}O_{30}$. The secondary mass spectrum reveals that these compounds generate major fragment ions at *m*/*z* 601 $[M–H–HHDP–glucose]^-$ and *m*/*z* 300 $[ellagic\ acid–H]^-$, indicating significant neutral losses of HHDP and glucose. Peaks **19** and **20** also exhibit characteristic fragments at *m*/*z* 451 [Flavogallonic acid–H–H_2O]$^-$. Peak **19** is identified as Punicacortein C and peak 20 as Punicacortein D based on the literature comparison. Peaks **31** and **43** are characterized by *m*/*z* 781 $[M–H–HHDP]^-$ fragments. The retention time and fragmentation pathway of peaks **31** and **43** were confirmed to correspond to Punicalagin-α and Punicalagin-β; the fragmentation pathway of Punicalagin is showed in Figure 8. Peak **84** is also characterized by *m*/*z* 449 [M–H–HHDP–glucose–galloyl]$^-$, identified as *T. chebula* based on its retention time. The ions at *m*/*z* 542 and 1085 are observed in the primary mass spectrum, while *m*/*z* 542 disappears in the secondary spectrum. These are identified as $[M–2H]^{2-}$ and *m*/*z* 1085 as $[M–H]^-$ with the molecular formula $C_{48}H_{30}O_{30}$. Through the loss of HHDP and galloyl groups, the fragment ions *m*/*z* 783 $[M–H–HHDP]^-$, *m*/*z* 631 [M–H–HHDP–galloyl]$^-$, and characteristic fragments at *m*/*z* 451 [Flavogallonic acid–H–H_2O]$^-$ are generated. Based on ClogP values (Table S1) and the literature, peak **34** was identified as Rhoipteleanin G and peak **46** as Terflavin A. The ions at *m*/*z* 494 and 989 are observed in the primary mass spectrum, while *m*/*z* 494 disappears in the secondary spectrum. These ions are identified as $[M–2H]^{2-}$ and *m*/*z* 989 as $[M–H]^-$, with the molecular formula $C_{41}H_{34}O_{29}$. The compound first loses a neoche group, forming the *m*/*z* 651 [M–H–neoche]$^-$ fragment, and then loses a gallic acid to generate the *m*/*z* 481 [M–H–neoche–gallic acid]$^-$ fragment. The characteristic fragment at *m*/*z* 337 [neochebulic acid–H–H_2O]$^-$ is generated, and peak **58** is identified as Carpinusnin.

Peaks **64**, **67**, **70**, **74**, **79**, **87**, **101**, **109**, **118**, and **121** display ions at *m*/*z* 485 and 971 in the primary mass spectrum. The *m*/*z* 485 ion disappears in the secondary mass spectrum, while *m*/*z* 971 is identified as an $[M–H]^-$ ion with the molecular formula $C_{41}H_{32}O_{28}$. These compounds are detected in the secondary mass spectrum through the loss of groups such as gallic acid, HHDP, and neoche. The fragment ions observed include *m*/*z* 801 $[M\text{-}H\text{-}170]^-$, *m*/*z* 669 $[M–H–302]^-$, *m*/*z* 633 $[M–H–338]^-$, *m*/*z* 499 $[M–H–170–302]^-$, *m*/*z* 463 $[M–H–338–302]^-$, *m*/*z* 337 [neochebulic acid–H–H_2O]$^-$, and *m*/*z* 301 [ellagic acid–H]$^-$. Based on the chemical structure of components in *Terminalia chebula*, HHDP is likely attached to the 3,6, 4,6, or 2,3 hydroxyl groups of glucose, while neoche groups generally do not attach to the 1-hydroxyl group of glucose. Using ClogP values (Table S1) to assess polarity, the following isomer structures were inferred and confirmed: Peak **64** corresponds to 4-galloyl-6-neoche-2,3-HHDP-glucose and peak **67** to 1-galloyl-2-neoche-4,6-HHDP-glucose. Peak **70** is identified as 1-galloyl-3-neoche-4,6-HHDP-glucose and peak **74** as 1-galloyl-2-neoche-3,6-HHDP-glucose. Peak **79** corresponds to 1-galloyl-4-neoche-3,6-HHDP-glucose and peak **87** to 1-galloyl-4-neoche-2,3-HHDP-glucose. Peak **101** is identified as 2-galloyl-3-neoche-4,6-HHDP-glucose and peak **109** as 1-galloyl-6-neoche-2,3-HHDP-glucose. Peak **118** corresponds to 2-galloyl-4-neoche-3,6-HHDP-glucose and peak **121** to 6-galloyl-4-neoche-2,3-HHDP-glucose.

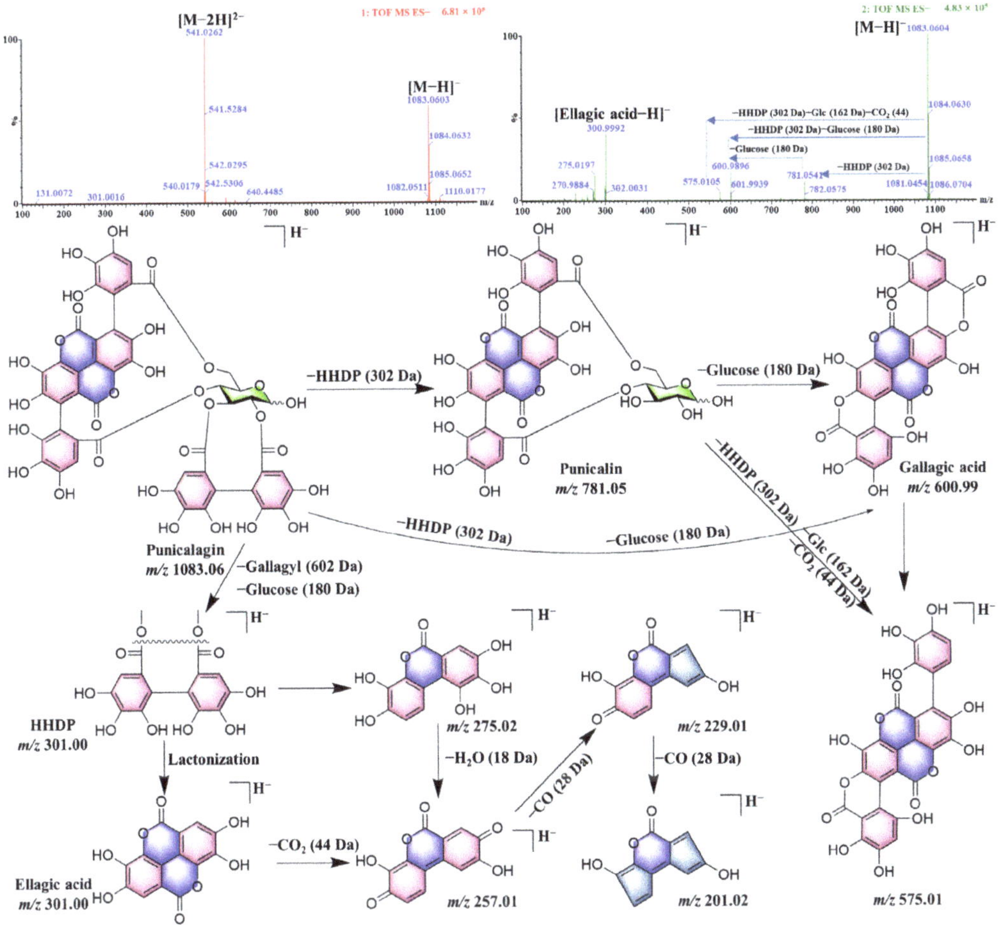

Figure 8. The putative fragmentation pathway of Punicalagin.

In the primary mass spectrum, the ion signal at *m*/*z* 953 was observed in peaks **86**, **90**, **95**, **105**, **119**, and **131**, while the ion at *m*/*z* 476 was detected in peaks **86**, **90**, **95**, **105**, **119**, and **123**, but not at peak **131**. In the secondary mass spectrum, the signal at *m*/*z* 476 disappeared. Thus, *m*/*z* 476 was inferred to be the $[M–2H]^{2-}$ ion peak, and *m*/*z* 953 was identified as the $[M–H]^-$ ion peak. The molecular formula was determined to be $C_{41}H_{30}O_{27}$. Further analysis of the secondary mass spectrum indicates that the fragment patterns of peaks **86**, **90**, **95**, **105**, **119**, and **123** are very similar. The main fragment ions observed include *m*/*z* 783 $[M–H–170]^-$, *m*/*z* 651 $[M–H–302]^-$, *m*/*z* 633 $[M–H–320]^-$, *m*/*z* 481 $[M–H–302–170]^-$, *m*/*z* 463 $[M–H–320–170]^-$, *m*/*z* 337 $[\text{chebuloyl–H}]^-$, *m*/*z* 331 $[M–H–320–302]^-$, *m*/*z* 319 $[M–H–302–170–162]^-$, and *m*/*z* 301 $[\text{ellagic acid–H}]^-$. These fragment ions primarily result from the loss of groups such as gallic acid, chebuloyl, HHDP, and glucose. Based on the fragment ion information, six isomers were identified. Peak **105** was identified as chebulagic acid by comparing its retention time and intensity with those of a reference standard; the fragmentation pathway of chebulagic acid is showed in Figure 9. By assessing the polarity of each peak using ClogP values (Table S1), the following structures were assigned: Peak **86** corresponds to 1-*O*-galloyl-3,4-chebuloyl-2,6-HHDP-D-glucose, and peak **90** corresponds to 1-*O*-galloyl-3,6-chebuloyl-2,4-HHDP-D-glucose.

Peak **95** was assigned as 1-*O*-galloyl-4,6-chebuloyl-3,3-HHDP-D-glucose and peak **119** as 1-*O*-galloyl-2,6-chebuloyl-3,4-HHDP-D-glucose. Peak **123** was assigned as 1-*O*-galloyl-2,3-chebuloyl-4,6-HHDP-D-glucose. The secondary fragments of peak **131** mainly include *m*/*z* 935.0799 [M–H–H_2O]$^-$, *m*/*z* 917.0695 [M–H–2H_2O]$^-$, *m*/*z* 635.0896 [M–H–DHHDP]$^-$, *m*/*z* 617.0781 [M–H–H_2O– DHHDP]$^-$, *m*/*z* 465.0671 [M–H–DHHDP–gallic acid]$^-$, and *m*/*z* 316.9932 [DHHDP–H]$^-$. The loss of DHHDP (318 Da) is the main neutral loss characteristic of the compound, suggesting that peak **131** is terchebin.

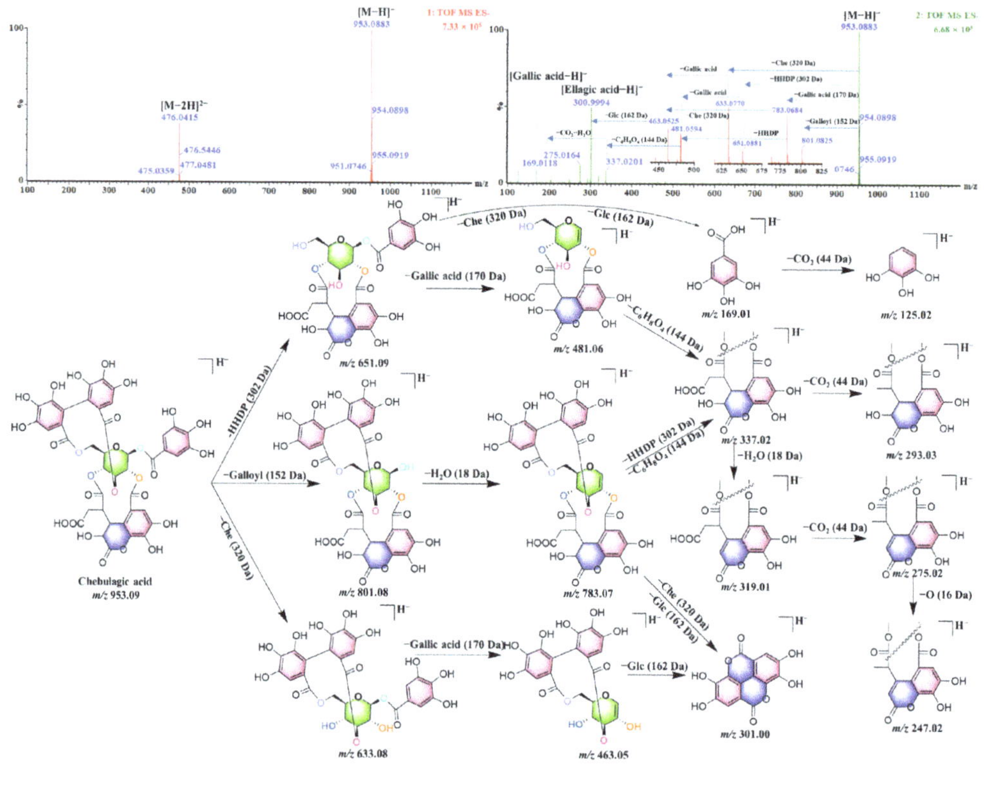

Figure 9. The putative fragmentation pathway of chebulagic acid.

Peaks **113** and **115** display identical ions at *m*/*z* 492 and 985 in the primary mass spectrum. The *m*/*z* 492 ion disappears in the secondary mass spectrum, indicating that *m*/*z* 492 corresponds to the [M–2H]$^{2-}$ ion, while *m*/*z* 985 is the [M–H]$^-$ ion, with the molecular formula determined as $C_{42}H_{34}O_{28}$. In the secondary mass spectrum, fragment ions were observed at *m*/*z* 815 [M–H–170]$^-$, *m*/*z* 683 [M–H–302]$^-$, *m*/*z* 633 [M–H–352]$^-$, *m*/*z* 513 [M–H–302–170]$^-$, *m*/*z* 463 [M–H–352–170]$^-$, *m*/*z* 351 [6′-*O*-methyl neochebulic acid-H_2O-H]$^-$, and *m*/*z* 301 [ellagic acid–H]$^-$. These ions are generated through the sequential loss of gallic acid, HHDP, and 6′-*O*-methyl neochebuloyl. Based on the location of the galloyl group, two isomers were deduced, and their polarity was determined using ClogP values (Table S1). Peak **113** corresponds to 1-*O*-galloyl-3,6-HHDP-4-6′-methyl neochebuloyl-glucose, while peak **115** corresponds to 2-*O*-galloyl-3,6-HHDP-4-6′-methyl neochebuloyl-glucose.

Peaks **89**, **93**, **96**, **103**, **116**, and **133** display identical ions at *m*/*z* 462 and 925 in the primary mass spectrum. The ion at *m*/*z* 462 disappears in the secondary mass spectrum, indicating that it corresponds to the [M–2H]$^{2-}$ ion, while *m*/*z* 925 is assigned to the [M–H]$^-$ ion. The molecular formula is determined to be $C_{40}H_{30}O_{26}$. In the secondary mass spectrum, fragment ions were observed at *m/z* 773 [M–H–152]$^-$, *m/z* 633 [M–H–292]$^-$, *m/z* 481 [M–H–292–152]$^-$, and *m/z* 465 [M–H–292–170]$^-$, resulting from the sequential loss of galloyl, THDP, and other groups. Based on the positions of the galloyl and THDP groups, six isomers were deduced, and their polarity was determined using ClogP values (Table S1). Peak **89** was identified as 1-*O*-galloyl-3,4-THDP-2,6-HHDP-D-glucose. Peak **93** was identified as 1-*O*-galloyl-2,4-THDP-6,6-HHDP-D-glucose (Phyllanthusiin C) and peak **96** as 1-*O*-galloyl-3,6-THDP-2,4-HHDP-D-glucose. Peak **103** was identified as 1-*O*-galloyl-4,6-HHDP-2,3-HHDP-D-glucose, while peak **116** was identified as 1-*O*-galloyl-2,3-HHDP-4,6-HHDP-D-glucose. Peak **133** was identified as 1-*O*-galloyl-2,6-HHDP-3,4-HHDP-D-glucose.

2.3.3. Terpenoids

Terpenoids generate abundant deprotonated molecular ions [M–H]$^-$ in primary mass spectrometry, with some also forming [2M–H]$^-$ ions. These characteristics facilitate the identification of excimer ions and the determination of their molecular formulas. The saponins in *Terminalia chebula* are typically linked to glucose (Glc), galloyl, and glucoheptonic acid through various substitution patterns. Aglycone fragments are relatively stable, primarily losing sugar and galloyl fragments (152, 162, 208 Da) under normal voltages. By analyzing the fragment ions of saponins, aglycones can be classified into 503 Da and 487 Da groups, each with multiple core configurations.

Peaks **139**, **149**, and **153** exhibit a high-intensity ion signal at *m/z* 711 in the primary mass spectrum, identified as the [M–H]$^-$ ion peak corresponding to the molecular formula $C_{37}H_{60}O_{13}$. In the secondary mass spectrum, the fragment at *m/z* 503 [M–H–208]$^-$ displayed a high-intensity signal, suggesting that it resulted from the loss of glucoheptonic acid. Based on the saponin characteristics, the fragment at *m/z* 503 is attributed to the loss of glucoheptonic acid. Peak **139** was identified as Arjungenin-24-*O*-glucoheptonic acid, and peak 149 as Madecassic acid-24-*O*-glucoheptonic acid. Peak **153** was identified as Terminolic acid-24-*O*-glucoheptonic acid. Peaks **141**, **145**, and **147** exhibit strong ion signals at *m/z* 817 in the primary mass spectrum. Combined with secondary mass spectrometry, *m/z* 817 is identified as the [M–H]$^-$ ion peak, corresponding to the molecular formula $C_{43}H_{62}O_{15}$. The secondary mass spectrum reveals intense fragment ions at *m/z* 655 [M–H–162]$^-$ and *m/z* 503 [M–H–162–152]$^-$, likely resulting from the loss of glucose and galloyl groups, as indicated by the characteristic fragments at *m/z* 503. **Peak 141** was identified as Quercotriterpenoside I (Arjungenin-24-galloyl-28-glucose), **peak 145** as Madecassic acid-24-galloyl-28-glucose, and peak **147** as Terminolic acid-24-galloyl-28-glucose. Peaks **158**, **159**, and **161** exhibit similar ion signals at *m/z* 503, 549, and 1007 in the primary mass spectrum. The *m/z* 1007 peak is identified as the [2M–H]$^-$ ion, while the *m/z* 549 peak corresponds to the [M+HCOO]$^-$ ion, with a molecular formula of $C_{30}H_{48}O_6$. Peak **158** was identified as Arjungenin, peak **159** as Madecassic acid, and peak **161** as Terminolic acid.

Peaks **148**, **151**, and **155** appear as [M–H]$^-$ ion peaks at *m/z* 801, corresponding to the molecular formula $C_{43}H_{62}O_{14}$. The intense fragments at *m/z* 639 [M–H–162]$^-$ and *m/z* 487 [M–H–162–152]$^-$ in the secondary mass spectrum suggest that the characteristic fragments at *m/z* 487 result from the loss of glucose and gallic acid groups. Peak **148** was identified as Rotundic acid-24-galloyl-28-glucose and peak **151** as Asiatic acid-24-galloyl-28-glucose. Peak **155** was identified as Arjunolic acid-24-galloyl-28-glucose. Peaks **150**, **154**, and **156** all appear as [M–H]$^-$ ion peaks at *m/z* 695, with the molecular formula identified as $C_{37}H_{60}O_{12}$. In the secondary mass spectrum, an intense fragment at *m/z* 487 [M–H–208]$^-$ was observed, which was attributed to the loss of glucoheptonic acid. Peak **150** was identified as Rotundic acid-24-*O*-glucoheptonic acid. Peak **154** was identified as Asiatic acid-24-*O*-glucoheptonic acid and peak 156 as Arjunolic acid-24-*O*-

glucoheptonic acid. Peak **160** exhibits a strong ion signal at *m*/*z* 639, corresponding to the molecular formula $C_{37}H_{52}O_9$. In the secondary mass spectrum, a fragment at *m*/*z* 487 $[M–H–152]^-$ was detected, attributed to the loss of a galloyl group. Peak **160** was identified as 23-galloyl-arjunolic acid. Peaks **162, 163**, and **164** show ion peaks at *m*/*z* 487 and 975 in the primary mass spectrum. The *m*/*z* 975 peak is identified as the $[2M–H]^-$ ion, while *m*/*z* 487 corresponds to the $[M–H]^-$ ion, with a molecular formula of $C_{30}H_{48}O_5$. Peak **162** was identified as Rotundic acid. Peak **163** was identified as Asiatic acid and peak **164** as Arjunolic acid.

2.3.4. Other Components

Besides the previously mentioned ingredients, four additional compounds (**2, 16, 47, 54, 69**) have been identified in *T. chebula*. Details regarding their chemical compositions are provided in the accompanying table.

2.4. The Applicability of the KID-NLF Strategy

The KID-NLF strategy has established identification methods for various medicinal plants by summarizing the research of Moilanen et al. [60] and examining the mass spectrometry rules in *T. chebula*. Employing this strategy allows for the identification of gallic acid derivatives, ellagitannins, and triterpenoids by precisely recognizing quasi-molecular ion peaks and their related secondary characteristic fragments. In addition to *T. chebula*, plants like *Phyllanthus emblica* [61] and *Punica granatum* [62] (rich in tannins) as well as *Panax ginseng*, *P. quinquefolium*, and *P. notoginseng* (containing triterpenoids) [63] also adhere to this identification rule. Further research has shown that iridoid glycosides, phenolic acids, and flavonoids are compatible with this strategy. For instance, iridoid glycosides dipsanosides A and B demonstrate specific mass spectrometry features, such as *m*/*z* 1519.519 $[M+HCOO]^-$, 1473.514 $[M–H]^-$, 759.252 $[M+2HCOO]^{2-}$, and 736.249 $[M–2H]^{2-}$ as adduct ions. Similarly, the phenolic acids isochlorogenic acid A, B, and C display peaks at *m*/*z* 1029.229 $[2M–H]^-$, 537.100 $[M+Na–2H]^-$, and 515.118 $[M–H]^-$, and the flavonoids hyperoside and isoquercitrin show ions at *m*/*z* 949.164 $[2M+Na–2H]^-$, 927.182 $[2M–H]^-$, and 463.089 $[M–H]^-$ [64,65]. Notably, alkaloids generate primarily $[M+H]^+$ type quasi-molecular ions and lack distinctive adduct patterns, making them incompatible with the KID-NLF strategy's rule system for mass spectrometry. This finding indicates the need for methodological validation based on component type when applying this strategy. Consequently, it is advised to undertake methodological validation according to component type when implementing this strategy and to formulate a specific identification strategy for alkaloids.

3. Experimental

3.1. Chemicals, Reagents, and Plant Materials

Deionized water was prepared using a Millipore Q purification system (Rephile, Shanghai, China). HPLC-grade acetonitrile, methanol, and formic acid were obtained from Sigma-Aldrich (Milwaukee, WI, USA). The crude medicinal materials derived from dried, pitted mature fruits of *T. chebula* were procured from Chengdu in 2023. The author identified these materials, and the specimens were deposited in the laboratory of the author. Reference products including shikimic acid (**2**), gallic acid (**8**), punicalagin α (**31**), punicalagin β (**43**), corilagin (**71**), ellagic acid (**88**), chebulagic acid (**105**), chebulinic acid (**127**), and 1,2,3,4,6-penta-*O*-galloyl-β-D-glucose (**130**) were purchased from Yuanye (Shanghai, China), with purity exceeding 98% as determined by HPLC analysis.

3.2. Sample Preparation

Approximately 1 g of pitted and dried fruit powder of *T. chebula* powder, sieved through a No. 3 mesh, was accurately weighed and transferred into a stoppered conical flask. Subsequently, 50 mL of 70% (*v*/*v*) methanol was added with precision. The sealed flask was weighed and then subjected to ultrasonic treatment at 250 W and 40 kHz for 30 min. After being allowed to cool, the flask was reweighed and any weight loss compensated with 70% methanol. The supernatant was centrifuged at 13,000 rpm for 15 min and filtered through a 0.22 μm microporous filter membrane.

3.3. UHPLC Analysis

UHPLC analysis was conducted using a Shimadzu LC-40 system (Shimadzu Corporation, Kyoto, Japan), which was equipped with an SPD-M40 detector. A Shim-pack GIST-C18 column (2.1 × 100 mm, 2 μm) was utilized. The mobile phase consisted of 0.2% phosphoric acid aqueous solution (A) and acetonitrile (B). A gradient elution profile was employed as follows: 0–3 min, 3–6% B; 3–4 min, 6–6% B; 4–5 min, 6–7% B; 5–6 min, 7–7% B; 6–7 min, 7–15% B; 7–10 min, 15–15% B; 10–11 min, 15–21% B; 11–23 min, 21–21% B; 23–26 min, 21–33% B; 26–28 min, 33–35% B; 28–40 min, 35–35%. Detection wavelengths were monitored at 254 nm. The flow rate was set at 0.4 mL/min, injection volume at 2 μL, and column temperature at 30 °C.

3.4. UPLC-Q-TOF/MS Analysis

UPLC analysis was carried out using a 100 mm × 2.1 mm, 1.7 μm Waters Acquity UPLCR BEH C18 column (Waters Corporation, Milford, MA, USA). The mobile phase comprised acetonitrile (A) and water (B), each containing 0.1% formic acid. The linear gradient program was set as follows: from 2% to 4% B over 0–2 min; from 4% to 5% B over 2–4 min; from 5% to 6% B over 4–11 min; from 6% to 9% B over 11–17 min; from 9% to 13% B over 17–21 min; maintained at 13% B over 21–27 min; from 13% to 60% B over 27–32 min; from 60% to 99% B over 32–33 min; maintained at 99% B over 33–38 min; from 99% to 1% B over 38–40 min. The flow rate was set at 0.4 mL/min, injection volume at 1 μL, and column temperature at 40 °C.

The Waters SYNAPT G2HDMS system with an ion source was employed for electrospray ionization (ESI). Scanning was conducted in negative (ESI^-) ion mode, using nitrogen as the atomization and conical gas. The source temperature was set at 100 °C, and the cone gas flow rate was maintained at 40 L/h. The desolvation temperature was held at 350 °C, and the gas flow rate was 800 L/h. Further MS settings included a sampling cone voltage of 40 V, extraction cone voltage of 4 V, capillary voltage of 2.5 kV, scan time of 0.3 s, inter scan time of 0.02 s, and a mass-to-charge ratio (m/z) ranging from 100 to 1200. Leucine-enkephalin (200 pg/mL) flowing at 10 μL/min was employed to calibrate the mass number m/z 554.2615. Data processing was conducted using MassLynx V4.2 and UNIFI software (Version 1.9) from Waters Corporation, Milford, MA, USA.

4. Conclusions

In conclusion, a post-acquisition LC-MS data processing strategy, key ion diagnostics–neutral loss filtering (KID-NLF), can effectively identify the structure of the natural products responsible for the herbal extract. In this study, a total of 164 compounds were identified by UPLC-Q-TOF/MS technique and KID-NLF strategy screening in 33 min running time, 47 of which were reported for the first time. This study provides a powerful strategy for rapid profiling of chemical constituents of herbal medicines.

Supplementary Materials: The following supporting information can be downloaded at: https://www.mdpi.com/article/10.3390/molecules30112451/s1, Table S1: The LogP and CLogP values of the chemical components in *T. chebula* were identified by UPLC-Q-TOF/MS.

Author Contributions: J.Y.: writing—original draft preparation, drawing preparation, table arrangement; X.Z.: drawing preparation, table arrangement; Y.H.: drawing preparation, table arrangement; Y.Z.: funding acquisition, conceptualization, supervision. C.T.: conceptualization, writing—review and editing, supervision. All authors have read and agreed to the published version of the manuscript.

Funding: The authors gratefully acknowledge the financial support from the National Key Research and Development Program of China (No. 2023YFC3504400, 2023YFC3504401, 2023YFC3504402), the Natural Science Foundation of Sichuan Province (2025ZNSFSC1821), and the "Xinglin Scholars" Program of Chengdu University of TCM (CCYB2022009).

Institutional Review Board Statement: Not applicable.

Informed Consent Statement: Not applicable.

Data Availability Statement: The datasets used and/or analyzed during the current study are available from the corresponding author on reasonable request.

Conflicts of Interest: The authors declare that they have no competing financial interests or personal relationships that may have influenced the work reported in this study.

References

1. Klein-Junior, L.C.; de Souza, M.R.; Viaene, J.; Bresolin, T.M.B.; de Gasper, A.L.; Henriques, A.T.; Heyden, Y.V. Quality Control of Herbal Medicines: From Traditional Techniques to State-of-the-art Approaches. *Planta Med.* **2021**, *87*, 964–988. [CrossRef] [PubMed]
2. Li, X.; Wen, D.; He, Y.; Liu, Y.; Han, F.; Su, J.; Lai, S.; Zhuang, M.; Gao, F.; Li, Z. Progresses and Prospects on Glucosinolate Detection in Cruciferous Plants. *Foods* **2024**, *13*, 4141. [CrossRef] [PubMed]
3. Ye, C.X.; Huang, W.J.; Qin, S.Y.; Mai, Z.Y.; Wang, L.Q.; Li, S.S.; Xu, H.R.; Wang, D.H.; Yan, S.K.; Xiao, X. Application Progress of 2D-LC in Analysis and Quality Control of Traditional Chinese Medicine. *Chem. Reag.* **2022**, *44*, 1103–1110. [CrossRef]
4. Hong, L.L.; Wang, H.D.; Xu, X.Y.; Hu, W.D.; Liu, J.Y.; Wang, X.Y.; Gao, X.M.; Yang, W.Z. Systematic characterization and identification of the chemical constituents of the *Schisandra chinensis* decoction based on a hybrid scanning technique of UHPLC/IM-QTOF-MS. *Acta Pharm. Sin.* **2024**, *59*, 678–692. [CrossRef]
5. Qiao, X.; Li, R.; Song, W.; Miao, W.J.; Liu, J.; Chen, H.B.; Guo, D.A.; Ye, M. A targeted strategy to analyze untargeted mass spectral data: Rapid chemical profiling of Scutellaria baicalensis using ultra-high performance liquid chromatography coupled with hybrid quadrupole orbitrap mass spectrometry and key ion filtering. *J. Chromatogr. A* **2016**, *1441*, 83–95. [CrossRef]
6. Chen, J.; Wang, X.; Xin, X.; Zheng, Y.; Hou, F.; Li, C.; Guo, L.; Wang, L. Comprehensive comparison of two colour varieties of *Perillae folium* by UHPLC-Q-TOF/MS analysis combining with feature-based molecular networking. *Food Chem.* **2025**, *463*, 141293. [CrossRef]
7. Schmid, R.; Petras, D.; Nothias, L.F.; Wang, M.X.; Aron, A.T.; Jagels, A.; Tsugawa, H.; Rainer, J.; Garcia-Aloy, M.; Dührkop, K.; et al. Ion identity molecular networking for mass spectrometry-based metabolomics in the GNPS environment. *Nat. Commun.* **2021**, *12*, 3832. [CrossRef]
8. Duhrkop, K.; Fleischauer, M.; Ludwig, M.; Aksenov, A.A.; Melnik, A.V.; Meusel, M.; Dorrestein, P.C.; Rousu, J.; Bocker, S. SIRIUS 4: A rapid tool for turning tandem mass spectra into metabolite structure information. *Nat. Methods* **2019**, *16*, 299–302. [CrossRef]
9. Wang, C.; Zhang, H.; Wang, X.; Wang, X.; Li, X.; Li, C.; Wang, Y.; Zhang, M. Comprehensive Review on Fruit of *Terminalia chebula*: Traditional Uses, Phytochemistry, Pharmacology, Toxicity, and Pharmacokinetics. *Molecules* **2024**, *29*, 5547. [CrossRef]
10. Dhar, A.; Maurya, S.K.; Mishra, A.; Singh, G.K.; Singh, M.K.; Seth, A. Preliminary Screening of a Classical Ayurvedic Formulation for Anticonvulsant Activity. *Anc. Sci. Life* **2016**, *36*, 28–34. [CrossRef]
11. Nigam, M.; Mishra, A.P.; Adhikari-Devkota, A.; Dirar, A.I.; Hassan, M.M.; Adhikari, A.; Belwal, T.; Devkota, H.P. Fruits of *Terminalia chebula* Retz.: A review on traditional uses, bioactive chemical constituents and pharmacological activities. *Phytother. Res.* **2020**, *34*, 2518–2533. [CrossRef] [PubMed]
12. Gao, H.; Lu, H.; Fang, N.; Su, J.; Li, R.; Wang, W.; Zhang, Y. The potential of *Terminalia chebula* in alleviating mild cognitive impairment: A review. *Front. Pharmacol.* **2024**, *15*, 1484040. [CrossRef]
13. Kim, H.L.; Choi, B.K.; Yang, S.H. *Terminalia chebula* Medicinal Uses: A Review of in vitro and in vivo Studies. *Biotechnol. Bioprocess Eng.* **2022**, *27*, 703–713. [CrossRef]

14. Zhang, Y.X.; Zhang, Y.S.; Zhang, G.X.; Sun, Y.; Yang, W.L.; Zhang, C.Y. Research Progress and Application Prospects of Medicinal and Edible Tibetan Medicine *Terminalia chebula*. *Chin. Wild Plant Res.* **2024**, *43*, 49–59. [CrossRef]
15. Tiwari, M.; Barooah, M.S. A comprehensive review on the ethno-medicinal and pharmacological properties of *Terminalia chebula* fruit. *Phytochem. Rev.* **2024**, *23*, 125–145. [CrossRef]
16. Singh, A.; Bajpai, V.; Kumar, S.; Sharma, K.R.; Kumar, B. Profiling of Gallic and Ellagic Acid Derivatives in Different Plant Parts of *Terminalia arjuna* by HPLC-ESI-QTOF-MS/MS. *Nat. Prod. Commun.* **2016**, *11*, 239–244. [CrossRef]
17. Kaunda, J.S.; Zhang, Y.J. Two New Phenolic Constituents from the Stems of *Euphorbia griffithii*. *Nat. Prod. Bioprospect.* **2019**, *9*, 405–410. [CrossRef]
18. Li, K.; Han, X.; Li, R.; Xu, Z.; Pan, T.; Liu, J.; Li, B.; Wang, S.; Diao, Y.; Liu, X. Composition, Antivirulence Activity, and Active Property Distribution of the Fruit of *Terminalia chebula* Retz. *J. Food Sci.* **2019**, *84*, 1721–1729. [CrossRef]
19. Ding, G.; Liu, Y.Z.; Yan, Z.; Lu, Y.R.; Sheng, L.S. Two new isomers of chebulic acid from Terminalia chebula. *J. Chin Pharma Univ.* **2000**, *32*, 333–335.
20. Lee, D.Y.; Kim, H.W.; Yang, H.; Sung, S.H. Hydrolyzable tannins from the fruits of *Terminalia chebula* Retz and their α-glucosidase inhibitory activities. *Phytochemistry* **2017**, *137*, 109–116. [CrossRef]
21. Bouchet, N.; Levesque, J.; Blond, A.; Bodo, B.; Poisset, J.L. 1,3-di-*O*-galloylquinic acid from *Guiera senegalensis*. *Phytochemistry* **1996**, *42*, 189–190. [CrossRef]
22. Ahn, M.J.; Kim, C.Y.; Lee, J.S.; Kim, T.G.; Kim, S.H.; Lee, C.K.; Lee, B.B.; Shin, C.G.; Huh, H.; Kim, J. Inhibition of HIV-1 integrase by galloyl glucoses from *Terminalia chebula* and flavonol glycoside gallates from *Euphorbia pekinensis*. *Planta Med.* **2002**, *68*, 457–459. [CrossRef] [PubMed]
23. Li, Y.J.; He, F.Q.; Zhao, H.H.; Li, Y.; Chen, J. Screening and identification of acetylcholinesterase inhibitors from *Terminalia chebula* fruits by immobilized enzyme on cellulose filter paper coupled with ultra-performance liquid chromatography-quadrupole time-of-flight mass spectrometry and molecular docking. *J. Chromatogr. A* **2022**, *1663*, 462784. [CrossRef] [PubMed]
24. Liang, C.C.; Zhang, F.Q.; Chen, J. Screening and characterization of cosmetic efficacy components of Terminalia chebula based on biological activity-guided methodology. *Biomed. Chromatogr.* **2024**, *38*, e5974. [CrossRef]
25. Tang, Y.; Yin, L.; Zhang, Y.; Huang, X.; Zhao, F.; Cui, X.; Shi, L.; Xu, L. Study on anti-inflammatory efficacy and correlative ingredients with pharmacodynamics detected in acute inflammation rat model serum from *Caulis Lonicerae* japonicae. *Phytomedicine* **2016**, *23*, 597–610. [CrossRef]
26. Saha, S.; Imran, I.B.B. Sensitive lanthanide label array method for rapid fingerprint analysis of plant polyphenols based on time-resolved luminescence. *Anal. Methods* **2019**, *11*, 5044–5054. [CrossRef]
27. Ajala, O.S.; Jukov, A.; Ma, C.M. Hepatitis C virus inhibitory hydrolysable tannins from the fruits of *Terminalia chebula*. *Fitoterapia* **2014**, *99*, 117–123. [CrossRef]
28. Chen, H.F.; Zhang, C.; Yao, Y.; Li, J.M.; Du, W.D.; Li, M.L.; Wu, B.; Yang, S.L.; Feng, Y.L.; Zhang, W.G. Study on anti-hyperuricemia effects and active ingredients of traditional Tibetan medicine TongFengTangSan (TFTS) by ultra-high-performance liquid chromatography coupled with quadrupole time-of-flight mass spectrometry. *J. Pharm. Biomed. Anal.* **2019**, *165*, 213–223. [CrossRef]
29. Wang, Z.; Zhang, Y.; Fu, Z.; Jin, T.; Zhao, C.; Zhao, M. A comprehensive strategy for quality evaluation of Changan powder by fingerprinting combined with rapid qualitative and quantitative multi-ingredients profiling. *Phytochem. Anal.* **2024**, *35*, 840–859. [CrossRef]
30. Han, N.; Ye, Q.; Guo, Z.; Liang, X. Metabolomics analysis of differential chemical constituents and α-glucosidase inhibiting activity of *Phyllanthus urinaria* L. root, stem, leaf and fruit. *Nat. Prod. Res.* **2023**, *37*, 642–645. [CrossRef]
31. Tanaka, T.; Jiang, Z.H.; Kouno, I. Structures and biogenesis of rhoipteleanins, ellagitannins formed by stereospecific intermolecular C-C oxidative coupling, isolated from *Rhoiptelea chiliantha*. *Chem. Pharm. Bull.* **1997**, *45*, 1915–1921. [CrossRef]
32. Jha, A.K.; Sit, N. Effect of ultrasound, microwave, and enzymatically pre-treated *Terminalia chebula* pulp on extraction of bioactive compounds using supercritical CO_2. *Sustain. Chem. Pharm.* **2023**, *33*, 101098. [CrossRef]
33. Li, C.W.; Dong, H.J.; Cui, C.B. The synthesis and antitumor activity of twelve galloyl glucosides. *Molecules* **2015**, *20*, 2034–2060. [CrossRef] [PubMed]
34. de Oliveira, F.M.G.; Lyrio, M.V.V.; Filgueiras, P.R.; de Castro, E.V.R.; Kuster, R.M. ESI(-)FT-ICR MS for the determination of best conditions for producing extract abundant in phenolic compounds from leaves of *E. uniflora* and FTIR-PCA as a sample screening method. *Anal. Methods* **2024**, *16*, 3701–3713. [CrossRef] [PubMed]
35. Zheng, Y.; Zhao, L.; Yi, J.; Cai, S. Effects and Mechanisms of *Rhus chinensis* Mill. Fruits on Suppressing RANKL-Induced Osteoclastogenesis by Network Pharmacology and Validation in RAW264.7 Cells. *Nutrients* **2022**, *14*, 1020. [CrossRef]
36. Zheng, W.; Zhang, M.; Zhao, Y.; Miao, K.; Jiang, H. NMR-based metabonomic analysis on effect of light on production of antioxidant phenolic compounds in submerged cultures of Inonotus obliquus. *Bioresour. Technol.* **2009**, *100*, 4481–4487. [CrossRef]

37. Li, M.; Zhang, Y.; Yan, J.; Ding, F.; Chen, C.; Zhong, S.; Li, M.; Zhu, Y.; Yue, P.; Li, P.; et al. Comparative Metabolomic Analysis Reveals the Differences in Nonvolatile and Volatile Metabolites and Their Quality Characteristics in Beauty Tea with Different Extents of Punctured Leaves by Tea Green Leafhopper. *J. Agric. Food Chem.* **2023**, *71*, 16233–16247. [CrossRef]
38. Xu, M.; Zhu, H.T.; Cheng, R.R.; Wang, D.; Yang, C.R.; Tanaka, T.; Kouno, I.; Zhang, Y.J. Antioxidant and hyaluronidase inhibitory activities of diverse phenolics in *Phyllanthus emblica*. *Nat. Prod. Res.* **2016**, *30*, 2726–2729. [CrossRef]
39. Chang, C.W.; Lin, M.T.; Lee, S.S.; Liu, K.C.; Hsu, F.L.; Lin, J.Y. Differential inhibition of reverse transcriptase and cellular DNA polymerase-alpha activities by lignans isolated from Chinese herbs, *Phyllanthus myrtifolius* Moon, and tannins from *Lonicera japonica* Thunb and *Castanopsis hystrix*. *Antiviral. Res.* **1995**, *27*, 367–374. [CrossRef]
40. Zhang, X.R.; Qiao, Y.J.; Zhu, H.T.; Kong, Q.H.; Wang, D.; Yang, C.R.; Zhang, Y.J. Multiple in vitro biological effects of phenolic compounds from *Terminalia chebula* var. tomentella. *J. Ethnopharmacol.* **2021**, *275*, 114135. [CrossRef]
41. Przybylska, D.; Kucharska, A.Z.; Cybulska, I.; Sozański, T.; Piórecki, N.; Fecka, I. *Cornus mas* L. Stones: A Valuable By-Product as an Ellagitannin Source with High Antioxidant Potential. *Molecules* **2020**, *25*, 4646. [CrossRef] [PubMed]
42. Shimozu, Y.; Kuroda, T.; Tsuchiya, T.; Hatano, T. Structures and Antibacterial Properties of Isorugosins H-J, Oligomeric Ellagitannins from *Liquidambar formosana* with Characteristic Bridging Groups between Sugar Moieties. *J. Nat. Prod.* **2017**, *80*, 2723–2733. [CrossRef] [PubMed]
43. Liu, J.B.; Ding, Y.S.; Zhang, Y.; Chen, J.B.; Cui, B.S.; Bai, J.Y.; Lin, M.B.; Hou, Q.; Zhang, P.C.; Li, S. Anti-inflammatory Hydrolyzable Tannins from *Myricaria bracteata*. *J. Nat. Prod.* **2015**, *78*, 1015–1025. [CrossRef]
44. Sehgal, R.; Sharma, A.K.; Singh, B.J.; Saini, R.V.; Saini, A.K.; Beniwal, V. Augmenting the antioxidant, anti-bacterial and anti-carcinogenic potential of *Terminalia chebula* and *Terminalia bellirica* after tannin acyl hydrolase mediated biotransformation. *Biocatal. Agric. Biotechnol.* **2024**, *56*, 103045. [CrossRef]
45. Avula, B.; Wang, Y.H.; Isaac, G.; Yuk, J.; Wrona, M.; Yu, K.; Khan, I.A. Metabolic Profiling of Hoodia, Chamomile, Terminalia Species and Evaluation of Commercial Preparations Using Ultrahigh-Performance Liquid Chromatography Quadrupole-Time-of-Flight Mass Spectrometry. *Planta Med.* **2017**, *83*, 1297–1308. [CrossRef]
46. Takemoto, M.; Kawamura, Y.; Hirohama, M.; Yamaguchi, Y.; Handa, H.; Saitoh, H.; Nakao, Y.; Kawada, M.; Khalid, K.; Koshino, H.; et al. Inhibition of protein SUMOylation by davidiin, an ellagitannin from *Davidia involucrata*. *J. Antibiot.* **2014**, *67*, 335–338. [CrossRef]
47. Haramiishi, R.; Okuyama, S.; Yoshimura, M.; Nakajima, M.; Furukawa, Y.; Ito, H.; Amakura, Y. Identification of the characteristic components in walnut and anti-inflammatory effect of glansreginin A as an indicator for quality evaluation. *Biosci. Biotechnol. Biochem.* **2020**, *84*, 187–197. [CrossRef]
48. Spinelli, L.V.; Anzanello, M.J.; Areze da Silva Santos, R.; Carboni Martins, C.; Freo Saggin, J.; Aparecida Silva Da Silva, M.; Rodrigues, E. Uncovering the phenolic diversity of Guabiju fruit: LC-MS/MS-based targeted metabolomics approach. *Food Res. Int.* **2023**, *173*, 113236. [CrossRef]
49. Pfundstein, B.; El Desouky, S.K.; Hull, W.E.; Haubner, R.; Erben, G.; Owen, R.W. Polyphenolic compounds in the fruits of Egyptian medicinal plants (*Terminalia bellerica*, *Terminalia chebula* and *Terminalia horrida*): Characterization, quantitation and determination of antioxidant capacities. *Phytochemistry* **2010**, *71*, 1132–1148. [CrossRef]
50. Engström, M.T.; Arvola, J.; Nenonen, S.; Virtanen, V.T.J.; Leppä, M.M.; Tähtinen, P.; Salminen, J.P. Structural Features of Hydrolyzable Tannins Determine Their Ability to Form Insoluble Complexes with Bovine Serum Albumin. *J. Agric. Food Chem.* **2019**, *67*, 6798–6808. [CrossRef]
51. Liu, L.; Jia, W.; Jiang, S.; Zhang, G.; Zhao, J.; Xu, J.; Wang, L.; Wu, D.; Tao, J.; Yue, H.; et al. Inhibitory activities and rules of plant gallotannins with different numbers of galloyl moieties on sucrase, maltase and α-amylase in vitro and in vivo. *Phytomedicine* **2023**, *120*, 155063. [CrossRef]
52. Pradeep, S.; Patil, S.M.; Dharmashekara, C.; Jain, A.; Ramu, R.; Shirahatti, P.S.; Mandal, S.P.; Reddy, P.; Srinivasa, C.; Patil, S.S.; et al. Molecular insights into the in silico discovery of corilagin from *Terminalia chebula* as a potential dual inhibitor of SARS-CoV-2 structural proteins. *J. Biomol. Struct. Dyn.* **2023**, *41*, 10869–10884. [CrossRef] [PubMed]
53. Kashiwada, Y.; Nonaka, G.; Nishioka, I.; Lee, K.J.; Bori, I.; Fukushima, Y.; Bastow, K.F.; Lee, K.H. Tannins as potent inhibitors of DNA topoisomerase II in vitro. *J. Pharm. Sci.* **1993**, *82*, 487–492. [CrossRef] [PubMed]
54. Quan, S.; Wen, M.; Xu, P.; Chu, C.; Zhang, H.; Yang, K.; Tong, S. Efficient screening of pancreatic lipase inhibitors from *Rheum palmatum* by affinity ultrafiltration-high-performance liquid chromatography combined with high-resolution inhibition profiling. *Phytochem. Anal.* **2024**, *35*, 540–551. [CrossRef] [PubMed]
55. Lee, D.Y.; Yang, H.; Kim, H.W.; Sung, S.H. New polyhydroxytriterpenoid derivatives from fruits of *Terminalia chebula* Retz. and their α-glucosidase and α-amylase inhibitory activity. *Bioorg. Med. Chem. Lett.* **2017**, *27*, 34–39. [CrossRef]
56. Xu, P.; Xiang, H.; Wen, W.; Quan, S.; Qiu, H.; Chu, C.; Tong, S. Application of two-dimensional reversed phase countercurrent chromatography × high-performance liquid chromatography to bioactivity-guided screening and isolation of α-glucosidase inhibitors from *Rheum palmatum* L. *J. Chromatogr. A* **2024**, *1717*, 464667. [CrossRef]

57. Xu, Y.; Zhang, L.; Wang, Q.; Luo, G.; Gao, X. An integrated strategy based on characteristic fragment filter supplemented by multivariate statistical analysis in multi-stage mass spectrometry chromatograms for the large-scale detection and identification of natural plant-derived components in rat: The rhubarb case. *J. Pharm. Biomed. Anal.* **2019**, *174*, 89–103. [CrossRef]
58. Kim, M.S.; Lee, D.Y.; Sung, S.H.; Jeon, W.K. Anti-cholinesterase activities of hydrolysable tannins and polyhydroxytriterpenoid derivatives from *Terminalia chebula* Retz. fruit. *Rec. Nat. Prod.* **2018**, *13*, 284–289. [CrossRef]
59. Wang, Z.F.; Sun, W.Y.; Yu, D.H.; Zhao, Y.; Xu, H.M.; He, Y.F.; Li, H.J. Rotundic acid enhances the impact of radiological toxicity on MCF-7 cells through the ATM/p53 pathway. *Int. J. Oncol.* **2018**, *53*, 2269–2277. [CrossRef]
60. Moilanen, J.; Sinkkonen, J.; Salminen, J.P. Characterization of bioactive plant ellagitannins by chromatographic, spectroscopic and mass spectrometric methods. *Chemoecology* **2013**, *23*, 165–179. [CrossRef]
61. Avula, B.; Wang, Y.H.; Wang, M.; Shen, Y.H.; Khan, I.A. Simultaneous determination and characterization of tannins and triterpene saponins from the fruits of various species of Terminalia and Phyllantus emblica using a UHPLC-UV-MS method: Application to triphala. *Planta Med.* **2013**, *79*, 181–188. [CrossRef] [PubMed]
62. Hernández-Corroto, E.; Marina, L.; García, C. Extraction and identification by high resolution mass spectrometry of bioactive substances in different extracts obtained from pomegranate peel. *J. Chromatogr. A* **2019**, *1594*, 82–92. [CrossRef] [PubMed]
63. Mi, Y.G.; Xu, X.Y.; Hong, L.L.; Jiang, M.T.; Chen, B.X.; Li, X.H.; Wang, H.D.; Zou, Y.D.; Zhao, X.; Li, Z.; et al. Comparative Characterization of the Ginsenosides from *Six Panax* Herbal Extracts and Their In Vitro Rat Gut Microbial Metabolites by Advanced Liquid Chromatography-Mass Spectrometry Approaches. *J. Agric. Food Chem.* **2023**, *71*, 9391–9403. [CrossRef] [PubMed]
64. Tang, C.; Li, H.J.; Fan, G.; Kuang, T.T.; Meng, X.L.; Zou, Z.M.; Zhang, Y. Network pharmacology and UPLC-Q-TOF/MS studies on the anti-arthritic mechanism of *Pterocephalus hookeri*. *Trop. J. Pharm. Res.* **2018**, *17*, 1095–1110. [CrossRef]
65. Geng, Z.; Wang, J.; Zhu, L.; Yu, X.; Zhang, Q.; Li, M.; Hu, B.; Yang, X. Metabolomics provide a novel interpretation of the changes in flavonoids during sea buckthorn (*Hippophae rhamnoides* L.) drying. *Food. Chem.* **2023**, *413*, 135598. [CrossRef] [PubMed]

molecules

MDPI

Article

Geochemical Study of Bitumen Residues on Potsherds from the al-Qusur Monastery (7th–9th c. CE): Composition and Origin

Jacques Connan [1,*], Julie Bonnéric [2], Rémi Perrogon [3], Michael H. Engel [4], Renaud Gley [5], Alex Zumberge [6] and Philippe Schaeffer [7]

[1] Université de Strasbourg, 23 rue Antoine de Saint-Exupéry, 64000 Pau, France
[2] French Research Centre in the Arabian Peninsula, USR3141, Kuwait City 92438, Kuwait; bonneric.julie@gmail.com
[3] LA3M-UMR 7298, Aix-Marseille University, 13100 Aix-en-Provence, France; remi.perrogon@gmail.com
[4] School of Geosciences, The University of Oklahoma, 100 East Boyd Street, Norman, OK 73019-1009, USA; ab1635@ou.edu
[5] Laboratoire Interdisciplinaire des Environnements Continentaux, UMR7360, Site Chamois, 15 Avenue de Chamois, 54500 Vandeouvre-les-Nancy, France; renaud.gley@univ-lorraine.fr
[6] GeoMark Research Inc., 9748 Whithorn Drive, Houston, TX 77095, USA; azumberge@geomarkresearch.com
[7] Institut Le Bel, UMR 7177, Université de Strasbourg, 4 rue Blaise Pascal, 67000 Strasbourg, France; p.schaef@unistra.fr
* Correspondence: connan.jacques@orange.fr

Abstract: Geochemical and isotopic analysis of bitumen lining potsherds from the al-Qusur monastery (second half of the 7th c. CE and the middle of the 9th c. CE), at the central part of Failaka Island (Kuwait Bay), confirms the presence of two distinct compositional categories that can be matched to contemporary sources from two different areas of Iran: the Kermanshah province on one side, and the Khuzestan–Fars–Busher provinces on the other side. Potsherds comprise different types: TORP-S amphorae, TORP-C amphorae, SPORC storage jar, turquoise alkaline-glazed jar (TURQ.T), and CREAC jar. There is no relationship between the type of potsherd and the origin of bitumen. The bitumen coating SPORC jar, first identified as a kind of juice strainer to filter the «garum-like juice», was examined in greater details to try to identify traces of fish sauce mentioned in the Arabic kitchen books as *'murri'*, and quite similar to the Roman *garum.* The mineralogical analysis exhibits the classical minerals of archaeological mixtures (quartz, calcite, dolomite) and no halite. Hydrocarbons, alcohols, and methyl esters show a typical biodegraded bitumen signature but no fatty acids and terpenoids. It seems that the bitumen matrix has not adsorbed any molecules from the presumed «*garum*» filtered in the basin.

Keywords: Kuwait; Failaka Island; al-Qusur; monastery; potsherds; bitumen; biomarkers; steranes; terpanes; carbon and hydrogen isotopes; geochemical analysis; oil seeps; Iran

1. Introduction

The bitumen samples originated from the site of al-Qusur, located in the middle of Failaka Island, c.19 km from the southern tip of Kuwait Bay. The central part of al-Qusur was identified by the French-Kuwaiti Archaelogical Mission in Failaka (MAFKF) as a monastery mainly occupied from the second half of the 7th c. CE to the middle of the 9th c. CE. The al-Qusur monastery, at the central part of Failaka Island, has been excavated by the French-Kuwaiti Archaeological Mission in Failaka (MAFKF) since 2012. The results of the past campaigns have been published [1–4], authenticating the monastic nature of the site and dating its main occupation from the second half of the 7th c. CE and the middle of the 9th c. CE. The site was first excavated by the Archaeological Mission in the Arabian Gulf in 1975–1976, then the French Mission in Kuwait in 1988–1989 and 2007–2009, the Kuwaiti-Polish Archaeological Mission in 2011 and 2013, and is still being studied today by the French-Kuwaiti Archaeological Mission in Failaka, since 2011, and the Kuwaiti-Slovak Archaeological Mission in 2007. In spite of the number of excavation seasons carried out on the site by many teams, there is always a debate about the interpretation of the whole site. If the center of al-Qusur is clearly a monastery, as evinced by the discovery of a monumental church (building A1), a refectory (building B23), and at least one monk's cell (building B16), it is still to be determined if the periphery of the site is part of the monastery or not (Figure 1). One of the current hypotheses is that the central part of the monastery (c. 0.78 ha) could be a *coenobium* surrounded by scattered cells accommodating monks wishing to live more isolated from the community. Following this hypothesis, the monastery would be covering c. 335 ha and be accurate with the rules of Abraham of Kashkar, promoting semi-cenobitic types of monasteries.

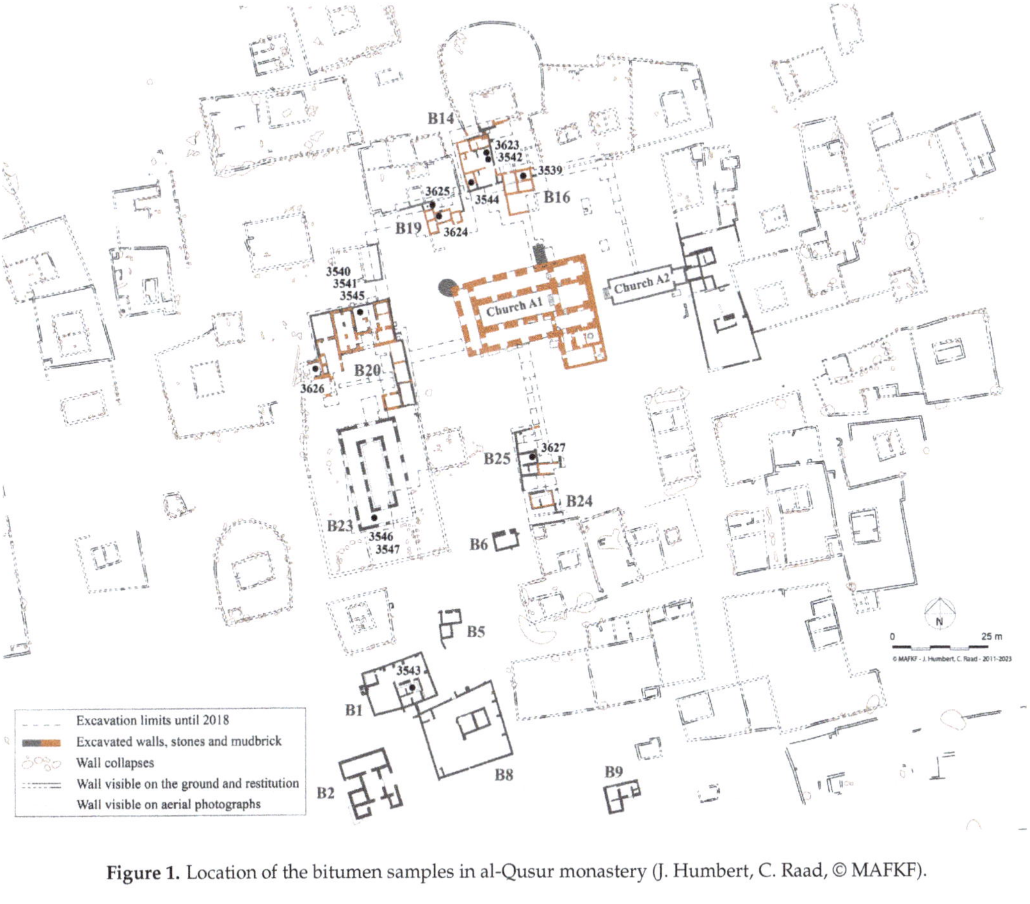

Figure 1. Location of the bitumen samples in al-Qusur monastery (J. Humbert, C. Raad, © MAFKF).

Al-Qusur was part of a network of monasteries associated with the Church of the East that spread across what is now Iraq, Iran, and the Gulf. Not only monks traveled between these areas but also goods. For instance, most of the pottery used in al-Qusur was made in Iraq and Iran. One of the aims of the French-Kuwaiti Archaeological Mission in Failaka is to study these exchanges and to understand how the monastery was supplied. For this reason, geochemical and isotopic analyses of nine samples of bituminous mixtures coating potteries were carried out. The purpose of this study was to document the composition of the bituminous mixture and to find the source of the bitumen.

2. Materials and Methods

2.1. Bitumen at al-Qusur

In al-Qusur, no bitumen had yet to be shown to be used for building construction. However, in building B20, a food processing building, bitumen was found to be associated with pottery in cooking areas. It mainly appears on pottery jars, as inside coating to waterproof them. One Syriac inscription on a sherd may also have been written with this material. The main documented use of bitumen is as coating on the inside of the pottery vessels. The jars do not appear to have been coated with bitumen on the site, but were probably imported already coated. However, it is worth mentioning that bitumen might be supplied by the nearby and contemporary settlement of al-Qurainiyah, on Failaka's northern coast, where the Kuwaiti-Italian Archaeological Mission reported remains that could be a bitumen production workshop [5]. The bitumen might have been delivered there in solid form before being heated and used in the architecture or maintenance of boats [5].

2.2. Archaeological Samples Analyzed

Thirteen bituminous samples were selected and a total of fifteen geochemical analyses (two samples were analyzed with two aliquots) were undertaken. Initial information (date, archaeological reference, context) including the type of potsherd is listed in Table 1. On potsherds, the coating of the interior face may be very thin as on the bottom of the torpedo jar No.3539, or more extensive as in samples No.3542 and No.3543 (Figure 2). The bituminous mass may also show some imprints (ropes?) as in sample No.3545 (Figure 2). The quantity of bituminous mixture, collected from a torpedo jar (TORP-C), suggests that it was a bituminous mixture stored in the jar. This sample was studied with several aliquots: No.3546, 3547 and 3547bis (Figure 2).

The ceramic categories that were sampled are part of the same group of sandy buff wares, sharing a similar clay composition with a rather fine calcareous ground mass, fine quartz elements, and small magmatic rocks. There are two categories of amphorae (Torpedo -S and -C) with an internal bitumen coating, known as *torpedo jars* because of their profiled shape. The samples No.3542, 3544, 3623, and 3624 were collected on sherds of TORP-S amphorae belonging to the Late Sassanian, Early Islamic assemblages (5th–8th c. CE). This type is found in abundance in the region, but is gradually replaced by lighter-paste amphorae (TORP-C) during the 8th century [6]. The samples No.3539, 3546, 3547, 3625, and 3626 are related to the latter type. It is contemporaneous to large storage jars (SPORC), such as sample No.3545. This type is quite rare in the area and has some parallels with material from the coastal site of Siraf [7]. The main centers of origin for these categories are supposedly located around the urban centers of central and southern Mesopotamia, from Susa (Kermanshah province, Iran) to al-Hira (Najaf district, Iraq) and Basra province (Iraq). A TORP-S production kiln has been identified in Telūl Hamediyāt, northeast of Baghdad (Hamrin basin, Diyala province, Iraq) and dated from the 6th–7th century [8,9]. This workshop was located alongside a wine-production facility and the amphorae were apparently potted on the site. This could seem to support the hypothesis that ceramics were produced close to the agricultural area in the Mesopotamian basin, instead of near the source of bitumen production in southwest Iran [10].

Table 1. Basic information on samples. The color refers to the type of potsherd.

Sample Number	Country	Lab Number	Location	Building	Context	Archaeological Reference	Latitude	Longitude	Date Range	Pottery Type
1	Kuwait	3539	al-Qusur	building B16 (monk's cell)	5313	QSR-12-A5/18	29°26′12″ N	48°20′52″ E	Mid. 7th–early 9th CE	TORP-C
2		3540		building B20 (food processing), room R8	6067	QSR-16-A4/25			Late 8th–10th c. CE	CREAC (HAR-CR17) ?/BUFF ?
3		3541		building B20, room R8	6067	QSR-16-A4/25			Late 8th–10th c. CE	CREAC (HAR-CR17) ./BUFF ?
4		3542		building B14, courtyard	7103	QSR22-A5/0002			Mid. 7th–early 9th c. CE	TORP-S
5		3543		building B1	4	QSR07/2100			Mid. 7th–mid. 8th c. CE	TURQ.T (TUR-CR7)
6		3544		building B14, room R1	7162	QRS2022-A5/0001			Mid. 7th–early 9th c. CE	TORP-S
7		3545		building B20 (food processing), room R8	6067	QRS16-A4/57			Late 8th–10th c. CE	SPORC
8		3546		building B23 (refectory)	5635	QSR15-A4/243			Mid. 8th–10th c. CE	TORP-C
9		3547		building B23 (refectory)	5635	QSR15-A4/243			Mid. 8th–10th c. CE	TORP-C
26		3623		building B14, courtyard	7177	QSR23-A5-B14-7177			Mid. 7th–early 9th c. CE	TORP-S
27		3624		building B19, room R2	6854	QSR19-A5-U.97-6854			Mid. 7th–early 9th c. CE	TORP-S
28		3625		building B19, courtyard	6862	QSR19-A5-U.21-22-6862			Mid. 8th–early 9th c. CE	TORP-C
29		3626		building B20 (food processing), room R5	/	QSR23-A4-B20-U.63 R15			Mid. 8th–early 9th c. CE	TORP-C
30		3627		building B25, room R4	7701	QSR23-A6-B25-u.140-7701			/	/

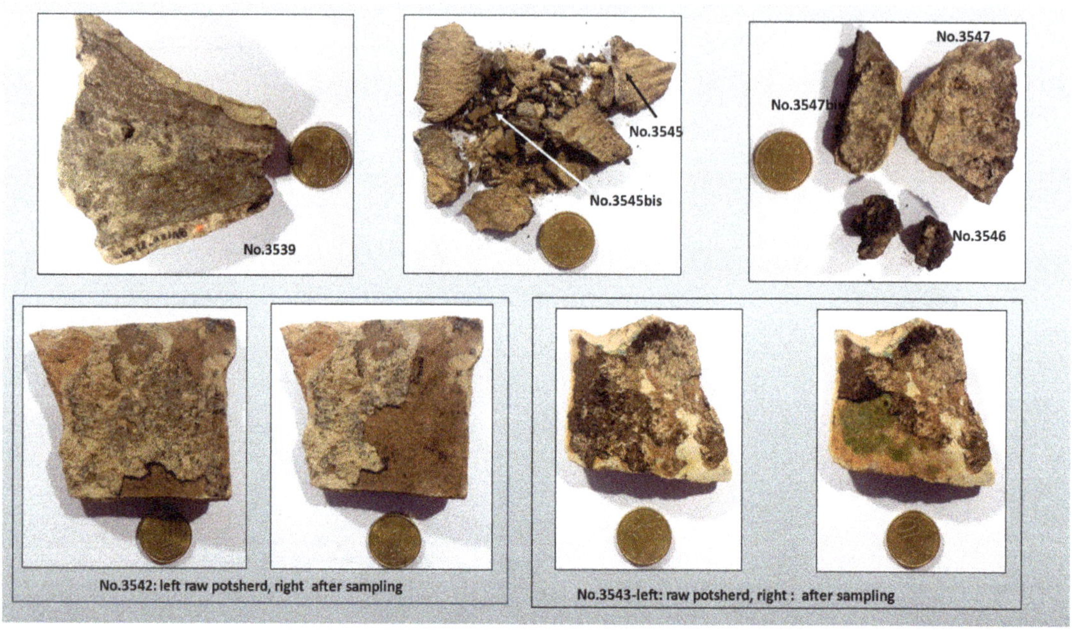

Figure 2. Picture of eight samples of al-Qusur showing various aspects of the bitumen coating the interior face of potsherds. Some pictures show the situation before and after sampling.

Sample No.3543 was recovered from a sherd of a turquoise alkaline-glazed small jar (TURQ.T). The layer of bitumen covers a large part of the internal and external surfaces and seems to have accumulated particularly on the shoulder and neck of the object. The various traces of dripping and the absence of repair marks suggest that this object was used to contain and handle bitumen in liquid form. Although turquoise ceramics were used over a long period of time, the forms found at the al-Qusur site correspond more closely to the types identified at the Sir Bani Yas site and dated mainly between the mid-7th and mid-8th century [11,12]. Based on the clay and glaze used, the area of origin of these products is probably in southern Mesopotamia, in the Basra region. Further archaeometrical analyses of the composition of the glazes should make it possible to identify even more precisely the regions of production [13].

Samples No.3540 and No.3541 are extracted from a ceramic category that is difficult to precisely identify. In terms of clay and surface treatment, this jar is similar to those made at Siraf on the Iranian coast, known as CREAC, and dated between the late 8th and the 10th century [12]. It is possibly another sandy buff ware belonging to the same group as TORP, TURQ, and SPORC, but with an overfired body, with a large number of magmatic rock inclusions within the clay matrix.

2.3. Analytical Procedures

2.3.1. Bitumen Analysis

All archaeological and geological samples were subjected to the same analytical procedure conducted at GeoMark Research Ltd. (Houston, TX, USA). The ground potsherd samples were Soxhlet extracted with a mixture of 372 mL dichloromethane and 28 mL methanol. The dried dichloromethane extracts were deasphalted using *n*-hexane. The deasphalted fraction was separated into saturated hydrocarbons, aromatic hydrocarbons, and resins using gravity flow column chromatography employing a 100–200 mesh silica gel support activated at 400 °C prior to use. Hexane was used to elute saturates, dichloromethane to elute the aromatic hydrocarbons, and dichloromethane/methanol (50:50 *v/v*) to elute the resin (NSO) fraction. Following solvent evaporation, the recovered fractions were quantified gravimetrically. The C_{15+} saturated hydrocarbon fraction was subjected to molecular sieve filtration (silicalite, Union Carbide S-115 powder) after the technique described by West et al. [14]. An aliquot of the total alkane fraction was not fractionated by silicalite in order to preserve access to the *n*-alkanes.

Gas Chromatography-Mass Spectrometry (GC-MS) analyses of C_{15+} branched/cyclic and aromatic hydrocarbon fractions (in order to determine sterane and terpane biomarker distributions and quantities) and aromatic hydrocarbon fractions were performed using an Agilent 7890A or 7890B GC interfaced to a 5975C or 5977A mass spectrometer (Wilmington, DE, USA). The J&W HP-5 column (50 m × 0.2 mm; 0.11 µm film thickness) is temperature programmed from 150 °C to 325 °C at 2 °C/min for the branched/cyclic fraction and 100 °C at 3 °C/min for the aromatic fraction. The mass spectrometer is run in the selected ion mode (SIM), monitoring ions m/z 177, 191, 205, 217, 218, 221, 231, and 259 for branched and cyclic alkanes, and m/z 133, 156, 170, 178, 184, 188, 192, 198, 231, 239, 245, and 253 for aromatics. In order to determine absolute concentrations of individual biomarkers, a deuterated internal standard (d4-C29 20R Ethylcholestane, Chiron Laboratories, Norway) is added to the C_{15+} branched/cyclic hydrocarbon fraction and a deuterated anthracene standard (d10) is added to the aromatic hydrocarbon fraction. Response factors (RF) were determined by comparing the mass spectral response at m/z 221 for the deuterated standard to hopane (m/z 191) and sterane (m/z 217) authentic standards. These response factors were found to be approximately 1.4 for terpanes and 1.0 for steranes. Concentrations of individual biomarkers on the branched/cyclic fraction were determined using the equation below:

$$\text{Conc. (ppm)} = [(\text{ht.biomarker})\text{ng standard})]/[(\text{ht.standard})\ (\text{RF})\ (\text{mg b/cy fraction})]$$

RF = response factor, ng = nanograms, for branched and cyclic alkanes (b/cy) peak heights (ht) are used, for aromatics the same equation is used with peak areas.

The C_{15+} saturates, C_{15+} aromatics, asphaltenes, and resins were analyzed for their respective carbon isotope ($\delta^{13}C$ in ‰/VPDB = Vienna Pee Dee Belemnite) compositions. The asphaltenes and resins were also analyzed for their stable hydrogen isotope compositions. Details for all of the methods for stable isotope analyses are described in Connan et al. [15].

2.3.2. Mineralogical Analysis

All samples were ground prior to the XRD analysis. The mineralogy was determined using a D8 Discover Bruker diffractometer (Bruker, Karlsruhe, Germany) equipped with an X-ray tube emitting Cu-Kα1 radiation (λ = 1.5406 Å) at 40 kV and 40 mA. X-ray diffraction patterns were recorded with a LynxEye detector (Bruker) under ambient conditions and from 2.5° to 65° 2θ, with a step size of 0.035° 2θ, and with a calculation time of 3 s per step. Analyses were performed using DIFFRAC.EVA Bruker software (Version Eva, Release 2025, 7.3.0.2 (64 Bits) access via Eva 3) and diffraction peaks were identified using the powder diffraction file database.

2.3.3. Complementary Organic Analyses

An aliquot of the sample No.3545 was extracted by a mixture of CH_2Cl_2/MeOH 1:1 (*v/v*) and then acetylated (Ac_2O/Pyr) and methylated (CH_2N_2). The mixture was subsequently fractionated by silica gel chromatography into two fractions: an apolar one (F1) containing the hydrocarbons and a more polar one (F2) containing the alcohols (as acetates) and the fatty acids (as methyl esters).

GC-FID

GC analyses were carried out on a Hewlett Packard 6890 gas chromatograph (Agilent Technologies Inc., Wilmington, DE, USA) equipped with an on-column injector used in the "track-oven" mode, a flame ionization detector, and an HP-5 fused silica capillary column (30 m × 0.25 mm; 0.25 µm film thickness). H_2 was used as carrier gas (constant flow mode, 2.5 mL/min), and the oven was programmed as follows: 70 °C (5 min), 70–200 °C (4 °C/min), 200–300 °C (10 °C/min), isothermal at 300 °C.

GC/MS

GC/MS analyses were performed on a Thermo GC Trace gas chromatograph (Thermo Fisher Scientific, Waltham, MA, USA) equipped with an HP5 MS column (30 m × 0.25 mm × 0.1 µm) and a temperature-programmable injector linked to a Thermo TSQ Quantum mass spectrometer (Thermo Fisher Scientific). Helium (constant flow, 1.2 mL/min) was used as carrier gas and the oven was programmed as follows: 70 °C (5 min), 70–200 °C (4 °C/min), 200–300 °C (10 °C/min), 40 min isothermal at 300 °C. The mass spectrometer was operating in electron impact (EI) mode at 70 eV with a scan range of *m/z* 50–700.

3. Results and Discussion

3.1. Gross Composition of Extractable Organic Matter

Gross composition data are listed in Table 2. The plot of % HC (sat + aro) vs. % resins (NSO's) vs. % asphaltenes (Figure 3b) and % saturates vs. % aromatics vs. % polars (Figure 3a) shows that bituminous mixtures of al-Qusur possess the characteristic composition of archaeological bitumens: high amounts of polars (resins + asphaltenes) and asphaltenes and low quantities of hydrocarbons and resins. One may notice some slight differences among samples in Figure 3b.

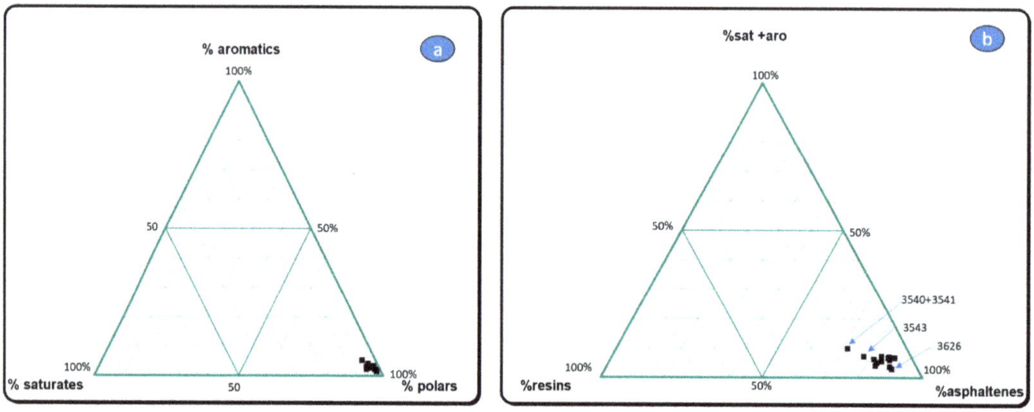

Figure 3. Gross composition of the dichloromethane extract in two ternary diagrams: (**a**) % saturates vs. % aromatics vs. % polars and (**b**) % sat + aromatics vs. % resins vs. % asphaltenes.

Table 2. Gross composition of the dichloromethane extract (EO in % by weight/sample) and isotopic data: $\delta^{13}C$ (‰/VPDB) and δD (‰/VSMOW = Vienna Standard Mean Ocean Water). Significance of abbreviations: sat = saturated hydrocarbons, aro = aromatic hydrocarbons, NSO= resins, asp = asphaltenes, pol = polars = resins + asphaltenes.

Sample Number	Country	Lab Number	Location	%EO	%Sat	%aro	%HC	%NSO	%ASP	%Pol	δ^{13}Csat	δ^{13}Caro	δ^{13}CNSO	δ^{13}Casp	δDNSO	δDasp
1	Kuwait	3539	al-Qusur	13.8	1.4	2	3.4	8.7	87.9	96.6	−28.5	−27.4	−28	−28	−84	−82
2		3540 + 3541		13.4	4.5	5.2	9.7	18.7	71.6	90.3	−27.4	−25.9	−28.1	−28.1	−91	−88
4		3542		53.5	3	4	7	9.5	83.5	93	−27.5	−26.4	−26.5	−27.1	−83	−74
5		3543		21	4	3.1	7.1	14.9	78	92.9	−27.1	−26	−26.1	−25.6	−87	−84
6		3544		15	1.7	3.2	4.9	12	83.1	95.1	−28.8	−27.3	−28.1	−28.2	−86	−71
7		3545		34.3	3.9	2.7	6.6	7.5	85.9	93.4	−27.2	−26.5	−26.9	−27.1	−85	−73
		3545bis		36.7	3.4	2.5	5.9	7.5	86.6	94.1	−27.6	−26.6	−26.8	−26.8	−82	−74
8		3546		22.1	3.3	3.5	6.8	7.1	86.1	93.2	−27	−25.9	−26.2	−25.8	−89	−69
9		3547		32.4	2.7	2.7	5.4	10.2	84.4	94.6	−27	−26	−26.3	−25.8	−108	−72
		3547bis		28.3	3.2	3	6.2	12.2	81.6	93.8	−27	−25.8	−26.5	−25.9	−101	−79
26	Kuwait	3623	al-Qusur	42.8	4.1	2.6	6.7	7.5	85.8	93.3	−27.5	−26.2	−26.8	−26.8	−80	−74
27		3624		41.5	4.5	2.2	6.7	5.3	88	93.3	−27.2	−26.3	−26.7	−26.9	−77	−77
28		3625		5.2	1.7	2.2	3.9	12.9	83.2	96.1	−27.2	−27.7	−28.2	−28	−103	−74
29		3626		1.7	1.4	1.4	2.8	8.4	88.8	97.2	−28.7	−27.8	−27.9	−27.9	−83	−72
30		3627		32.2	4.1	2.6	6.7	5.8	87.5	93.3	−27.4	−26.3	−26.8	−26.8	−81	−68

3.2. Isotopic Compositions

Isotope data are listed in Table 2. The plot of $\delta^{13}C_{sat}$ (‰/VPDB) vs. $\delta^{13}C_{aro}$ (‰/VPDB) and $\delta^{13}C_{asp}$ (‰/VPDB) vs. $\delta^{13}C_{NSO}$ (‰/VPDB) in Figure 4 reveals at least two main types of bitumen. One may notice a good match in duplicate samples (No.3545 and 3545bis and 3547 and 3547bis, Figure 4).

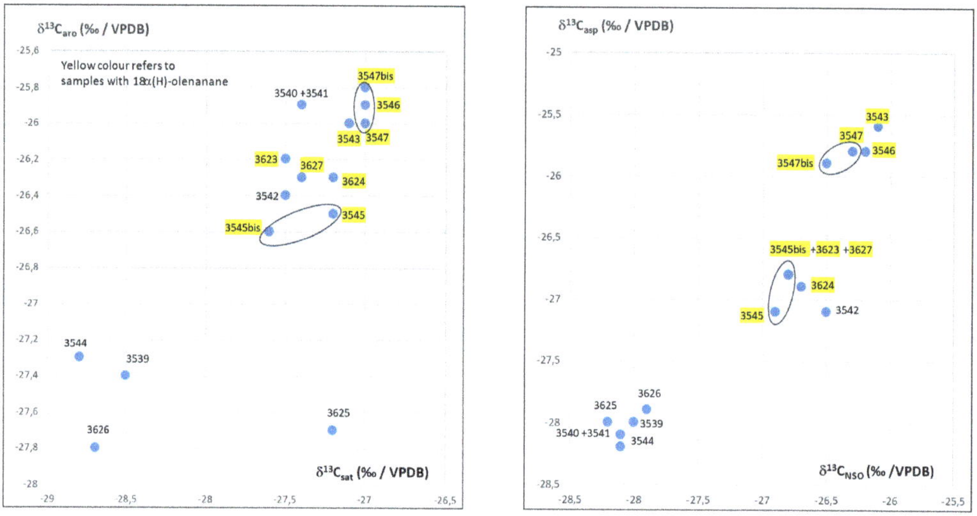

Figure 4. Plot of $\delta^{13}C$ (‰/VPDB) of aromatics vs. $\delta^{13}C$ (‰/VPDB) of saturates and $\delta^{13}C$ (‰/VPDB) of asphaltenes vs. $\delta^{13}C$ (‰/VPDB) of NSO for the al-Qusur samples.

Comparison of the al-Qusur data with data collected on oil seeps from Iraq (Figure 5) and Iran (Figure 6) shows that the bitumen of al-Qusur matches the oil seeps of Iran, but not those of Iraq.

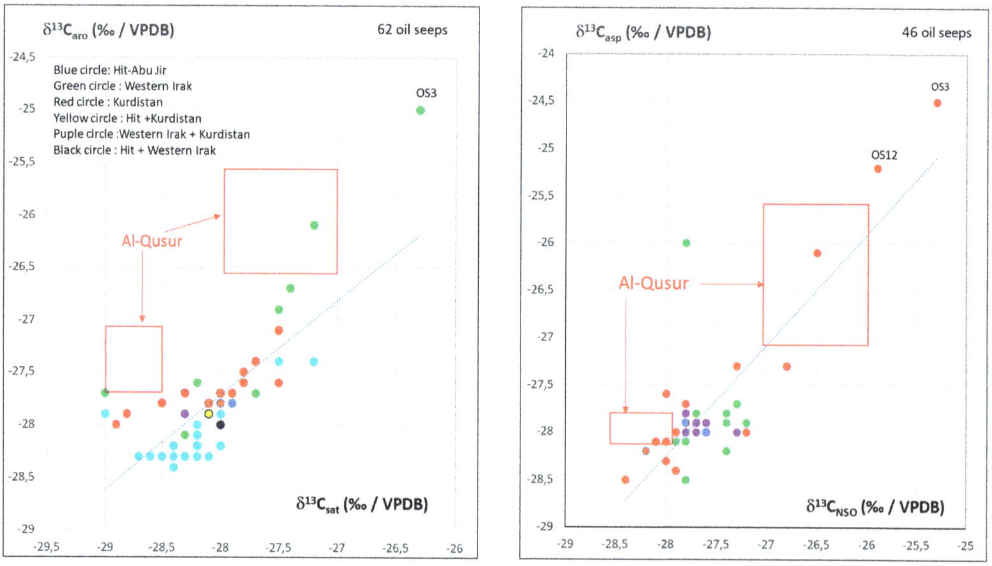

Figure 5. Plot of $\delta^{13}C$ (‰/VPDB) of aromatics vs. $\delta^{13}C$ (‰/VPDB) of saturates and $\delta^{13}C$ (‰/VPDB) of asphaltenes vs. $\delta^{13}C$ of NSO for the 62 and 46 oil seeps of Iraq: comparison with al-Qusur's data.

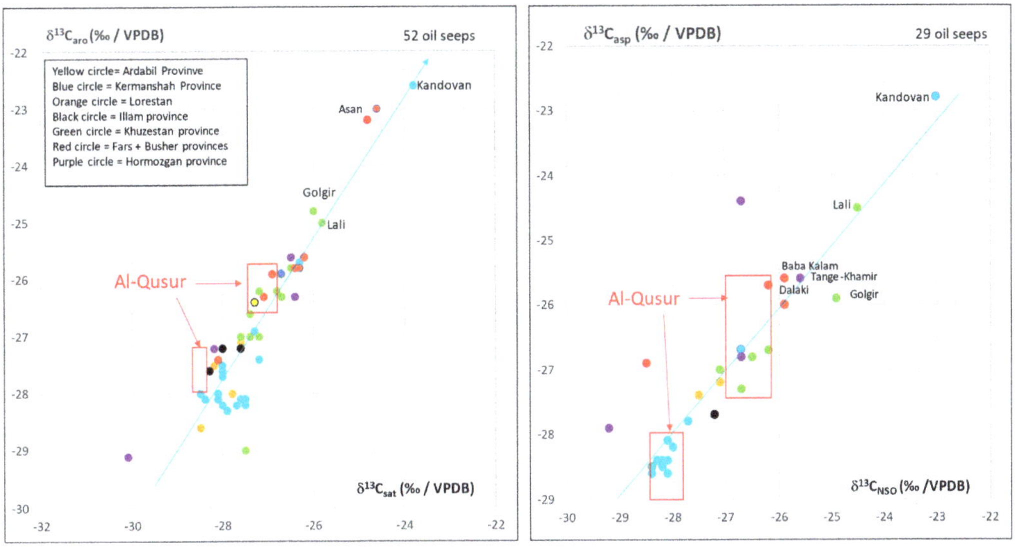

Figure 6. Plot of $\delta^{13}C$ (‰/VPDB) of aromatics vs. $\delta^{13}C$ (‰/VPDB) of saturates and $\delta^{13}C$ (‰/VPDB) of asphaltenes vs. $\delta^{13}C$ of NSO for 52 and 29 oil seeps of Iran: comparison with al-Qusur's data.

One group of bitumen is originating from Kermanshah province whereas the second group is likely coming from the Khuzestan–Busher–Fars provinces. The plot of δD_{asp} (‰/VSMOW) vs. δD_{NSO} (‰/VSMOW) in Figure 7 shows values (−90 to −70 ‰/VSMOW) for asphaltenes which are currently reported for archaeological bitumens. No relationship exists between the origin of bitumen and δD values (Figure 7). This situation is usual for δD parameter is not a source parameter but rather reflects the state of bitumen alteration which is not very intense here [15,16]. Values higher than −100 (‰/VSMOW) in resins of samples No.3547 and No.3547bis reflect even a well-preserved fraction [17]. One may note the good match between the values for samples No.3547 and No.3547bis as well as for samples No.3545 and No.3545bis (Figure 7) which documents the reliability of the δD values.

3.3. Biomarker Analysis

Mass fragmentograms of steranes (m/z 217) and terpanes (m/z 191) of two torpedo C potsherds are presented in Figure 8.

Terpanes of sample No.3539 show a high Tm/Ts ratio, a moderate gammacerane content, the complete hopane family, and the occurrence of the hexahydrobenzohopane series ($C_{32}/6$ to $C_{35}/6$). The tricyclopolyprenanes are missing (water washing? biodegradation? origin?). The steranes are biodegraded with a classical pattern ($C_{27} < C_{28} < C_{29}$). No selective degradation of $C_{29}\alpha\alpha\alpha R$ is seen.

Terpanes of sample No.3546 also have a high Tm/Ts ratio but the outstanding feature is the occurrence of 18(α)H-oleanane, characteristic of oils from Khuzestan, Busher and Fars [18]. In this sample, tricyclopolyprenanes ($C_{23}/3$ to $C_{29}/3$) and hexahydrobenzohopanes ($C_{32}/6$ to $C_{35}/6$) are present. Steranes are not biodegraded and as expected with this type of chemistry, contain diasteranes (C_{27}S and C_{27}R dia). The occurrence of 18α(H)-oleanane, identified on samples in Figures 4 and 7, confirm the existence of at least two main groups of bitumens as already expected on the basis of $\delta^{13}C$ data.

Biomarker data are listed in Table 3.

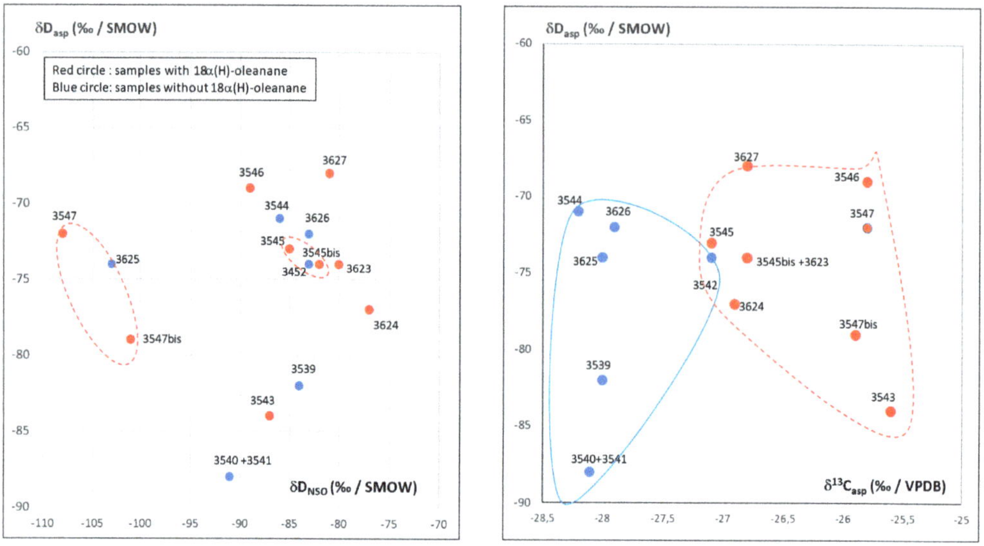

Figure 7. Plot of the δD (‰/VSMOW) of asphaltenes vs. δD (‰/VSMOW) of resins (NSO) and δD (‰/VSMOW) of asphaltenes vs. $\delta^{13}C$ (‰/VPDB) of asphaltenes with the identification of samples containing 18α(H)-oleanane.

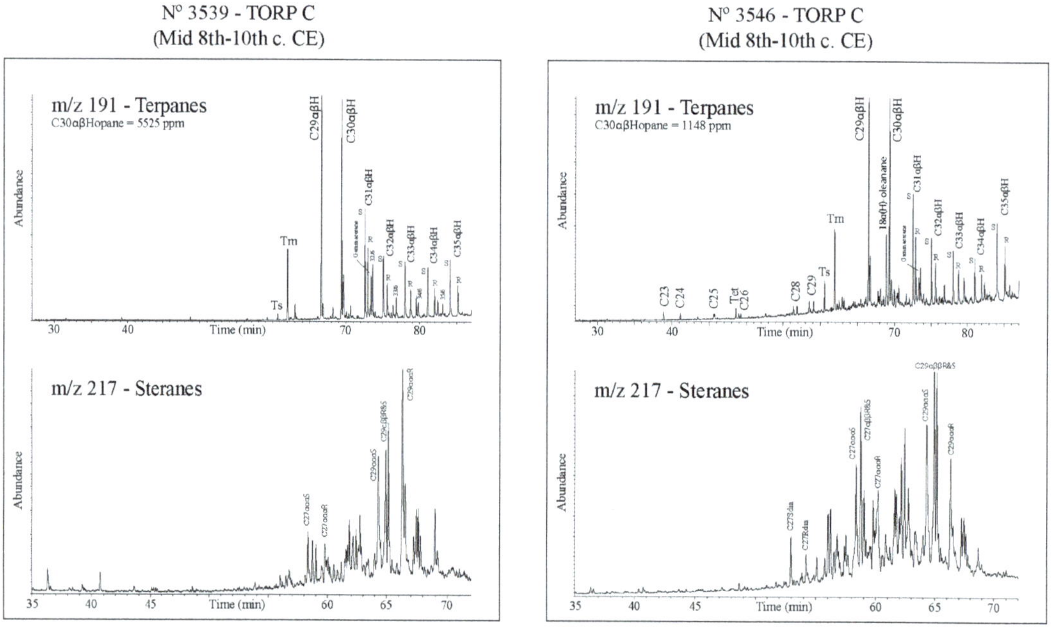

Figure 8. Mass fragmentograms of steranes (m/z 217) and terpanes (m/z 191) of sample No.3529 without 18α(H)-oleanane and sample No.3546 with 18α(H)-oleanane. Significance of abbreviations: $C_{29}\alpha\beta H$ = norhopane, $C_{30}\alpha\beta H$ = 17α,21β-hopane, $C_{31}\alpha\beta HR$ = 17α,21β,22R-30-homohopane, $C_{27}\alpha\beta\beta R{+}S$ = 5α,14β,17β-20R+20S-cholestane, $C_{28}\alpha\beta\beta R{+}S$ = 5α,14β,17β-20R+20S-24methylcholestane, $C_{29}\alpha\beta\beta R{+}S$ = 5α,14β,17β-20R+20S-24ethylcholestane, $C_{29}\alpha\alpha\alpha 20S$ = 5α,14α,17α-20S-24ethylcholestane, Ts = 18α-22,29,30-trisnorneohopane, Tm = 17α-22,29,30-trisnorhopane, $C_{34}\alpha\beta HS$ = C34-17α,21β-22S-extended hopane.

Table 3. Molecular data on steranes and terpanes. Significance of abbreviations: $C_{30}\alpha\beta$Hopane = 17α,21β-hopane, OL/H = 18α(H)-oleanane/hopane, GA/C_{31}R = Gammacerane/17α,21β,22R-30-homohopane, ster/terp = steranes/terpanes, Dia/reg = diasteranes/regular steranes, %$C_{27}\alpha\beta\beta$R+S = 5α,14β,17β-20R+20S-cholestane,%$C_{28}\alpha\beta\beta$R+S = 5α,14β,17β-20R+20S-24methylcholestane, %$C_{29}\alpha\beta\beta$R+S = 5α,14β,17β-20R+20S-24ethylcholestane, $C_{29}\alpha\alpha\alpha$20S/20R = 5α,14α,17α-20S-24ethylcholestane/5α,14α,17α-20R-24ethylcholestane, C_{29}H/C_{30}H = norhopane/hopane, C_{27}Ts/Tm = 18α-22,29,30-trisnorneohopane/17α-22,29,30-trisnorhopane, C_{35}S/C_{34}S = C_{34}-17α,21β-22S-extended hopane/C_{35}-17α,21β-22S-extended hopane.

Lab Number	C30αβHopane (ppm)	Tet/C23	C29/H	Ol/H	C31R/H	GA/ C31R	GA/ C30H	C35S/ C34S	Ster/ Terp	Dia/ Reg	%C27	%C28	%C29	C2920S/ 20R	C29αββS/ C29αααR	Ts/Tm	Tricyclics	Terpanes	Steranes	C27dia-steranes
3539	**5525**	1.27	1.02	0	0.32	0.59	0.19	1.14	0.04	0.12	15.5	18.8	65.7	0.55	0.68	0.09	almost absent	well preserved	biodegraded -C27 < C28 < C29	
3540 + 3541	**13,768**	3.41	1.19	0	0.23	0.56	0.13	0.87	0.08	0.08	4.3	15.5	80.2	0.38	0.81	0.1	almost absent	preserved?	severely biodegraded-C27	
3542	**3428**	4.15	1.35	0	0.3	0.55	0.17	1.03	0.13	0.19	19.5	31.2	49.3	0.71	0.99	0.33	almost absent	preserved?	biodegraded	
3543	**109**	0.53	1.65	3.72	0.43	1.07	0.46	3.7	0.72	0.28	4.8	30.9	64.4	11.38	2.6	1.37	absent	severely biode-graded	severely biodegraded-	
3544	**10,580**	3.44	0.94	0	0.27	0.58	0.16	1.08	0.09	0.04	22.3	21	56.8	0.65	1.06	0.09	absent	preserved	slightly biodegraded	
3545	**3563**	1.15	0.98	0.1	0.31	0.36	0.11	1.15	0.36	0.69	25.1	29.9	45	0.76	1.37	0.35	present low	preserved	preserved	present
3545bis	**3426**	1.79	1.01	0.11	0.33	0.35	0.12	1.25	0.33	0.86	21.1	30	48.9	1.05	1.52	0.38	present low	preserved	preserved	present
3546	**1148**	1.26	1.01	0.35	0.32	0.36	0.12	1.72	0.75	0.75	18.3	32	49.7	1.18	1.71	0.4	present low	preserved	preserved	present
3547	**1569**	1.24	0.94	0.2	0.38	0.29	0.11	1.26	0.36	0.92	15.9	31.8	52.2	1.19	1.71	0.49	present low	preserved	slightly biodegraded-C29αββR -C27αββR biodegraded	present
3447bis	**2027**	1.85	0.94	0.17	0.36	0.3	0.11	1.19	0.34	0.87	14.8	33.1	52.1	1.16	1.79	0.49	almost absent	preserved	slightly biodegraded-C29αββR -C27αββR biodegraded	present
3623	1464	1.1	1.19	0.16	0.34	0.34	0.12	1.3	0.82	1.09	24.1	32	43.9	1.28	2.05	0.46	almost absent	preserved	slightly biodegraded-C29αββR -C27αββR biodegraded	present
3624	4313	1.46	1.06	0.1	0.32	0.31	0.1	1.05	0.33	0.81	27.8	30	42.2	0.88	1.42	0.39	almost absent	preserved	preserved	abundant
3625	2786	2.14	1.04	0	0.35	0.73	0.26	1.18	0.04	0.59	17.5	21.8	60.7	0.71	0.74	0.13	absent	preserved	biodegraded-C27 < C28 < C29	present low
3626	21,510	3.62	1.13	0	0.25	0.54	0.14	0.9	0.08	0.05	24.9	22.9	52.2	0.47	0.77	0.1	absent	preserved	biodegraded-C27 < C28 < C29	absent
3627	1244	0.58	1.23	0.17	0.32	0.42	0.13	1.55	0.7	1.09	24.6	30.8	44.6	1.18	1.44	0.53	present	preserved?	preserved	abundant

The plot of Ts/Tm vs. δ^{13}Casp (‰/VPDB) and 18α(H)-oleanane vs. $C_{30}\alpha\beta$Hopane (ppm) in Figure 9 confirms the distribution of the samples into two main groups: samples No.3543, 3546, 3547, 3547bis, 3545, 3545bis, 3627, 3623, and 3624 on one side, samples No.3539, 3540 + 3541, 3542, 3544, 3625, and 3626 on the other side. The sample No.3543 shows extreme values which are due to the high degree of biodegradation of steranes and terpanes as documented by the mass fragmentograms shown in Figure 9. Reference to types of potsherd and amphorae in Figure 10 show that there is no relationship between the type of bitumen and the type of potsherd.

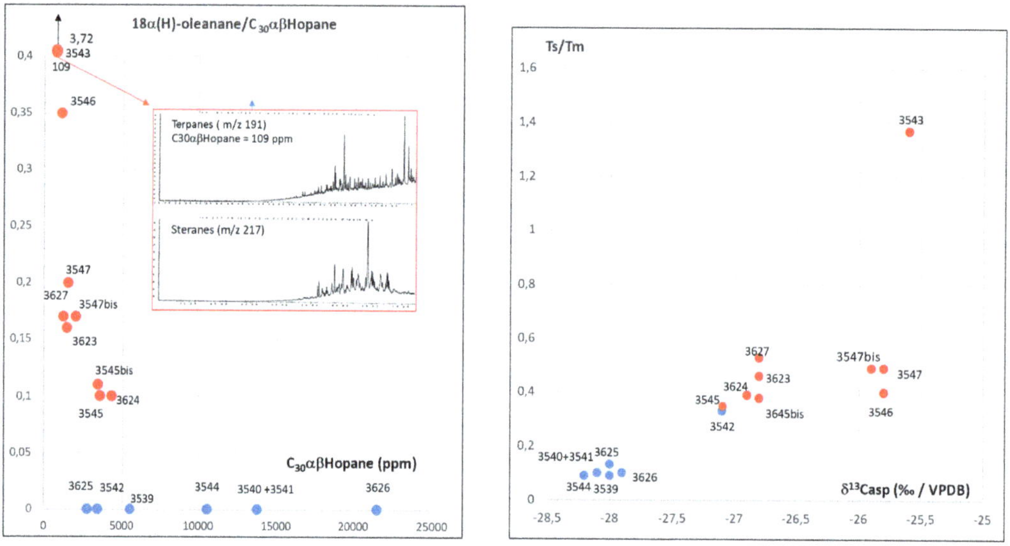

Figure 9. Plot of 18α(H)-oleanane/$C_{30}\alpha\beta$Hopane vs. $C_{30}\alpha\beta$Hopane (ppm) and Ts/Tm vs. δ^{13}Casp (‰/VPDB) with the al-Qusur samples.

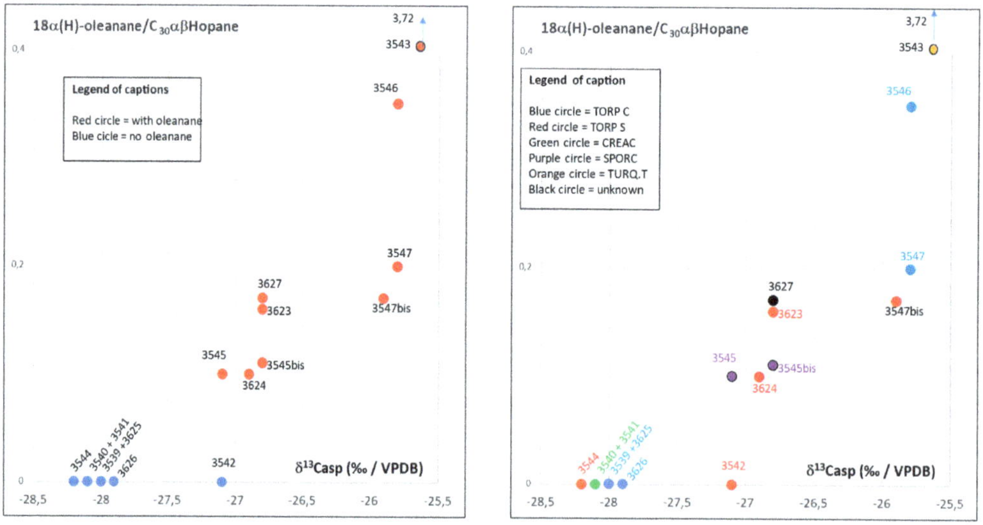

Figure 10. Plot of 18α(H)-oleanane/$C_{30}\alpha\beta$Hopane vs. δ^{13}Casp (‰/VPDB) in samples of al-Qusur with and without the identification of type of potsherds.

The plot of %$C_{27}\alpha\beta\beta$steranes vs. %$C_{28}\alpha\beta\beta$steranes vs. %$C_{29}\alpha\beta\beta$steranes in a ternary diagram (Figure 11), used to determine the origin of bitumens, confirms the two main groups of samples already defined: group I with samples No. 3539, 3540 + 3541, 3542, 3544, 3525, and 3526 originating from the Kermanshah province and group II with samples No.3543, 3545, 3545bis, 3546, 3547, 3547bis, 3623, 3624, and 3627 sourced from the Khuzestan and Fars provinces. Sample No.3542, which has a particular behavior, falls within group II.

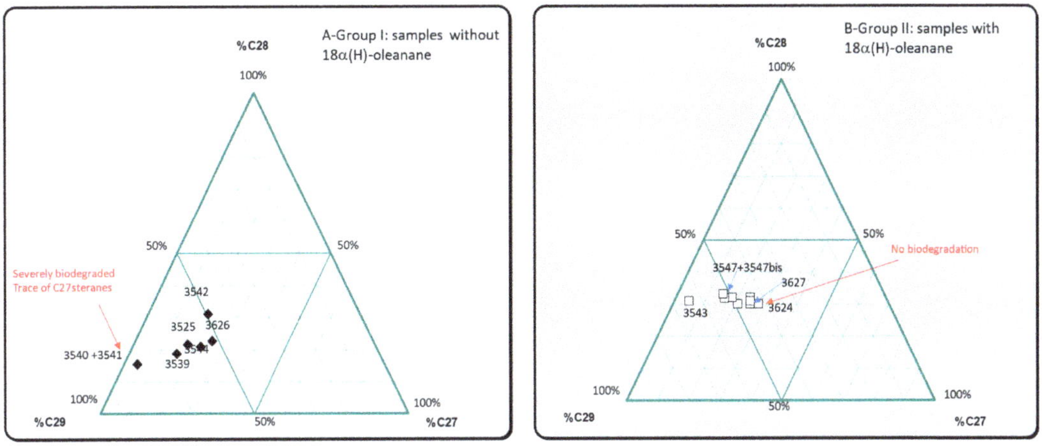

Figure 11. Plot of $\alpha\beta\beta$steranes in a ternary diagram: %C_{27} vs. %C_{28} vs. %C_{29}. (**A**) Group I: samples without 18α(H)-oleanane. (**B**) Group II: samples with 18α(H)-oleanane.

The plot of diasteranes/regular steranes vs. steranes/terpanes or vs. C_{24}tetracyclic terpane/C_{23}tricyclic terpanes confirms the identified two groups of samples.

The plot of Ts/Tm vs. $\delta^{13}C_{asp}$ (‰/VPDB) of al-Qusur data with other data already published [19] on other sites (F5, F6, and B6) from Failaka Island clearly differentiate al-Qusur from other sites (Figure 12). These sites covered other periods: Hellenistic (B6 and F5), Kassite (F6), and Early Dilmun (F6), and bitumens came from Iraq and Kuwait.

3.4. Detailed Geochemical Study of Samples No.3545 and 3545bis

Large bitumen coated basins (SPORC) were identified within room R8 of building B20, in the last occupation layer of this food-processing workshop, dated between the end of the 8th and the beginning of the 9th century (Figure 1). These jars stand out from the rest of the ceramic corpus in terms of their shape and size. A functional study was carried out by combining all the data available prior to the physico-chemical analysis. The presence of numerous ichthyological remains in the adjoining room, engravings in the shape of fish on the body of one of the jars, the presence of bitumen, and a base studded with perforations suggesting a filtration system all point to the manufacture of '*murri*'. This medieval recipe, similar to ancient *garum*, used small, dried fish, wine, and various herbs and spices. This hypothesis was based on textual sources linking the consumption of this condiment to Christian communities [3].

The sample that we analyzed in duplicate (No.3545 and 3545bis) came from the larger basin, which seems to be a kind of juice strainer to filter the «*garum-like juice*». The mineralogical analysis carried out by Renaud Gley at the University of Lorraine (France) did not find evidence for any halite (NaCl), but classical mineral additives of archaeological bituminous mixtures [18]: quartz, calcite, and dolomite as the dominant minerals and calcium sulfates (gypsum and anhydrite) were often present in carbonates lithologies (Figure 13). In addition, clays (illite) and plagioclases (albite) were also identified.

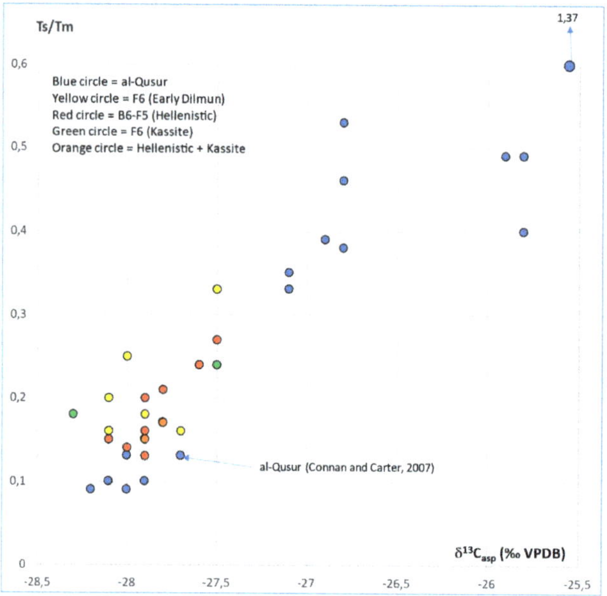

Figure 12. Plot of Ts/Tm vs. δ^{13}Casp (‰/VPDB): comparison of al-Qusur to other sites of Failaka Island.

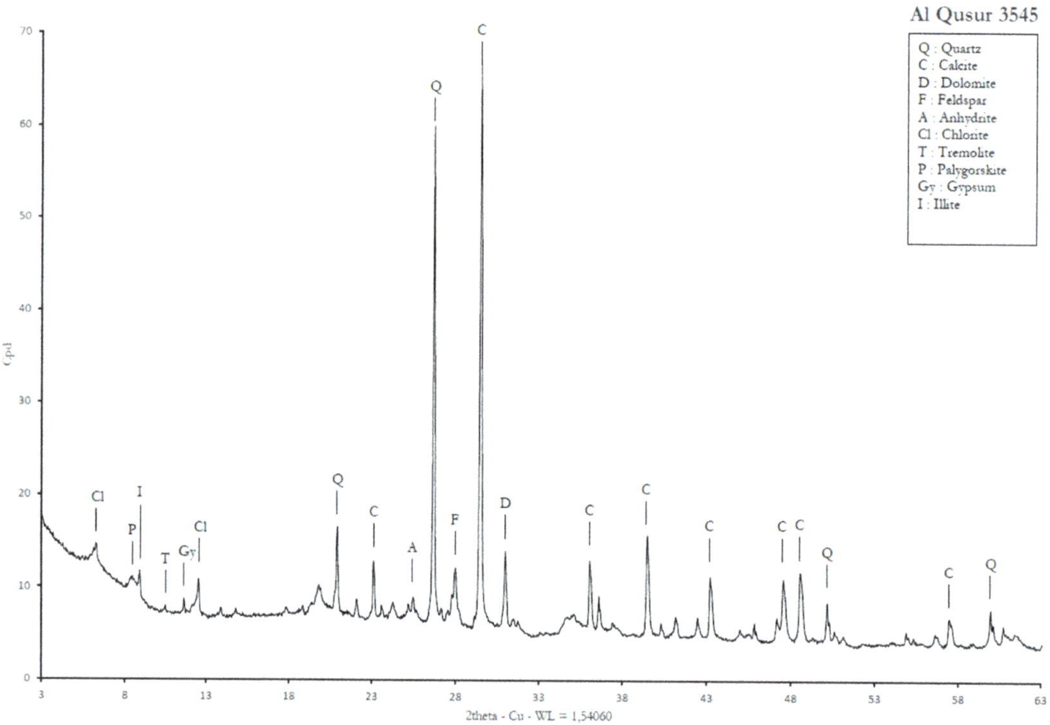

Figure 13. X-Ray diffraction diagram: identification of minerals in the sample No.3545.

The sample was analyzed by Philippe Schaeffer, University of Strasbourg, for complementary organic analyses.

The mixture has been subsequently fractionated into two fractions: hydrocarbons and then alcohols and methyl esters. The gas chromatograms of the two fractions are shown in Figure 14. The hydrocarbon fraction F1 shows a distribution typical for a biodegraded bitumen (abundant unresolved complex mixture, lack of *n*-alkanes and isoprenoids, abundant hopane family). The fraction F2 also shows an unresolved complex mixture associated with biodegraded bitumen. This fraction has also been analyzed by GC-MS in order to try to detect diagnostic compounds: fatty acids, terpenoids, for instance. However, no specific compounds could be identified. As a consequence, it seems that the bitumen matrix has not adsorbed any molecules from the presumed "*garum*" filtered in the basin.

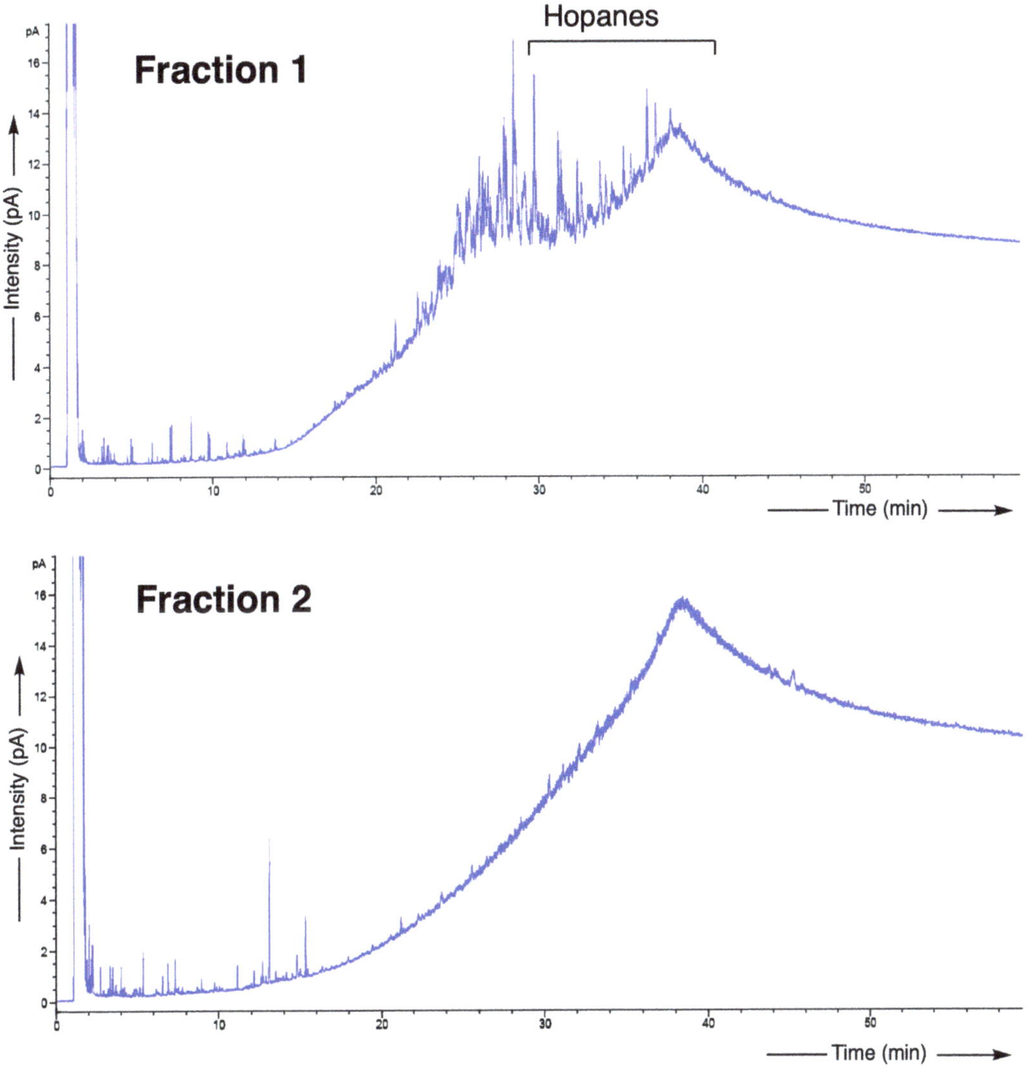

Figure 14. Gas chromatogram of the chromatographic fractions: F1 (hydrocarbons) and F2 (fatty acids and alcohols, analyzed as acetate and methyl ester derivatives, respectively) of the sample No.3545.

Although there are many arguments in favor of the production of '*murri*', the absence of organic markers linked to fish products means that an alternative should be carried out.

4. Conclusions

The geochemical analysis of the 13 samples has shown the occurrence of bitumen in all samples.

This bitumen has different origins: Kermanshah province and Khuzestan, Fars, and Busher provinces in Iran.

The coating of the big basin with numerous holes (QSR16-A4/57) is composed mainly of bitumen and minerals, a typical archaeological bituminous mixture. No trace of the "*garum*" has been found within this mixture. It is interesting to note that whilst the archaeological evidence indicates the production of '*murri*', the lack of organic biomarkers for fish opens the possibility for alternative uses.

Author Contributions: Validation, J.C. and A.Z.; Investigation, J.B., R.P., M.H.E., R.G., A.Z. and Philippe Schaeffer; Resources, J.B., R.P., M.H.E., R.G. and A.Z.; Data curation, J.C.; Writing—original draft, J.C., J.B., R.P., R.G. and P.S.; Writing—review & editing, M.H.E. All authors have read and agreed to the published version of the manuscript.

Funding: This work was partly done with resources from the Pôle de compétences en physico-chimie de l'environnement ANATELo, LIEC laboratory, UMR 7360 CNRS—Université de Lorraine.

Institutional Review Board Statement: Not applicable.

Informed Consent Statement: Not applicable.

Data Availability Statement: Data are contained within the article.

Conflicts of Interest: Author Alex Zumberge was employed by the company GeoMark Research Inc. The remaining authors declare that the research was conducted in the absence of any commercial or financial relationships that could be construed as a potential conflict of interest.

References

1. Bonnéric, J. Mission archéologique franco-koweïtienne de Faïlaka: Le monastère sassanido-islamique d'al-Qusur. In *Dossier «Prix Clio»*; Centre Français de Recherche de la Péninsule Arabique: Kuwait City, Kuwait, 2018; pp. 1–19. Available online: https://ssl.clio.fr/securefilesystem/Bonneric.pdf (accessed on 20 April 2025).
2. Bonnéric, J. Archaeological evidence of an early Islamic monastery in the centre of al-Qusur (Failaka Island, Kuwait). *Arab. Arch. Epig.* **2021**, *32*, 50–61. [CrossRef]
3. Bonnéric, J. Why islands? Understanding the Insular Location of Late Sasanian and Early Islamic Christian Monasteries in the Arab-Persian Gulf. In *Networked Spaces: The Spatiality of Networks in the Red Sea and Western Indian Ocean*; Durand, C., Marchand, J., Redon, B., Schneider, P., Eds.; MOM Éditions: Lyon, France, 2022; pp. 193–210.
4. Perrogon, R.; Bonnéric, J. A consideration on the interest of a pottery typology adapted to the late Sasanian and early Islamic monastery at Al-Qusur (Kuwait). *Arab. Arch. Epig.* **2021**, *32*, 70–82. [CrossRef]
5. Di Miceli, A. The site of Al-Qurainiyah: Topography and phases of an early Islamic coastal settlement on Failaka Island. *Arab. Arch. Epig.* **2021**, *32*, 62–69. [CrossRef]
6. Priestman, S. Bushehr, Dashtestan and Siraf: The Transformation of the Maritime Trade Network in the Middle Persian Gulf. In *Sasanian Archaeology: Settlements, Environment and Material Culture*; Simpson, S.-J., Ed.; Archaeopress Publishing Ltd.: Oxford, UK, 2022; pp. 153–177.
7. Priestman, S. *Ceramic Exchange and the Indian Ocean Economy (AD 400–1275). Volume 2: Indian Ocean Pottery Classification*; The British Museum: London, UK, 2021. [CrossRef]
8. Simpson, S.-J. Sasanian Pottery: Archaeological Evidence for Production, Circulation and Diachronic Change. In *Sasanian Archaeology: Settlements, Environment and Material Culture*; Simpson, S.-J., Ed.; Archaeopress Publishing Ltd.: Oxford, UK, 2022; pp. 255–374.
9. Kawamata, M. Telūl Hamediyāt near Tells Gubba and Songor: Part III. *Al-Rāfidān* **1991**, *12*, 249–260.
10. Carter, R.; Connan, J.; Priestman, S.; Tomber, R. Torpedo jars from Sir Bani Yas, Abu Dhabi. *Tribulus J. Emir. Nat. Hist. Group* **2011**, *19*, 162–163.

11. Carter, R. Christianity in the Gulf during the first centuries of Islam. *Arab. Arch. Epig* **2008**, *19*, 71–108. [CrossRef]
12. Priestman, S. *Ceramic Exchange and the Indian Ocean Economy (AD 400–1275). Volume 1: Analysis*; The British Museum: London, UK, 2021. [CrossRef]
13. Carvajal López, J.; Guérin, A.; Georgakopoulou, M. Petrographic analysis of ceramics from Murwab, an early Islamic site in Qatar. *Proc. Semin. Arab. Stud.* **2022**, *51*, 55–70.
14. West, N.; Alexander, R.; Kagi, R.I. The use of silicalite for rapid isolation of branched and cyclic fractions in petroleum. *Org. Geochem.* **1990**, *15*, 499–501. [CrossRef]
15. Connan, J.; Engel, M.H.; Jackson, R.; Priestman, S.; Vosmer, T.; Zumberge, A. Geochemical analysis of two Samples of bitumen from jars discovered on Muhut and Masirah islands (Oman). *Separations* **2021**, *8*, 182. [CrossRef]
16. Charrié-Duhaut, A.; Lemoine, S.; Adam, P.; Connan, J.; Albrecht, P. Abiotic oxidation of bitumens under natural conditions. *Org. Geochem.* **2000**, *31*, 977–1003. [CrossRef]
17. Connan, J.; Borrell, F.; Vardi, J.; Wolff, S.; Ortiz, S.M.; Engel, M.; Gley, R.; Zumberge, A. Geochemical analysis of bituminous samples from the Pre-Pottery Neolithic B site of Nahal Efe (Northern Negev, Israel): Earliest evidence in the region and an example of alteration of the Dead Sea bitumen. *Org. Geochem.* **2024**, *195*, 104844. [CrossRef]
18. Connan, J. *Le Bitume dans l'Antiquité*; Editions Errance: Arles, France, 2012; p. 272.
19. Connan, J.; Carter, R. A geochemical study of bituminous mixtures from Failaka and Umm an-Namel (Kuwait), from the Early Dilmun to the Early Islamic period. *Arab. Arch. Epig.* **2007**, *18*, 139–181. [CrossRef]

molecules

MDPI

Article

Differential Analysis of Anthocyanins in Red and Yellow Hawthorn (*Crataegus pinnatifida*) Peel Based on Ultra-High Performance Liquid Chromatography-Electrospray Ionization Tandem Mass Spectrometry

Dongsheng Wang [1,2,3], Beibei Cheng [2,3], Liyang Yu [1,2,3], Guomei Yuan [2,3], Yate Ma [2,3], Jijun Zhang [2,3] and Furong Lin [4,*]

1 Hebei Province Yanshan Agriculture Characteristic Industry Technology Research Institute, Hebei Normal University of Science and Technology, Qinhuangdao 066004, China
2 Hebei Key Laboratory of Horticultural Germplasm Excavation and Innovative Utilization, Hebei Normal University of Science and Technology, Qinhuangdao 066600, China
3 Hebei Higher Institute Application Technology Research and Development Center of Horticultural Plant Biological Breeding, Qinhuangdao 066004, China
4 State Key Laboratory of Tree Genetics and Breeding, Key Laboratory of Tree Breeding and Cultivation of State Forestry and Grassland Administration, Research Institute of Forestry, Chinese Academy of Forestry, Beijing 100091, China
* Correspondence: linfr@caf.ac.cn

Abstract: Anthocyanins constitute the primary pigment components in hawthorn (*Crataegus pinnatifida*) peel, yet their specific composition and concentration profiles remain poorly characterized. This study employed ultra-performance liquid chromatography–electrospray ionization–tandem mass spectrometry (UPLC-ESI-MS/MS)-based metabolomics to systematically compare anthocyanin profiles between red-peel (CPR) and yellow-peel (CPY) hawthorn cultivars. Our analysis identified 26 anthocyanin metabolites in CPR and 24 in CPY, with cyanidin-3-O-galactoside and cyanidin-3-O-arabinoside being the predominant compounds in both. Multivariate analysis revealed seven significantly differential metabolites, including cyanidin-3-O-galactoside, cyanidin-3-O-arabinoside, pelargonidin-3-O-galactoside, pelargonidin-3-O-glucoside, pelargonidin-3-O-arabinoside, and peonidin-3-O-galactoside. Notably, all the differential metabolites exhibited reductions in CPY compared to CPR. Chromatic analysis demonstrated that CPR possessed highly significantly lower hue angle values (h_{ab}) than CPY (47.7093 ± 4.1706, 83.6427 ± 1.4604, $p < 0.01$), showing strong negative correlations with key anthocyanins. These findings enhance the scientific understanding of anthocyanin biosynthesis in hawthorn peel and provide a certain reference for the development and utilization of anthocyanins in hawthorn peel.

Keywords: hawthorn; anthocyanins; peel; UHPLC-ESI-MS/MS

1. Introduction

The genus *Crataegus* (Rosaceae), commonly known as hawthorn, comprises approximately 1000 species distributed across Europe, North America, and East Asia [1]. This genus has gained recognition as a valuable source for natural pigment extraction and antioxidant-rich functional food development [2]. Among the 12 species and six varieties native to China, *C. pinnatifida* Bge. var. major N.E.Br. (Chinese hawthorn) dominates commercial cultivation. While primarily consumed fresh or processed, its peel exhibits significant anthocyanidin accumulation, with most cultivars displaying red pigmentation and rare occurrences of yellow/orange phenotypes [3,4].

Anthocyanidins, the glycosylated precursors of anthocyanins, encompass six major aglycones: cyanidin, delphinidin, peonidin, pelargonidin, malvidin, and petunidin [5,6]. These water-soluble pigments typically conjugate with glucose, arabinose, or galactose via O-glycosidic bonds, conferring both chromatic properties (cyanidin/peonidin: purple-red; pelargonidin: red-orange; delphinidin/malvidin/petunidin: blue-purple) and potent antioxidant capacities [7]. Despite their pivotal role in peel pigmentation, previous phytochemical analyses of Chinese hawthorn have prioritized organic acids, triterpenoids, and flavonoids [8–10], with limited attention to anthocyanin profiling. Current knowledge remains restricted to red-peel cultivars, while the anthocyanin composition of yellow-peel variants persists as a scientific enigma.

Modern metabolomics employs chromatography–mass spectrometry platforms to achieve high-resolution metabolite profiling, combining separation efficiency, analytical precision, and environmental sustainability [11–13]. One anthocyanin, cyanidin-3-O-galactoside, was identified when Liu et al. [14] analyzed the phenolic compounds in Chinese hawthorn fruits using HPLC-DAD-ESI/MS (high-performance liquid chromatography–diode array detector–mass spectrometry). Moreover, Wang, et al. [15] isolated and identified cyanidin-3-O-β-galactoside and cyanidin-3-O-α-arabinoside of *C. pinnatifida* Bge. var. major fruits. Subsequently, cyanidin-3-O-galactoside and cyanidin-3-O-arabinoside were extracted from Chinese hawthorn peel and analyzed using HPLC-MS [16]. Eight anthocyanins were separated and identified in the fruit of *C. pinnatifida* Bge. var. major; this is by far the most complete study of anthocyanins in the species [17]. Wang et al. [18] revealed two additional derivatives—cyanidin-3-xyloside and cyanidin-3-O-glucoside—with broad-spectrum metabolomics at five developmental stages. Ultra-high-performance liquid chromatography–tandem mass spectrometry (UHPLC-MS/MS) is becoming increasingly widely used in the analysis of metabolites [19,20].

Notwithstanding these advancements, the systematic profiling of hawthorn peel anthocyanins—particularly comparative analyses between chromatic phenotypes—remains unaddressed. This study employs UPLC-ESI-MS/MS-based metabolomics to qualitatively and quantitatively analyze the major anthocyanins of red- and yellow-skinned *C. pinnatifida* Bge. var. major and to identify phenotype-specific differential anthocyanins. Our findings are helpful in revealing the reason for the coloration of hawthorn peel and in understanding the mechanism of the color change of hawthorn peel.

2. Results

2.1. Quality Control (QC) Analysis of Samples

The total ion chromatogram (TIC) profiles exhibited comprehensive ion intensity distribution across the analytical time course (Figure S1a). A mixed solution was used as the quality control (QC) sample; the QC sample was inserted every 10 test samples during the analysis process. High overlap of the QC-derived TIC curves (Figure S1b) confirmed excellent system stability and method reproducibility. The target anthocyanins showed well-resolved peaks in extracted ion chromatograms (Figure S1c). The method validation demonstrated limits of detection (LOD) and quantification (LOQ) ranging from 0.1909 to 0.5342 ng/g and 0.6592 to 1.8445 ng/g, respectively, meeting the stringent sensitivity requirements for plant metabolite analysis.

2.2. Qualitative and Quantitative Analysis of Anthocyanin

A total of 26 anthocyanin metabolites in six categories were detected in CPR and CPY, including cyanidin derivatives (n = 4), delphinidin derivatives (n = 6), malvidin derivatives (n = 2), pelargonidin derivatives (n = 6), peonidin derivatives (n = 4), and petunidin derivatives (n = 4) (Table 1). Ten compounds were authenticated using reference standards: cyanidin-3-O-galactoside, cyanidin-3-O-arabinoside, cyanidin-3-O-(6-O-malonyl-β-D-glucoside), delphinidin-3-O-galactoside, malvidin-3-O-arabinoside, malvidin-3-O-galactoside, pelargonidin-3-O-glucoside, peonidin-3-O-glucoside, peonidin-3-O-arabinoside, and petunidin-3-O-glucoside. The remaining compounds were identified through spectral library matching (Maiwei Biotechnology Co., Ltd., Wuhan, China). The CPR-exclusive metabolites included pelargonidin-3-O-glucoside and pelargonidin-3-O-arabinoside. The total anthocyanin content differed significantly between CPR (2411.76 μg/g) and CPY (53.967 μg/g), with cyanidin-3-O-galactoside (45.3-fold higher in CPR) and cyanidin-3-O-arabinoside (51.0-fold higher) being predominant in both cultivars.

Table 1. Absolute quantification of anthocyanin results.

Anthocyanin Metabolite	Retention Time	Regression Equation	Precursor Ions *m/z*	Characteristic Fragment *m/z*	DP	CE	CPR (μg/g)	CPY (μg/g)	Limits of Detection (ng/g)	Limits of Quantification (ng/g)
Cyanidin-3-O-galactoside	6.91	$y = 5424.31085x + 1862.34714$ (0.99881)	449.1	287.1	50	32	2327.962 ± 435.075 ^{Aa}	51.406 ± 30.260 ^{Bb}	0.1909	0.6592
Cyanidin-3-O-arabinoside	7.87	$y = 2.06939 \times 10^{5}x + 8.27387 \times 10^{4}$ (0.99727)	419.1	287.1	45	25	40.410 ± 10.411 ^{Aa}	0.792 ± 0.225 ^{Bb}	0.3996	1.3798
Cyanidin-3-O-(6-O-malonyl-beta-D-glucoside)	10.20	$y = 1.42967 \times 10^{5}x + 23953.76856$ (0.99900)	535.1	287.1	100	32	0.007 ± 0.003 ^{Aa}	0.004 ± 0.002 ^{Aa}	0.2292	0.8291
Cyanidin-3-(6-O-p-caffeoyl)-glucoside	11.38	$y = 3.62073 \times 10^{4}x + 4600.17059$ (0.99612)	611.1	287.1	75	35	0.013 ± 0.004 ^{Aa}	0.009 ± 0.008 ^{Aa}	-	-
Delphinidin-3-O-arabinoside	6.96	$y = 3.62073 \times 10^{4}x + 4600.17059$ (0.99612)	435.5	303.1	100	30	0.038 ± 0.022 ^{Aa}	0.046 ± 0.009 ^{Aa}	-	-
Delphinidin-3-O-(6-O-p-coumaroyl)-glucoside	11.75	$y = 3.62073 \times 10^{4}x + 4600.17059$ (0.99612)	611.1	303.1	75	35	0.026 ± 0.008 ^{Aa}	0.033 ± 0.030 ^{Aa}	-	-
Delphinidin-3-O-(6-O-acetyl)-glucoside	10.97	$y = 3.62073 \times 10^{4}x + 4600.17059$ (0.99612)	507.1	303.1	75	35	0.180 ± 0.006 ^{Aa}	0.094 ± 0.093 ^{Aa}	-	-
Delphinidin-3-O-(6-O-malonyl-beta-D-glucoside)	9.49	$y = 3.62073 \times 10^{4}x + 4600.17059$ (0.99612)	551.1	303.1	100	32	1.240 ± 0.832 ^{Aa}	0.216 ± 0.217 ^{Aa}	-	-
Delphinidin-3-O-5-O-(6-O-coumaroyl)-diglucoside	10.93	$y = 3.62073 \times 10^{4}x + 4600.17059$ (0.99612)	773.2	303.1	75	35	0.030 ± 0.005 ^{Aa}	0.027 ± 0.016 ^{Aa}	-	-
Delphinidin-3-O-galactoside	6.01	$y = 1.32162 \times 10^{5}x - 5432.25388$ (0.99888)	465.1	303.1	60	31	0.332 ± 0.156 ^{Aa}	0.098 ± 0.030 ^{Aa}	0.4700	1.6230
Malvidin-3-O-arabinoside	9.77	$y = 1.52996 \times 10^{5}x + 1279.51381$ (0.99863)	463.3	331.1	120	27	0.161 ± 0.092 ^{Aa}	0.054 ± 0.028 ^{Aa}	0.2055	0.7095
Malvidin-3-O-galactoside	8.95	$y = 2.38083 \times 10^{5}x + 8974.76971$ (0.99770)	493.2	331.1	50	34	0.505 ± 0.310 ^{Aa}	0.272 ± 0.246 ^{Aa}	0.3676	1.2693
Pelargonidin-3-O-(6-O-p-coumaroyl)-glucoside	12.53	$y = 3.62073 \times 10^{4}x + 4600.17059$ (0.99612)	579.2	271.1	75	35	0.122 ± 0.035 ^{Aa}	0.105 ± 0.020 ^{Aa}	-	-
Pelargonidin-3-O-5-O-(6-O-coumaroyl)-diglucoside	12.04	$y = 3.62073 \times 10^{4}x + 4600.17059$ (0.99612)	741.2	271.1	75	35	0.026 ± 0.003 ^{Aa}	0.025 ± 0.004 ^{Aa}	-	-
Pelargonidin-3-O-galactoside	7.78	$y = 3.62073 \times 10^{4}x + 4600.17059$ (0.99612)	433.2	271.1	100	27	20.269 ± 8.060 ^{Aa}	0.194 ± 0.091 ^{Ab}	-	-
Pelargonidin-3-O-rutinoside-5-O-glucoside	8.92	$y = 3.62073 \times 10^{4}x + 4600.17059$ (0.99612)	741.2	271.1	75	35	0.036 ± 0.013 ^{Aa}	0.037 ± 0.005 ^{Aa}	-	-
Pelargonidin-3-O-glucoside	8.37	$y = 1.74307 \times 10^{5}x + 12033.58043$ (0.99738)	433.2	271.1	100	27	0.115 ± 0.025 ^{Aa}	- ^{Bb}	0.2673	0.9474
Pelargonidin-3-O-arabinoside	8.74	$y = 3.62073 \times 10^{4}x + 4600.17059$ (0.99612)	403.1	271.1	45	25	0.214 ± 0.0380 ^{Aa}	- ^{Bb}	-	-
Peonidin-3-O-(6-O-p-coumaroyl)-glucoside	12.60	$y = 3.62073 \times 10^{4}x + 4600.17059$ (0.99612)	609.2	301.1	75	35	0.017 ± 0.0010 ^{Aa}	0.019 ± 0.003 ^{Aa}	-	-
Peonidin-3-O-glucoside	8.96	$y = 3.26812 \times 10^{5}x + 1.63886 \times 10^{5}$ (0.99506)	463.3	301.1	100	30	0.093 ± 0.040 ^{Aa}	0.015 ± 0.003 ^{Ab}	0.5342	1.8445
Peonidin-3-O-galactoside	8.49	$y = 3.62073 \times 10^{4}x + 4600.17059$ (0.99612)	463.3	287.1	100	27	19.245 ± 2.193 ^{Aa}	0.316 ± 0.015 ^{Bb}	-	-
Peonidin-3-O-arabinoside	9.37	$y = 2.31273 \times 10^{5}x + 8.07477 \times 10^{4}$ (0.99769)	433.2	287.1	50	25	0.220 ± 0.036 ^{Aa}	0.004 ± 0.002 ^{Bb}	0.3559	1.2290
Petunidin-3-O-sophoroside	7.96	$y = 3.62073 \times 10^{4}x + 4600.17059$ (0.99612)	641.1	287.1	70	40	0.043 ± 0.023 ^{Aa}	0.014 ± 0.013 ^{Aa}	-	-

Table 1. *Cont.*

Anthocyanin Metabolite	Retention Time	Regression Equation	Precursor Ions *m/z*	Characteristic Fragment *m/z*	DP	CE	CPR (μg/g)	CPY (μg/g)	Limits of Detection (ng/g)	Limits of Quantification (ng/g)
Petunidin-3-O-glucoside	8.05	$y = 1.76301 \times 10^5x + 17143.60875(0.99872)$	479.1	287.1	100	30	0.024 ± 0.012 Aa	0.009 ± 0.008 Aa	0.2225	0.7888
Petunidin-3-O-galactoside	7.98	$y = 3.62073 \times 10^4x + 4600.17059(0.99612)$	479.1	303.1	100	30	0.129 ± 0.0623 Aa	0.047 ± 0.044 Aa	-	-
Petunidin-3-O-arabinoside	8.52	$y = 3.62073 \times 10^4x + 4600.17059(0.99612)$	449.1	303.1	45	25	0.303 ± 0.180 Aa	0.135 ± 0.027 Aa	-	-
Total							2411.760 ± 369.029 Aa	53.967 ± 24.424 Bb		

Note: '-' indicates that these compounds were not detected. Different capital letters represent highly significant differences between groups ($p < 0.01$); different lowercase letters represent significant differences between groups ($p < 0.05$).

2.3. Principal Component Analysis

Principal component analysis (PCA) is an unsupervised discriminant method that can evaluate the clustering relationship of red-peel (CPR) and yellow-peel (CPY) hawthorn samples. PCA analysis was performed on the anthocyanin metabolites of CPR and CPY (Figure 1). The samples in the same group were clustered together, and there were great differences between CPR and CPY, indicating that the samples within the group had good repeatability, the data of the samples were reliable, and the subsequent differential metabolite analysis could be performed. The scores of PC1 and PC2 were 70.27% and 13.67%, respectively. This result showed that the representative features of most metabolites were well aggregated.

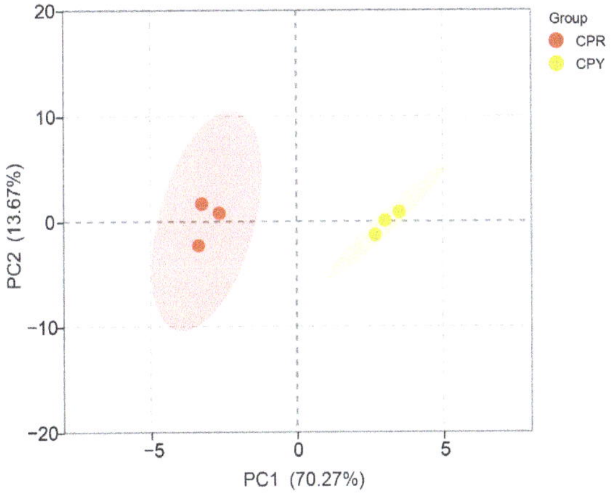

Figure 1. PCA scores plot of CPR and CPY. The shaded area represents the 95% confidence interval.

2.4. OPLS-DA Analysis and Permutation Test Analysis

Orthogonal partial least squares discriminant analysis (OPLS-DA) was also used to show the difference between CPR and CPY and to further screen differential metabolites. The results showed that the parameters R2X and R2Y of the OPLS-DA model were 0.698 and 0.999, respectively (Figure 2A), which indicated that the interpretation rate of the model was good. Q2 was 0.925, indicating that the prediction ability of the model was high. In the OPLS-DA model score plot, the T score of the principal component in the OSC (orthogonal signal correction) process was 48.2%, showing that the difference between the two groups was obvious. The orthogonal T score was 21.6%, indicating that the difference within the group was small (Figure 2B). The permutation test results showed that as the replacement retention decreased, R2 and Q2 decreased, and the regression line showed an upward trend, which indicated that the model did not have overfitting and was effective and repeatable (Figure 3).

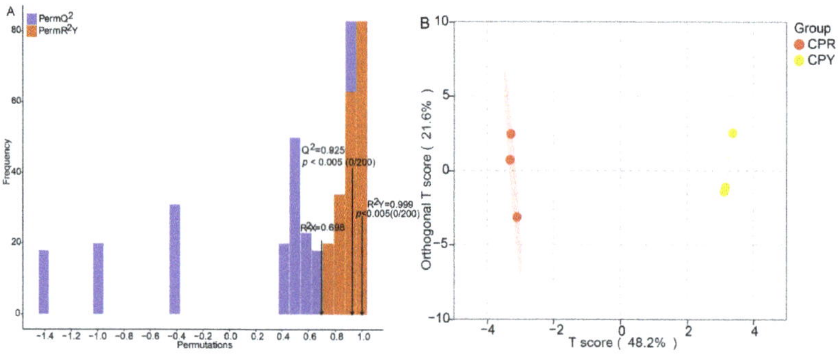

Figure 2. OPLS-DA analysis of CPR and CPY ((**A**) OPLS-DA verification plot of CPR and CPY; (**B**) OPLS-DA score plot of CPR and CPY, tthe shaded area represents the 95% confidence interval; ordinate and abscissa represent the scores of orthogonal and principal components in the OSC process, respectively).

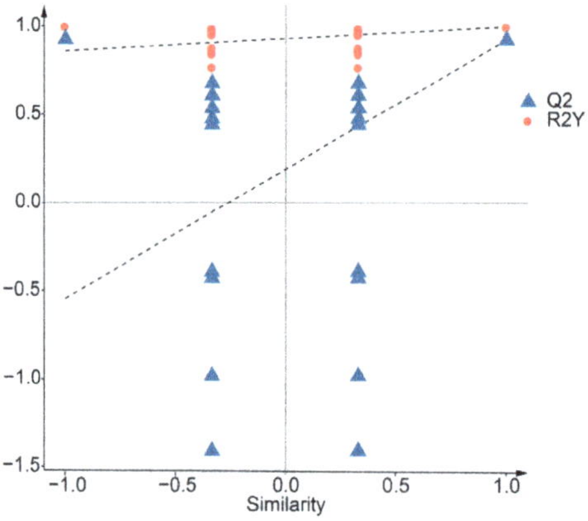

Figure 3. OPLS-DA permutation test of CPR and CPY. The X-axis represents the retention of permutation, while the Y-axis displays the values of R^2Y and Q^2. The two dashed lines correspond to the regression lines of R^2Y and Q^2, respectively.

2.5. Analysis of Differential Anthocyanin Metabolites

Differential anthocyanin metabolites were screened according to the following criteria: VIP (variable importance in project value) ≥ 1, *p*-value ≤ 0.05, and fold change (FC) ≥ 2 or fold change ≤ 0.5. Cyanidin-3-O-galactoside, cyanidin-3-O-arabinoside, pelargonidin-3-O-galactoside, pelargonidin-3-O-glucoside, pelargonidin-3-arabinoside, peonidin-3-O-galactoside and peonidin-3-O-arabinoside were identified as differential metabolites. Compared with CPR, all the differential metabolites in CPY showed a downward trend (Table 2).

Table 2. Analysis of differential anthocyanin metabolites between CPY and CPR.

Differential Anthocyanin Metabolites	VIP	*p*-Value	CPY vs. CPR Fold Change	Type
Cyanidin-3-O-galactoside	1.5141	0.0117	0.0221	down
Cyanidin-3-O-arabinoside	1.4857	0.0222	0.0196	down
Pelargonidin-3-O-galactoside	1.4037	0.0497	0.0095	down
Pelargonidin-3-O-glucoside	1.5081	0.0159	0.0000	down
Pelargonidin-3-O-arabinoside	1.5236	0.0103	0.0000	down
Peonidin-3-O-galactoside	1.5342	0.0044	0.0164	down
Peonidin-3-O-arabinoside	1.5196	0.0088	0.0174	down

2.6. Correlation Analysis Between Anthocyanin Differential Metabolites and Hue Angle

CPR displayed significantly lower hue angles (47.7093 ± 4.1706°) compared to CPY (83.6427 ± 1.4604°, $p < 0.01$), correlating with visual phenotypes (Table 3). Strong negative correlations (r = −0.9023 to −0.9955) between hue angle and differential anthocyanin content were observed (Table 4), confirming that anthocyanin accumulation drives red pigmentation intensity.

Table 3. Calculation of a*, b*, and hue angle.

	a*	b*	h_{ab}
CPR	29.1311 ± 2.6221 ^{A}	32.4199 ± 2.7185 ^{B}	47.7093 ± 4.1706 ^{B}
CPY	6.5816 ± 1.6875 ^{B}	59.4201 ± 2.9749 ^{A}	83.6427 ± 1.4604 ^{A}

Note: Different capital letters represent highly significant differences between CPR and CPY ($p < 0.01$).

Table 4. Correlation analysis between anthocyanin differential metabolites and hue angle in CPR and CPY.

Color Index	Differential Anthocyanin Metabolites	Correlation	*p*-Value
Hue angle	Cyanidin-3-O-galactoside	−0.9747	0.0010
Hue angle	Cyanidin-3-O-arabinoside	−0.9512	0.0035
Hue angle	Pelargonidin-3-O-galactoside	−0.9577	0.0026
Hue angle	Pelargonidin-3-O-glucoside	−0.9023	0.0139
Hue angle	Pelargonidin-3-O-arabinoside	−0.9259	0.0080
Hue angle	Peonidin-3-O-galactoside	−0.9918	0.0001
Hue angle	Peonidin-3-O-arabinoside	−0.99550	0.00003

3. Discussion

Traditional approaches for identifying bioactive components in *Crataegus* species have primarily relied on HPLC systems coupled with detectors such as DAD, MS, and photodiode arrays (PAD) [21–24]. Anthocyanins, which are critical pigments responsible for the coloration of hawthorn fruit peel, remain under-characterized in these studies, with significant discrepancies existing across reported identifications. For instance, Zorzi et al. [21] identified delphinidin-hexoside in *Crataegus monogyna* Jacq. using HPLC-DAD-ESI-HRMS, while Simirgiotis [23] detected four anthocyanin derivatives in the same species via HPLC-DAD-MS. To date, the most comprehensive profiling was achieved using HPLC-DAD-ESI/MS in *C. pinnatifida* Bge. var. *major* N.E.Br., revealing eight distinct anthocyanins [17], the highest number reported to date.

UHPLC-ESI-MS/MS has emerged as a cornerstone technology for targeted metabolomic profiling, particularly in fruit phytochemical characterization [18,25,26]. Recent applications of this platform include the identification of 17 anthocyanins in *Kadsura coccinea* (Lem.) A. C. Smith [27] and 26 distinct anthocyanin derivatives in cashew apple (*Anacardium occidentale*) [28]. Aligning with these advances, our LC-MS analysis revealed 26 anthocyanin metabolites in *C. pinnatifida* red-fruited varieties (CPR) and 24 in their yellow-fruited counterparts (CPY). Notably, 24 anthocyanins were common to both varieties, with pelargonidin-3-O-glucoside and pelargonidin-3-O-arabinoside being uniquely detected in CPR.

Previous studies on *C. pinnatifida* have identified diverse anthocyanin profiles. Liu et al. [14] quantified cyanidin-3-O-galactoside in red hawthorn dry weight (0.06–0.66 mg/g), while Liu et al. [17] isolated eight anthocyanins, including cyanidin-3,5-dihexoside, pelargonidin-3-rutinoside, cyanidin-3-galactoside, cyanidin-3-glucoside, cyanidin-3-rutinoside, cyanidin-3-arabinoside, pelargonidin-3-glucoside, and malvidin-3-glucoside. Recent work by Wang et al. [18] further confirmed cyanidin-3-xyloside and cyanidin-3-O-glucoside. Some anthocyanins were also identified in other hawthorn fruits. *C. monogyna* Jacq. contains acylated derivatives, such as malvidin-3-O-(4‴coumaroyl)-rutinose-5-O-glucose [23], while *C. monogyna* fruits exhibit cyanidin-O-hexoxide and cyanidin-3-O-glucoside and peonidin-O-hexoxide [24]. Notably, this study represents the first comprehensive anthocyanin analysis of yellow Chinese hawthorn, revealing only two undetected anthocyanins compared to the red varieties. Furthermore, delphinidin and petunidin derivatives were newly identified in Chinese hawthorn.

Cyanidin-3-O-galactoside emerged as the predominant anthocyanin in both CPR and CPY, constituting 98.2% and 96.7% of total anthocyanins when combined with cyanidin-3-O-arabinoside. This aligns with Liu et al.'s [17] findings in *C. pinnatifida* and parallels observations in red-skinned *C. monogyna*, where cyanidin-3-O-glucoside predominates [24]. Anthocyanin absence in yellow *C. azarolus* [29] and distinct profiles in black-skinned *C. maximowiczii* [30] further underscore the species-specific patterns. The cyanidin dominance (80% total anthocyanins) in apples [31] suggests conserved metabolic pathways across Rosaceae.

Quantitative analysis revealed that CPR's total anthocyanin content (TAC) exceeded that of CPY by 44.69-fold, which is consistent with Qi et al.'s [32] report of ≥20-fold differences between red and yellow hawthorn cultivars. It has been reported that the reduction in anthocyanins leads to a change in the color of the flowers from coral to pink, then to yellow [33,34]. The positive values of a* indicate red, and the positive values of b* indicate yellow. For hue angle, zero degrees (0°) represents red and 90° represents yellow [35,36]. Thus, this TAC disparity directly correlates with peel coloration, as evidenced by the colorimetric parameters: CPR exhibited 4.43× higher a* (redness) and 1.83× lower b* (yellowness) versus CPY. Hue angle (h_{ab}) values further confirmed phenotypic differences (CPR: 47.7 vs. CPY: 83.6), with significant negative correlations between h_{ab} and differential anthocyanin metabolites (Table 4). These findings align with apricot coloration studies [37] and anthocyanin degradation patterns in strawberries [38].

Multivariate analysis (PCA/OPLS-DA) effectively distinguished CPR and CPY metabolomes (Figures 1 and 2B). The screening criteria were slightly different in the studies. Differential metabolites were mined by $p < 0.05$ and VIP > 1 [39]. In another report, $p \leq 0.05$ and fold change ≥ 2 were applied [40]. Differential metabolites were screened using stringent criteria (VIP ≥ 1, $p \leq 0.05$, FC ≤ 0.5), identifying seven key anthocyanins: cyanidin-3-O-galactoside, cyanidin-3-O-arabinoside, pelargonidin-3-O-galactoside, pelargonidin-3-O-glucoside, pelargonid-in-3-O-arabinoside, peonidin-3-O-galactoside, and peonidin-3-O-arabinoside. All the metabolites showed reduced accumulation in CPY. Notably, despite cyanidin-3-O-galactoside predominance in CPY, its chromatic expression may be masked by co-occurring flavonoids [35,41] through co-pigmentation effects. Additional factors, including pH fluctuations, metal ion interactions [42], and environmental conditions [43], likely contribute to coloration differences, highlighting the need for process/storage condition studies.

4. Materials and Methods

4.1. Plant Materials and Color Parameters Determination

Six elite hybrid progeny with red peel (designated CPR1, CPR2, CPR3) and yellow peel (CPY1, CPY2, CPY3) were selected from crosses between *C. pinnatifida* Bge. var. major 'Damianqiu' and *C. pinnatifida* Bge. var. major 'Dahuangmianzha' (Figure 4). These seedlings were uniformly grafted onto 'Damianqiu' rootstocks in 2017 and initiated fruiting by 2021. All the grafted plants were cultivated in identical plots under standardized horticultural management. At 140 days after anthesis in 2022, ten fully illuminated mature fruits were harvested from the upper canopy of each tree for color parameter analysis and anthocyanin quantification. Five fruits per tree were subjected to color measurements, with three technical replicates per fruit. The remaining five fruits per plant were pooled for anthocyanin extraction.

Figure 4. Phenotypic characterization of CPR (red peel) and CPY (yellow peel) accessions. **Upper** panel (**left** to **right**): CPR1, CPR2, CPR3; **lower** panel (**left** to **right**): CPY1, CPY2, CPY3.

Peel color indices (a* and b*) at the equatorial region were recorded using an X-Rite SP60 portable spectrophotometer under ambient natural illumination. Hue angle (h_{ab}) was calculated as h_{ab} = arctan (b*/a*) [44].

4.2. Extraction of Hawthorn Peel Anthocyanin

The fruits were peeled and frozen in liquid nitrogen for the determination of anthocyanin metabolites. First, freeze-dried peel was ground into powder by a Mixer mill mm 400 (30 Hz, 1.5 min, Retsch, Haan, Germany); then, 50mg of peel powder was dissolved in 500 μL of extracting solution (methanol/water/hydrochloric acid = 500:500:1 (*V*/*V*/*V*)). The extract was sonicated at 4 °C for 5 min (KQ5200E, Kunshan Shumei, Kunshan, China) and centrifuged at 12,000 rpm for 3 min (5424R, Eppendorf, Hamburg, Germany). The supernatant was filtered through 0.22 μm nylon membranes (Doudian, Shenzhen, China) prior to LC-MS/MS analysis.

4.3. UHPLC and ESI-MS/MS Conditions

The extracts were detected and analyzed using an UHPLC-ESI-MS/MS system (UPLC, ExionLC™ AD, Framingham, MA, USA; MS, SCIEX OTRAP 6500+, Framingham, MA, USA). An Acquity beh C18 chromatographic column (1.7 μm, 2.1 mm × 100 mm) was used. Ultrapure water containing 0.1% formic acid (Sigma-Aldrich, St Louis, MO, USA) and methanol with 0.1% formic acid (Merck, Darmstadt, Germany) were used as mobile phases A and B, respectively. Elution gradient: At 0.00 min, the proportion of phase B was 5%, increased to 50% at 6.00 min, further increased to 95% at 12.00 min, maintained for 2 min, decreased to 5% at 14 min, and equilibrated for 2 min. The column temperature was 40 °C, and the injection volume was 2 μL.

ESI+ was used as an ion source with a positive ion mode; the temperature was set at 550 °C, the ion spray voltage (IS) was 5500 V, and the curtain gas was 35 psi. Anthocyanins were quantified via multiple reaction monitoring (MRM), with optimized declustering potentials (DP) and collision energies (CE) for target ion pairs (Table 1). Parent ions were fragmented in the collision cell, and characteristic product ions were selected for quantification.

4.4. Anthocyanin Content Calculation and Statistical Analysis

There were 11 standards in total, including cyanidin-3-O-galactoside, cyanidin-3-O-arabinoside, cyanidin-3-O-(6-O-malonyl-beta-D-glucoside), delphinidin-3-O-galactoside, malvidin-3-O-arabinoside, malvidin-3-O-galactoside, pelargonidin-3-O-glucoside, peonidin-3-O-glucoside, peonidin-3-O-arabinoside, petunidin-3-O-glucoside, and delphinidin-3,5-O-diglucoside. All the standards (isoReag, Shanghai, China) were first prepared into a 1mg/mL stock solution with 50% methanol and stored at −20 °C. Before analysis, the working solution was diluted with 50% methanol to 0.01 ng/mL, 0.02 ng/mL, 0.05 ng/mL, 0.1 ng/mL, 0.5 ng/mL, 1 ng/mL, 5 ng/mL, 10 ng/mL, 50 ng/mL, 100 ng/mL, 500 ng/mL, 1000 ng/mL, 2000 ng/mL, and 5000 ng/mL. Then, the standard curve was drawn with the standard concentration as the abscissa and the peak area as the ordinate. With reference to the retention time of the standard, the mass spectrometry data were qualitatively analyzed based on the self-built database (Maiwei Biotechnology Co., Ltd., Wuhan, China). Analyst 1.6.3 (Sciex, Framingham, MA, USA) was used for data analysis and, MultiQuant 3.0.3 (Sciex) further performed integral correction on the mass spectrometry peaks to ensure qualitative and quantitative accuracy. Afterwards, it was combined with the standard curve, and the peak area of the component was used to obtain the concentration; then, the formula for quantitative analysis was as follows: (μg/g) = C × V/1,000,000/m. (C: concentration value (μg/mL); V: the volume of the solution used for extraction (mL); m: sample mass (g)).

The limit of quantification (LOQ) and the limit of detection (LOD) were calculated using the standard. Their formulas were LOD = $k\sigma$ and LOQ = 10σ [45], respectively. The standards were prepared into a 1ng/mL solution, and detection was repeated 10 times. Then, the standard deviation (σ) of the actual measured concentration was calculated. According to the degree of freedom (n) = 9 and p (1) = 0.01 in the t-test critical value table, the k value was 2.821.

Principal component analysis (PCA) and orthogonal partial least-squares discriminant analysis (OPLS-DA) were implemented using Stats 3.5.1 and Metabo Analyst R 1.0.1 in R 3.5.1. Model validity was confirmed through 200 permutation tests. VIP, p-value, and fold change were combined to select the differential metabolites.

5. Conclusions

A total of 26 anthocyanin metabolites were identified through UHPLC-ESI-MS/MS analysis, with complete detection (26/26) in CPR and partial detection (24/26) in CPY. Both accessions exhibited identical dominant anthocyanin profiles, with cyanidin-3-O-galactoside and cyanidin-3-O-arabinoside constituting the most abundant components in peel tissues. Comparative analysis revealed seven differentially accumulated metabolites: cyanidin-3-O-galactoside, cyanidin-3-O-arabinoside, pelargonidin-3-O-galactoside, pelargonidin-3-O-glucoside, pelargonidin-3-O-arabinoside, and peonidin-3-O-galactoside. Notably, all the differential metabolites displayed downregulation in CPY compared to CPR. This study expands the documented anthocyanin repertoire of *C. pinnatifida* fruits. However, comprehensive characterization of hawthorn peel anthocyanins requires further investigation with expanded sample sizes and enhanced metabolite coverage.

Supplementary Materials: The following supporting information can be downloaded at: https://www.mdpi.com/article/10.3390/molecules30051149/s1, Figure S1: Ion chromatogram. (a). The total ion chromatogram (TIC) of quality control (QC) sample. (b). The overlap diagram of the TIC plot from QC mass spectrometry at different times. (c). The multipeak detection plot shows the extracted ion chromatogram of anthocyanins. The abscissa represents the retention time (Time, min), and the ordinate represents the ion current intensity (Intensity, cps).

Author Contributions: D.W. and F.L. contributed to the study conceptualization. Data curation and investigation were performed by B.C., G.Y. and Y.M. Validation was performed by L.Y. Study materials were provided by J.Z. The original draft of the manuscript was written by D.W., and all authors commented on previous versions of the manuscript. All authors have read and agreed to the published version of the manuscript.

Funding: This work was funded by National Forestry and Grassland Genetic Resource Center (2005DKA21003); Hebei Province Yanshan Agriculture Characteristic Industry Technology Research Institute (PT2024-10); Hebei Agriculture Research System (HBCT2024190202).

Institutional Review Board Statement: Not applicable.

Informed Consent Statement: Not applicable.

Data Availability Statement: The datasets used and/or analyzed during the current study are available from the corresponding author on reasonable request.

Conflicts of Interest: The authors declare no conflicts of interest.

References

1. Edwards, J.E.; Brown, P.N.; Talent, N.; Dickinson, T.A.; Shipley, P.R. A review of the chemistry of the genus Crataegus. *Phytochemistry* **2012**, *79*, 5–26. [CrossRef]
2. Martinelli, F.; Perrone, A.; Yousefi, S.; Papini, A.; Castiglione, S.; Guarino, F.; Cicatelli, A.; Aelaei, M.; Arad, N.; Gholami, M.; et al. Botanical, Phytochemical, Anti-Microbial and Pharmaceutical Characteristics of Hawthorn (*Crataegus monogyna* Jacq.), Rosaceae. *Molecules* **2021**, *26*, 7266. [CrossRef]
3. Özcan, M.; Hacıseferoğulları, H.; Marakoğlu, T.; Arslan, D. Hawthorn (*Crataegus* spp.) fruit: Some physical and chemical properties. *J. Food Eng.* **2005**, *69*, 409–413. [CrossRef]
4. Zhang, S.-Y.; Sun, X.-L.; Yang, X.-L.; Shi, P.-L.; Xu, L.-C.; Guo, Q.-M. Botany, traditional uses, phytochemistry and pharmacological activity of *Crataegus pinnatifida* (Chinese hawthorn): A review. *J. Pharm. Pharmacol.* **2022**, *74*, 1507–1545. [CrossRef]
5. Tanaka, Y.; Sasaki, N.; Ohmiya, A. Biosynthesis of plant pigments: Anthocyanins, betalains and carotenoids. *Plant J.* **2008**, *54*, 733–749. [CrossRef]
6. Alappat, B.; Alappat, J. Anthocyanin Pigments: Beyond Aesthetics. *Molecules* **2020**, *25*, 5500. [CrossRef]
7. Glover, B.J.; Martin, C. Anthocyanins. *Curr. Biol.* **2012**, *22*, R147–R150. [CrossRef]
8. Li, C.-R.; Hou, X.-H.; Xu, Y.-Y.; Gao, W.; Li, P.; Yang, H. Manual annotation combined with untargeted metabolomics for chemical characterization and discrimination of two major crataegus species based on liquid chromatography quadrupole time-of-flight mass spectrometry. *J. Chromatogr. A* **2020**, *1612*, 460628. [CrossRef]
9. Wu, J.; Peng, W.; Qin, R.; Zhou, H. *Crataegus pinnatifida*: Chemical constituents, pharmacology, and potential applications. *Molecules* **2014**, *19*, 1685–1712. [CrossRef]
10. Cui, T.; Li, J.Z.; Kayahara, H.; Ma, L.; Wu, L.X.; Nakamura, K. Quantification of the polyphenols and triterpene acids in chinese hawthorn fruit by high-performance liquid chromatography. *J. Agric. Food Chem.* **2006**, *54*, 4574–4581. [CrossRef]
11. Allwood, J.W.; De Vos, R.C.; Moing, A.; Deborde, C.; Erban, A.; Kopka, J.; Goodacre, R.; Hall, R.D. Plant metabolomics and its potential for systems biology research background concepts, technology, and methodology. *Methods Enzymol.* **2011**, *500*, 299–336. [CrossRef]
12. Aharoni, A.; Goodacre, R.; Fernie, A.R. Plant and microbial sciences as key drivers in the development of metabolomics research. *Proc. Natl. Acad. Sci. USA* **2023**, *120*, e2217383120. [CrossRef]
13. Chacko, S.; Haseeb, Y.B.; Haseeb, S. Metabolomics Work Flow and Analytics in Systems Biology. *Curr. Mol. Med.* **2022**, *22*, 870–881. [CrossRef]
14. Liu, P.; Yang, B.; Kallio, H. Characterization of phenolic compounds in Chinese hawthorn (*Crataegus pinnatifida* Bge. var. major) fruit by high performance liquid chromatography-electrospray ionization mass spectrometry. *Food Chem.* **2010**, *121*, 1188–1197. [CrossRef]
15. Wang, J.; Yan, R.; Yang, B. Anthocyanins from the fruit of *Crataegus pinnatifida* Bge. var. major N. E. Br. *Lishizhen Med. Mater. Medica Res.* **2015**, *26*, 42–43. [CrossRef]
16. Wu, P. Study on Preparation and Biological Activities of Anthocyanin Extract from the Peel of Hawthorn Fruit. Master's Thesis, Anhui University of Chinese Medicine, Hefei, China, 2018.
17. Liu, S.; Chang, X.; Liu, X.; Shen, Z. Effects of pretreatments on anthocyanin composition, phenolics contents and antioxidant capacities during fermentation of hawthorn (*Crataegus pinnatifida*) drink. *Food Chem.* **2016**, *212*, 87–95. [CrossRef]
18. Wang, Y.; Hao, R.; Guo, R.; Nong, H.; Qin, Y.; Dong, N. Integrative Analysis of Metabolome and Transcriptome Reveals Molecular Insight into Metabolomic Variations during Hawthorn Fruit Development. *Metabolites* **2023**, *13*, 423. [CrossRef]
19. Di Paola-Naranjo, R.D.; Sánchez-Sánchez, J.; González-Paramás, A.M.; Rivas-Gonzalo, J.C. Liquid chromatographic-mass spectrometric analysis of anthocyanin composition of dark blue bee pollen from Echium plantagineum. *J. Chromatogr. A* **2004**, *1054*, 205–210. [CrossRef]
20. Swartz, M.E. UPLC™: An Introduction and Review. *J. Liq. Chromatogr. Relat. Technol.* **2005**, *28*, 1253–1263. [CrossRef]
21. Zorzi, M.; Gai, F.; Medana, C.; Aigotti, R.; Peiretti, P.G. Identification of Polyphenolic Compounds in Edible Wild Fruits Grown in the North-West of Italy by Means of HPLC-DAD-ESI HRMS. *Plant Foods Hum. Nutr.* **2020**, *75*, 420–426. [CrossRef]

22. Wu, P.; Li, F.; Zhang, J.; Yang, B.; Ji, Z.; Chen, W. Phytochemical compositions of extract from peel of hawthorn fruit, and its antioxidant capacity, cell growth inhibition, and acetylcholinesterase inhibitory activity. *BMC Complement. Altern. Med.* **2017**, *17*, 151. [CrossRef]
23. Simirgiotis, M.J. Antioxidant Capacity and HPLC-DAD-MS Profiling of Chilean Peumo (*Cryptocarya alba*) Fruits and Comparison with German Peumo (*Crataegus monogyna*) from Southern Chile. *Molecules* **2013**, *18*, 2061–2080. [CrossRef]
24. Tamayo-Vives, C.; García-Herrera, P.; Sánchez-Mata, M.C.; Cámara-Hurtado, R.M.; Pérez-Rodríguez, M.L.; Aceituno, L.; Pardo-de-Santayana, M.; Días, M.I.; Barros, L.; Morales, P. Wild Fruits of *Crataegus monogyna* Jacq. and *Sorbus aria* (L.) Crantz: From Traditional Foods to Innovative Sources of Pigments and Antioxidant Ingredients for Food Products. *Foods* **2023**, *12*, 2427. [CrossRef]
25. Ni, M.; Chen, J.; Fu, M.; Li, H.; Bu, S.; Hao, X.; Gu, W. UPLC-ESI-MS/MS-Based Analysis of Various Edible Rosa Fruits Concerning Secondary Metabolites and Evaluation of Their Antioxidant Activities. *Foods* **2024**, *13*, 796. [CrossRef]
26. Wang, F.; Chen, L.; Chen, S.; Chen, H.; Liu, Y. Characterization of two closely related citrus cultivars using UPLC-ESI-MS/MS-based widely targeted metabolomics. *PLoS ONE* **2021**, *16*, e0254759. [CrossRef]
27. Huang, D.; Ming, R.; Yao, S.; Li, L.; Huang, R.; Tan, Y. Identification of anthocyanins in the fruits of Kadsura coccinea using UPLC-MS/MS-based metabolomics. *Biochem. Syst. Ecol.* **2021**, *98*, 104324. [CrossRef]
28. Huang, H.; Zhao, L.; Zhang, B.; Huang, W.; Zhang, Z.; An, B. Integrated analysis of the metabolome and transcriptome provides insights into anthocyanin biosynthesis of cashew apple. *Food Res. Int.* **2024**, *175*, 113711. [CrossRef]
29. Mraihi, F.; Hidalgo, M.; de Pascual-Teresa, S.; Trabelsi-Ayadi, M.; Chérif, J.-K. Wild grown red and yellow hawthorn fruits from Tunisia as source of antioxidants. *Arab. J. Chem.* **2015**, *8*, 570–578. [CrossRef]
30. Zhang, X.; Wang, J.; Li, P.; Sun, C.; Dong, W. Integrative metabolome and transcriptome analyses reveals the black fruit coloring mechanism of *Crataegus maximowiczii* C. K. Schneid. *Plant Physiol. Biochem.* **2023**, *194*, 111–121. [CrossRef]
31. Treutter, D. Biosynthesis of phenolic compounds and its regulation in apple. *Plant Growth Regul.* **2001**, *34*, 71–89. [CrossRef]
32. Qi, X.; Li, Z.; Xu, S. Relationship between soluble sugars and peel pigments in hawthorn fruit. *J. Fruit Sci.* **2005**, *22*, 81–83. [CrossRef]
33. Guo, L.; Wang, Y.; da Silva, J.A.T.; Fan, Y.; Yu, X. Transcriptome and chemical analysis reveal putative genes involved in flower color change in Paeonia 'Coral Sunset'. *Plant Physiol. Biochem.* **2019**, *138*, 130–139. [CrossRef]
34. Luo, X.; Sun, D.; Wang, S.; Luo, S.; Fu, Y.; Niu, L.; Shi, Q.; Zhang, Y. Integrating full-length transcriptomics and metabolomics reveals the regulatory mechanisms underlying yellow pigmentation in tree peony (*Paeonia suffruticosa* Andr.) flowers. *Hortic. Res.* **2021**, *8*, 235. [CrossRef]
35. Nguyen, L.L.P.; Baranyai, L.; Nagy, D.; Mahajan, P.V.; Zsom-Muha, V.; Zsom, T. Color analysis of horticultural produces using hue spectra fingerprinting. *MethodsX* **2021**, *8*, 101594. [CrossRef]
36. McGuire, R.G. Reporting of Objective Color Measurements. *HortScience* **1992**, *27*, 1254–1255. [CrossRef]
37. Ruiz, D.; Egea, J.; Gil, M.I.; Tomás-Barberán, F.A. Characterization and quantitation of phenolic compounds in new apricot (*Prunus armeniaca* L.) varieties. *J. Agric. Food Chem.* **2005**, *53*, 9544–9552. [CrossRef]
38. Taghavi, T.; Patel, H.; Rafie, R. Extraction Solvents Affect Anthocyanin Yield, Color, and Profile of Strawberries. *Plants* **2023**, *12*, 1833. [CrossRef]
39. Lan, L.; Huang, W.; Zhou, H.; Yuan, J.; Miao, S.; Mao, X.; Hu, Q.; Ji, S. Integrated Metabolome and Lipidome Strategy to Reveal the Action Pattern of Paclobutrazol, a Plant Growth Retardant, in Varying the Chemical Constituents of Platycodon Root. *Molecules* **2022**, *27*, 6902. [CrossRef]
40. Du, S.; Wan, H.; Luo, J.; Duan, X.; Zou, Z. Metabolic profiling of Citrus maxima L. seedlings in response to cadmium stress using UPLC-QTOF-MS. *Plant Physiol. Biochem.* **2024**, *214*, 108920. [CrossRef]
41. Zhang, Z.; Chang, Q.; Zhu, M.; Huang, Y.; Ho, W.K.; Chen, Z. Characterization of antioxidants present in hawthorn fruits. *J. Nutr. Biochem.* **2001**, *12*, 144–152. [CrossRef]
42. Hong, Y.; Tang, X.; Huang, H.; Zhang, Y.; Dai, S. Transcriptomic analyses reveal species-specific light-induced anthocyanin biosynthesis in chrysanthemum. *BMC Genom.* **2015**, *16*, 202. [CrossRef]
43. Yoshida, K.; Mori, M.; Kondo, T. Blue flower color development by anthocyanins: From chemical structure to cell physiology. *Nat. Prod. Rep.* **2009**, *26*, 884–915. [CrossRef]
44. Yuan, B.; Guo, Y.; Liu, Z. The influence of light path length on the color of synthetic ruby. *Sci. Rep.* **2022**, *12*, 5943. [CrossRef]
45. Li, W.; Morgan, M.K.; Graham, S.E.; Starr, J.M. Measurement of pyrethroids and their environmental degradation products in fresh fruits and vegetables using a modification of the quick easy cheap effective rugged safe (QuEChERS) method. *Talanta* **2016**, *151*, 42–50. [CrossRef]

MDPI AG
Grosspeteranlage 5
4052 Basel
Switzerland
Tel.: +41 61 683 77 34

MDPI Books Editorial Office
E-mail: books@mdpi.com
www.mdpi.com/books

www.ingramcontent.com/pod-product-compliance
Lightning Source LLC
LaVergne TN
LVHW070455061126
851858LV00011B/242
* 9 7 8 3 7 2 5 8 8 4 5 7 5 *